Critical acclaim

"… a lucid explanation of the w   …… ……u reactors. It is must reading for anyone interested in our energy future."
   Leon Cooper, Brown University physicist and 1972 Nobel laureate for superconductivity

"… a wealth of information that I've never seen anywhere. Very informative and insightful."
   Steve Kirsch, San Jose entrepreneur and philanthropist

"The book describes mankind's hope for a sustainable and prosperous future: high-temperature thorium-based reactors. The writing is clear and factual, and the book will helpful to anyone interested in energy choices."
   Meredith Angwin, Director of Energy Education for the Ethan Allen Institute

"Robert Hargraves is right: for nuclear energy to rapidly displace fossil fuels, it must become safer, cleaner, and cheaper — and next generation thorium technologies are leading candidates for that mission. In his excellent new book, *Thorium: Energy Cheaper than Coal*, Hargraves makes a compelling case that thorium holds great potential for powering a world of seven going on 10 billion people, all yearning to live modern, energy-rich lives. But *Thorium* goes beyond advocacy to offer a detailed description of the tough technical challenges that face thorium, and the innovation policies needed to accelerate this potentially world-changing energy technology. In offering a vision of environmental sustainability achieved not through raising the cost of energy but lowering it, Hargraves' *Thorium* is a valuable public service."
   Ted Nordhaus and Michael Shellenberger, co-founders Breakthrough Institute, and co-authors of *Break Through: From the Death of Environmentalism to the Politics of Possibility*

"As our energy future is essential I can strongly recommend the book for everybody interested in this most significant topic."
   George Olah, 1994 Nobel laureate for carbon chemistry

"A terrific book-length description of the need for energy solutions for this century, leading the reader to the advantages of thorium fissioning in a fluid of of molten salt. He explains the technical basis for how such a power plant works and why it can be cheaper than making power from coal -- the dominant fuel for power plants today. This book will be a valuable aid for the many people who will take this demonstrated technology of the 1960s at the Oak Ridge National Laboratory in Tennessee through the rebirth phase and into deployment in this century possibly to dominate the power plants by the later part of the 21st century. "
   Ralph Moir, retired Lawrence Livermore Laboratory physicist, expert in fusion and molten salt reactors

"Anything from Robert Hargraves is worth reading....this one is a must as it wipes out all of the opposition to nuclear power. Read the book and understand how we all have been brainwashed by the fossil industry. There is no existential substitute for nuclear and Hargraves shows us how cheap it can be."
   Reese Palley, Author: THE ANSWER: Why Only Inherently Safe Mini Nuclear Power Plants Can Save The World

THORIUM
energy cheaper than coal

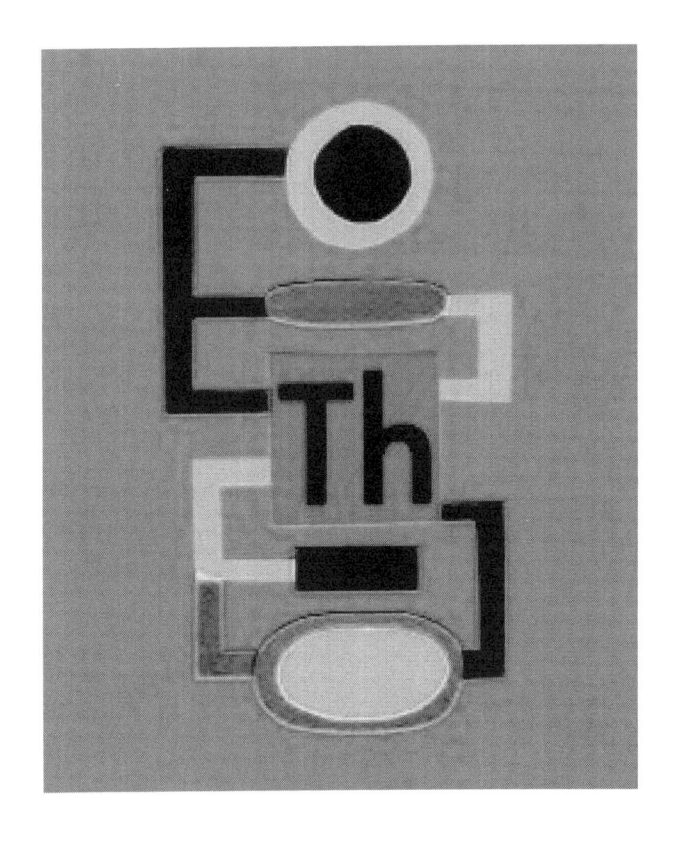

Robert Hargraves

Robert Hargraves has written articles and made presentations about the liquid fluoride thorium reactor and energy cheaper than coal – the only realistic way to dissuade nations from burning fossil fuels. His presentation "Aim High" about the technology and social benefits of the liquid fluoride thorium reactor has been presented to audiences at Dartmouth ILEAD, Thayer School of Engineering, Brown University, Columbia Earth Institute, Williams College, Royal Institution, the Thorium Energy Alliance, the International Thorium Energy Association, Google, the American Nuclear Society, and the Presidents Blue Ribbon Commission of America's Nuclear Future.

With coauthor Ralph Moir he has written articles for the American Physical Society Forum on Physics and Society: Liquid Fuel Nuclear Reactors (Jan 2011) and American Scientist: Liquid Fluoride Thorium Reactors (July 2010).

Robert Hargraves is a study leader for energy policy at Dartmouth ILEAD. He was chief information officer at Boston Scientific Corporation and previously a senior consultant with Arthur D. Little. He founded a computer software firm, DTSS Incorporated while at Dartmouth College where he was assistant professor of mathematics and associate director of the computation center.

He graduated from Brown University (PhD Physics 1967) and Dartmouth College (AB Mathematics and Physics 1961).

Copyright 2012 by Robert Hargraves, Hanover NH 03755
robert.hargraves@gmail.com

Website: http://www.thoriumenergycheaperthancoal.com

Cover design and graphic abstraction of the liquid fluoride thorium reactor by Suzanne Hobbs Baker of PopAtomic Studios
http://www.popatomic.org/

20130313

Chapters

1 Introduction: an introduction to world crises related to energy and the environment, and the potential for good solutions.

2 Energy and civilization: the relationship between energy, life, and human civilization, easy energy science, life's dependence on energy flows, civilization's progress with the energy of the Industrial Revolution, and the 21st century crises of global warming and energy consumption.

3 An unsustainable world: global warming and its terrifying implications for water, agriculture, food, and civilization; depletion of economical petroleum reserves, deadly air pollution from burning coal, increased competition for natural resources from a growing population, and the solution of new energy technology, cheaper than coal.

4 Energy sources: the character and cost of current and principal emerging energy sources: coal, oil, natural gas, hydropower, solar, wind, biomass, and nuclear.

5 Liquid fluoride thorium reactor (LFTR): the history and technology of liquid fuel nuclear reactors, the Oak Ridge demonstration molten salt reactors, thorium, LFTR, the denatured molten salt reactor (DMSR), builders, and possible contenders for energy cheaper than coal.

6 Safety: the safety of molten salt reactors, comparisons to alternative energy sources, radiation risks, waste, weapons, and fear.

7 A sustainable world: environmental benefits of thorium energy cheaper than coal: reduced CO_2 emissions, reduced petroleum consumption, synthetic fuels for vehicles, hydrogen power, water conservation, desalination.

8 Energy policy: current confused policies, failure to reduce CO_2 emissions, subsidies, recommendations, leadership.

TABLE OF CONTENTS

GLOSSARY 21

FOREWORD 23

1 INTRODUCTION 25
THE STORY OF FIRE 25
ASTONISHING BENEFITS OF NUCLEAR ENERGY 26
ENERGY AND ENVIRONMENT ISSUES ARE HARSH 27
Global warming is harming us all. 27
Increasing population stresses natural resources. 27
Cheap oil is ending. 28
Air pollution kills millions. 28
Energy insecurity leads to conflict. 28
Carbon taxes increase contention between rich and poor. 29
Less, more expensive food will not feed more, poor people. 29
A MARKET-BASED ENVIRONMENTAL SOLUTION 30

2 ENERGY AND CIVILIZATION 31
ENERGY 31
Kinetic energy is mass in motion. 31
Potential energy can be stored by the force of gravity. 33
Gravitational potential energy powers a water wheel. 34
Chemical bonds between atoms store energy. 34
Elastic potential energy is stored in a spring. 36
Energy can change solid, liquid, and gas states of matter. 36
Electrostatic fields can store energy. 37
A magnetic field stores energy. 38
Electromagnetic radiation is energy at the speed of light. 38
Mass is a dense form of energy. 39
Thermal energy is the kinetic energy of many molecules. 40
Thermal energy radiates away. 40
ELECTRICITY 42
Power is measured in watts, energy in watt-hours. 42
Electricity is the flow of electrons. 43
Electric energy is fleeting. 43
WORK AND HEAT 44

Work is force applied over a distance. 44
Energy flows from the Big Bang to Heat Death. 45
It's easy to make thermal energy. 45
It's harder to use thermal energy. 46
Thermal to kinetic energy conversion efficiency is always < 1. 47
Electric power is more valuable than thermal power. 48
Heat pumps are heat engines run in reverse. 49
LIFE **51**
Energy is the key to life. 51
Big fish eat little fish. 52
HUMANS **53**
Cooking with fire energy shaped human evolution. 53
Humans and animals consume energy to do work. 53
CIVILIZATION **54**
With the invention of fire came the need for fuel. 54
Energy from farming surpassed hunting and gathering. 55
Water power provided energy to mill the grain. 56
Wind kinetic energy was captured by horizontal windmills. 57
Coal energy enabled the Industrial Revolution. 58
Energy and the industrial revolution transformed the world. 59
Developing countries will consume more energy. 61
Fossil energy use increased world atmospheric CO_2. 62
Energy related CO_2 emissions will rise to over 30 Gt per year. 63
ENERGY AND CIVILIZATION SUMMARY **64**

3 AN UNSUSTAINABLE WORLD 65

The earth's resources are finite. 65
Limits to growth arise from finite resources. 65
Resource depletion may be more severe than climate change. 68
Population is stable in developed nations. 69
Impoverished countries birth the most children. 70
Prosperity stabilizes population. 71
Prosperity depends on energy. 72
Energy use is growing rapidly in developing nations. 74
Coal burning is increasing sharply in developing nations. 75
Global carbon dioxide emissions are rising. 75
GLOBAL WARMING **76**

Global temperatures are rising.	76
Carbon dioxide emissions increase global warming.	76
IPCC climate modeling projects a warming earth.	79
Unchecked global warming will end life as we know it.	80
Fossil fuel burning kills 34,000 US citizens per year.	82
Shipping emits more air pollution than all the world's cars.	83
The US is addicted to imported oil.	84
Reduced resources and increased population spark conflict.	86
Current world energy flows will not sustain civilization.	86
Carbon taxes are not a global solution.	88
CO2 emissions growth accelerated in 2012.	89
NEW ENERGY TECHNOLOGY	**91**
New energy technology can solve our environmental issues.	91
New technology makes clean energy, cheaper than coal.	92
Stopping particulate air pollution will save million of lives.	92
Lowering energy costs will increase economic productivity.	93
Ending energy poverty leads to a sustainable population.	93
Reducing CO2 emissions will check global warming.	94
LIQUID FLUORIDE THORIUM REACTOR	**95**
LFTR energy technology is cheaper and better.	97
LFTR produces electricity cheaper than from coal	97
Energy from LFTR is virtually inexhaustible.	97
Thorium fuel provides energy security to all nations.	97
LFTR produces little waste.	98
LFTR is affordable to developing nations.	98
LFTR can synthesize vehicle fuels.	98
LFTR is walk-away safe.	98
LFTR can zero world coal power plant emissions.	99
The benefits of LFTR will be well worth the cost.	99

4 ENERGY SOURCES   101

ENERGY DEMAND	**102**
2010 US energy demand was 98 quad BTU.	104
GENERATION CAPACITY	**105**
Average US electric power is 45% of 1,100 GW total capacity.	105
EIA calculates electric power generation capacity factors.	106
EIA estimates capital costs of electric power generation.	107
COAL	**108**

Coal burning is the world's largest CO2 source.	108
More efficient coal plants could use less coal.	109
"Clean Coal" is a marketing achievement.	110
Carbon capture and sequestration projects are tiny.	111
Electricity from coal is inexpensive.	113
Indirect pollution damage adds to the cost of coal power.	115
Energy cheaper than coal must be < 5.6 cents/kWh.	115
NATURAL GAS	**116**
Natural gas will dominate the near-term energy scene.	116
Natural gas burns twice as cleanly as coal.	116
Natural gas can burn more efficiently than coal.	117
Natural gas combustion turbines supply peak power.	118
Combined cycle gas turbines are most efficient.	119
CCGT CO2 emissions are much less than those of coal.	119
CCGTs emit less CO2 than proposed clean coal with CCS.	120
CCGT generation depletes natural gas reserves at half speed.	120
Hydraulic fracturing taps natural gas trapped in shale.	121
Environmental concerns with fracking will be addressed.	121
The Marcellus shale alone contains 55% of US natural gas.	123
The US natural gas pipeline network is diverse and growing.	124
The US has abundant natural gas reserves.	124
US natural gas reserves will not be depleted for decades.	125
Natural gas arbitrage opportunities beckon.	126
Post-Fukushima LNG imports ended Japan's trade surplus.	127
CNG vehicles may also increase natural gas demand.	127
US natural gas prices ($/MBTU) are historically volatile.	128
EIA projects natural gas prices will rise.	128
Electricity from natural gas is inexpensive.	129
Electricity cheaper than natural gas must be < 4.8 cents/kWh.	130
Cheap natural gas electricity has its drawbacks.	131
WIND	**132**
Wind provided 3% of US electricity in 2011.	132
Offshore wind power has a higher capacity factor.	134
Intermittent wind power requires backup power sources.	135
Hydroelectric power can back up wind power.	136
With coal backup power, wind does not reduce CO2.	136
Natural gas turbines commonly back up wind turbines.	137

Introducing wind turbines can increase CO2 emissions. 137
EPA proposes 454 g/kWh CO2 emissions limits. 138
In the future, wind with CCGT backup may emit less CO2. 140
Gas turbines running at less than full power emit more CO2. 140
Diversified wind turbines may reduce gas ramping penalties. 140
Field studies show minor CO2 reductions from wind power. 142
Electricity cheaper than from wind must be < 18 cents/kWh. 142
SOLAR **144**
Passive solar heating absorbs sunlight within a building. 144
Solar hot water readily displaces fossil fuel CO2. 145
Photovoltaic cells convert sunlight directly to electricity. 146
Concentrated solar power creates electricity from heat. 147
Concentrated solar power plants can store energy as heat. 148
Electricity cheaper than from solar must be < 24 cents/kWh. 148
INTERMITTENT WIND AND SOLAR POWER **150**
Intermittent power can raise emissions and costs. 150
SOLID BIOFUELS **151**
About 20% of forest mass is carbon. 151
Forests absorb carbon dioxide from air until maturity. 151
Wood-fired electric power plants are not sustainable. 152
Energy cheaper than burning wood must be < 10 cents/kWh. 152
Energy from renewables costs more than from fossil fuels. 154
LIQUID BIOFUELS **155**
Substituting ethanol for gasoline *may* reduce CO2 emissions. 155
US corn ethanol return on invested energy is poor. 156
Cellulosic ethanol is not yet economically feasible. 156
Farming biomass for fuel raises food prices. 158
Biomass to ethanol energy conversion efficiency is < 32%. 158
ENERGY STORAGE **159**
Rechargeable storage batteries use chemical energy. 159
Batteries meet some special electric utility needs. 161
Flywheels can store electric energy. 161
Pumped storage hydroelectricity requires two reservoirs. 161
Compressed air energy storage also uses natural gas. 163
Energy storage costs vary with technology. 164
Energy storage adds to the cost of electricity. 165
Batteries are a very expensive solution to intermittent power. 165
Siemens proposes using hydrogen to store electric energy. 166

Molten salt can store pre-electric thermal energy.	167
HYDROELECTRIC POWER	**168**
Hydro electric power opportunities are limited.	168
ENERGY CONSERVATION	**170**
Conservation and efficiency are not enough.	171
Food choices impact energy consumption.	172
OTHER ELECTRICITY SOURCES	**174**
Oil powers cogeneration of electricity and desalinated water.	174
Nuclear power can generate more, clean, safe electricity.	174

5 LIQUID FLUORIDE THORIUM REACTOR 176

President John F Kennedy to the Atomic Energy Commission	176
AEC Chairman Glenn T Seaborg to President Kennedy	176
LFTR technology still answers Kennedy's request.	177
A supernova created uranium and thorium energy.	178
PRESSURIZED WATER REACTORS	**180**
Today US nuclear power reactors use solid fuel.	180
The solid fuel form limits energy production.	183
Spent fuel rods contain long-lived radioactive transuranics.	184
LIQUID-FUEL NUCLEAR REACTORS	**185**
Transuranics could continue to burn in a fluid fuel reactor.	185
Fermi started up the first of the fluid-fuel nuclear reactors.	186
Los Alamos operated a molten plutonium metal reactor.	187
Oak Ridge scientists conceived molten salt reactors.	187
Thorium is a mildly radioactive element, and a possible fuel.	189
Thorium was first burned in solid-fuel reactors.	189
The molten salt reactor realizes thorium's true potential.	191
ORNL's Molten Salt Reactor Experiment was a success.	192
LFTR makes its own fissile uranium from thorium fuel.	193
LFTR molten salts can be continuously reprocessed.	194
The uranium separator moves new U-233 to the core salt.	195
The waste separator uses chemistry and physical properties.	196
LFTR has inherent safety.	196
LFTR gains its efficiency from high temperature.	198
Brayton cycle power conversion is efficient and compact.	198
LFTR high temperature allows dry air cooling.	199
The Nixon administration stopped LFTR development.	199

LFTR ADVANTAGES AND FLEXIBILITY	**201**
LFTRs can be started with U-233, U-235, or Pu-239.	201
Fast MSRs can convert LWR waste to U-233 for LFTRs.	202
Fusion reactors might someday produce startup U-233.	203
A handful of thorium can provide a lifetime of energy.	203
LFTR energy from thorium is inexhaustible.	204
LFTR produces < 1% of long-lived radiotoxic waste of LWRs.	206
A single fluid LFTR has simpler plumbing.	207
DENATURED MOLTEN SALT REACTOR	**208**
DMSR fuel additions are 75% thorium and 25% uranium.	209
DMSR fuel salt can be reprocessed after 30 years.	209
DMSR electricity will be cheaper than coal.	210
DMSR can recycle LWR spent fuel.	210
The Denatured Molten Salt Reactor will be first to market.	211
PEBBLE BED MOLTEN-SALT-COOLED REACTOR	**212**
PB-AHTR is a molten-salt-cooled solid-fuel reactor.	212
PB-AHTR has many of the advantages of LFTR.	214
ENERGY CHEAPER THAN COAL	**216**
LFTR will provide energy cheaper than coal.	216
Molten salt reactor cost estimates have been about $2/watt.	217
The compact LFTR operates at atmospheric pressure.	218
Inherent thermal stability lowers control costs.	218
Decay heat removal systems are passive.	218
LFTR's high temperature increases its efficiency.	219
The high heat capacity of molten salt reduces size.	219
New power conversion systems are smaller.	219
Waste disposal costs are smaller.	219
Small modular LFTRs can be mass produced.	220
Ongoing research will lead to lower LFTR costs.	222
Initial fissile material quantities and costs are low.	222
Thorium fuel is plentiful and inexpensive.	222
Uranium enrichment costs are low.	222
Fuel fabrication costs are low.	223
New control system technologies can reduce labor costs.	223
Transmission line costs are less with distributed LFTRs.	223
The program *objective* must be energy cheaper than coal.	224
Cost challenges can be met at the R&D stage.	224
LFTR DEVELOPMENT ENGINEERING	**225**

LFTR DEVELOPMENT TASKS	**227**
Build a LFTR technology reference database.	227
Develop the program plan, budget, and schedule.	227
Design an appropriate neutron economy.	228
Control reactivity and power output.	229
Control molten salt chemistry.	230
Remove noble fission products.	232
Neutron-irradiated graphite swells, then shrinks.	233
Metals must withstand heat, irradiation, and corrosion.	234
Carbon composites might replace metal materials.	235
Heat exchangers isolate fluids, at a temperature loss.	235
Lithium-6 must first be removed from Flibe molten salt.	236
Tritium must be continually removed.	237
Select a high-temperature power conversion turbine.	238
Implement passive waste heat dissipation.	240
Design a safe, maintainable plant.	241
Develop nuclear materials safeguard systems.	241
Separate and immobilize waste.	243
Integrate licensing process with design process.	245
Manage for success.	245
Maintain cost priorities.	246
Shorter projects cost less and have less cancellation risk.	247
DEVELOPERS	**248**
UNITED STATES	**248**
US scientists rejuvenated 21st century LFTR interest.	248
The US is destroying its U-233, valuable to LFTR R&D.	249
US National Laboratories are capable of developing LFTR.	250
US Energy Secretary Chu discounted LFTR potential.	252
Flibe Energy is preparing to develop LFTR for the US military.	253
Transatomic Power features MSR waste burning capability.	254
Thorenco has a fast neutron LFTR design.	255
CHINA	**258**
China bases its nuclear expansion on Generation III LWRs.	258
China is building commercial pebble bed reactors.	259
Russia is selling two fast reactors to China.	260
China is undertaking a LFTR R&D project.	260
FRANCE	**264**

Grenoble scientists are designing fast neutron thorium MSRs. 264
OTHER EMERGING LFTR DEVELOPERS — 265
Czech Republic and Australia may develop LFTR. 265
Canada ventures are examining thorium MSR opportunities. 266
Dr. Kazuo Furukawa founded IThEMS to build the FUJI MSR. 267
CONTENDERS — 270
NGNP — 271
NGNP is US DOE's choice for next generation nuclear power. 271
WESTINGHOUSE AP1000 — 274
AP1000 design evolved from Westinghouse PWR experience. 274
Advances in computing and engineering enable new designs. 274
Westinghouse built the first PWRs. 275
Westinghouse's new AP1000 has fewer costly components. 275
The AP1000 uses new modular construction techniques. 276
AP1000 shutdown decay heat is passively removed. 277
The AP1000 may generate electricity cheaper than coal. 278
SMALL MODULAR REACTORS — 279
Babcock & Wilcox applies naval reactor expertise to SMRs. 280
NuScale's SMR evolved from INL and Oregon State R&D. 281
Holtec plans its first 140 MW SMR at Savannah River. 282
Westinghouse is designing a 225 MW SMR. 283
Gen4 Energy, née Hyperion, is designing a 25 MW SMR. 284
LIQUID METAL FAST BREEDER REACTORS — 285
Experimental Breeder Reactor II used metal uranium fuel. 286
The integral fast reactor is based on EBR-II. 288
GE-Hitachi's S-PRISM based on EBR-II and IFR development. 289
Russia's SVBR-100 is based on Alfa submarine experience. 290
Russia's BN-600 LMFBR has operated since 1980. 290
Bill Gates backs TerraPower's traveling wave reactor. 291
TerraPower's TWR-D shuffles its fuel pins internally. 292
LFTR advantages require a longer R&D path than LMFBRs. 295
ACCELERATOR-DRIVEN SUBCRITICAL REACTOR — 296
An accelerator-driven reactor is subcritical. 296
The ADSR can be switched off. 297
Large proton accelerators are expensive and unreliable. 298
ADSRs need reliable control rods. 298
Starting up an ADSR with no fissile material is impractical. 299
Britain's ThorEA promotes ADSR research. 299

ADSRs have been studied by several other start-ups.	300
ADSR has no advantage over LFTR.	300
LFTR ADVANTAGES	**302**

6 SAFETY   305

ACCIDENTS	**305**
22 energy disasters killed 608 people in 2010.	305
US NRC studied consequences of severe nuclear accidents.	307
IONIZING RADIATION	**309**
0.18% of ionizing radiation comes from nuclear power.	309
Four ionizing particles come from four sources.	310
Ionizing radiation can damage cells.	311
Radioactivity is measured by counting decays.	311
Radiation dose is measured in energy units.	312
LNT theory warns that any radiation is dangerous.	315
Everyday life activities bring a risk of death.	317
EPA tries to balance regulatory costs and values.	318
Fear saves lives.	318
Fear sells.	319
Fear causes flight to the perceived safety of nature.	320
The Linear No Threshold theory is disputed.	321
Leukemia deaths were not affected by radiation < 200 mSv.	322
Solid cancer deaths were not affected by radiation < 100 mSv.	323
Residents of Fukushima will not exhibit extra cancers.	324
Low-dose radiation research programs contradict LNT theory.	325
Cobalt-60 radiation reduced cancer in Taiwan.	327
Low dose radiation hormesis may protect against high doses.	328
Radiophobia is harmful.	328
Unreasonably low radiation limits injure people.	329
American Nuclear Society documents LNT fallacies.	330
WASTE	**331**
Nature safely buried its own nuclear reactor waste in Gabon.	331
US military nuclear waste is safely buried underground.	331
Long-lived waste may be sequestered in deep boreholes.	332
Less geological storage is needed for LFTR waste.	334
WEAPONS PROLIFERATION	**336**
Weapons arose from political ambitions, not nuclear power.	336

Advanced nuclear power must be proliferation resistant. 337
The liquid fluoride thorium reactor is proliferation resistant. 338
The single-fluid DMSR is highly proliferation resistant. 342
There are easier paths than U-233 to make nuclear weapons. 342
LFTR reduces existing weapons proliferation risks. 343

7 A SUSTAINABLE WORLD 344

COAL POWER REPLACEMENT 345
LFTR can zero world coal power plant emissions. 345
SHIPPING 346
LFTR can power commercial ships. 346
OIL 347
Postponing peak oil lowers EROI and raises CO_2 emissions. 347
The US has more oil than mankind has ever pumped. 348
Surface mining of oil shale is environmentally harsh. 350
In situ shale oil extraction has less environmental impact. 350
LFTR can supply cheap energy for shale oil extraction. 353
LFTR heat can extract crude from Canadian tar sands. 354
SYNTHETIC LIQUID VEHICLE FUELS 355
Carbonaceous fuels have valuable high energy density. 355
The world economy depends on petroleum for transport. 356
Hydrocarbon fuels can use LFTR-produced hydrogen. 356
Hydrogen can be combined with coal to make synfuels. 357
LFTR energy can combine coal and natural gas for synfuels. 359
Synfuels could be carbon-neutral by recycling CO_2. 360
More carbon sources are air, vegetation, and cement. 360
Green Freedom proposes extraction of CO_2 from air. 360
Nuclear heat and hydrogen can make biomass into synfuel. 362
LFTR-energized biomass fuels might supply US needs. 363
AMMONIA 365
Ammonia can transport much of hydrogen's energy. 365
Ammonia energy density is higher than that of hydrogen. 365
Ammonia is a common industrial chemical. 366
Ammonia can fuel internal combustion engines. 367
Ammonia fuel cells can generate vehicle electricity directly. 368
Solid state ammonia synthesis cuts ammonia costs. 368
The energy cost of nuclear ammonia is 1/3 that of gasoline. 369
Ammonia can be handled safely. 370

NUCLEAR CEMENT	**373**
Cement curing in concrete absorbs CO2.	373
Cement can be created with LFTR, rather than fossil fuel.	374
HYDROGEN	**376**
A hydrogen-economy infrastructure does not yet exist.	376
A hydrogen-fueled car is available in California.	378
Hydrogen can power airplanes.	378
WATER AND DESALINIZATION	**380**
World water resources are stressed.	380
LFTR power can reduce global water stress.	380
Water desalination is becoming more efficient.	381
POPULATION STABILITY	**382**
Ending energy poverty leads to a sustainable population.	382

8　ENERGY POLICY　　　　　　　　　　　　383

US spends $21 billion on federal tax preferences for energy.	383
DOE spends 3% of its budget on advanced nuclear power.	385
2009 risk transfers were $31 billion in loans and guarantees.	386
The 50 US states have 50 additional energy policies.	387
Existing energy policies are failing.	389
ENERGY POLICY RECOMMENDATIONS	**391**
Lead energy policy at the federal level, not the state level.	391
Audit energy policy with neutral experts.	391
End subsidy-based energy policy.	392
Reduce energy costs.	392
Reduce CO2 emissions.	392
End renewable-energy source favoritism.	393
Invest in energy cheaper than coal.	393
Invest in innovative nuclear power R&D.	393
Invest in thorium energy cheaper than coal R&D.	394
Invest in power conversion technology R&D.	394
Invest in high-temperature irradiated materials R&D.	394
Invest in high-temperature hydrogen production.	395
Invest in hydrogen-energized synfuel pilot plants.	395
Facilitate corporate development of advanced nuclear power.	395
Fund the NRC to learn about advanced nuclear power.	396
Invest in low-level radiation safety research.	396

Educate the public about nuclear power.	396
Prepare to compete with other nations.	397
Export LFTR nuclear power plants.	397
Lead!	398
THORIUM: ENERGY CHEAPER THAN COAL	**400**

REFERENCES 401

FRONT MATTER, FOREWORD, INTRODUCTION	**401**
ENERGY AND CIVILIZATION	**401**
AN UNSUSTAINABLE WORLD	**402**
ENERGY SOURCES	**404**
Coal	404
Gas	405
Wind	405
Solar	406
Biofuels	407
Energy storage	407
Conservation	408
Other	408
LIQUID FLUORIDE THORIUM REACTOR	**408**
Denatured Molten Salt Reactor (DMSR)	409
Pebble bed advanced high-temperature reactor (PB-AHTR)	410
LFTR energy cheaper than coal	410
Development tasks	411
Builders	412
Contenders	413
Small modular reactors	414
Liquid metal cooled fast breeder reactors	414
Accelerator-driven subcritical reactors	414
SAFETY	**415**
A SUSTAINABLE WORLD	**417**
Coal	417
Oil	417
Ammonia	418
Nuclear cement	419
Hydrogen	419
Water	419
ENERGY POLICY	**420**

APPENDICES	**420**

APPENDIX A — 421

AMERICAN SCIENTIST JUNE/JULY 2010 LIQUID FLUORIDE THORIUM REACTOR	**421**
An old idea in nuclear power gets re-examined.	421
The Choice	423
Thorium	425
Advantages of Liquid Fuel	430
Waste Not	434
Safety First	436
Cost Wise	438
Nonproliferation	442
Prospects	443
References	446

APPENDIX B — 449

UNDERGROUND POWER PLANT BASED ON MOLTEN SALT TECHNOLOGY	**449**
1. POWER PLANT DESIGN	450
II. WASTE FORM: SUBSTITUTED FLUORAPATITE	457
III. SAFETY	458
V. ALTERNATIVE FUEL CYCLE	460
VI. ECONOMIC COMPETITIVENESS	462
VIII. DEVELOPMENT REQUIREMENTS AND CONCLUSIONS	463
ACKNOWLEDGMENT	464
FOOTNOTES	464
REFERENCES	466

BIBLIOGRAPHY	**468**
INDEX	**470**
ACKNOWLEDGEMENTS	**481**

Glossary

actinide: an element of atomic number 89 (actinium) or higher

BTU: British Thermal Unit; 3412 BTU = 1 kWh of energy

BWR: boiling water reactor, LWR at ~60 atmospheres pressure

C: Celsius, temperature relative to water freezing

CANDU: Canadian deuterium uranium reactor

CCGT: combined cycle gas turbine

DMSR: denatured molten salt reactor; contains U-238

DOE: US Department of Energy

EIA: DOE Energy Information Agency

Flibe: fluoride salts of lithium and beryllium

FHR: fluoride high temperature reactor

G: giga, prefix meaning 1,000,000,000

GDP: gross domestic product, annual value of goods and services

Gt: gigatonne, 1,000,000,000 tonnes

GW: gigawatt, 1,000,000,000 watts

ha: hectare, area of a 100 x 100 meter square

HEU: highly enriched uranium, over 20% U-235 or 12% U-233

INL: Idaho National Laboratory

J: joule, a unit of energy = 1 watt-second

K: Kelvin, unit of temperature, relative to absolute zero, °C + 273

k: kilo, prefix meaning 1,000

kW: kilowatt, 1,000 watts, = 3412 BTU

kWh: kilowatt-hour, the energy of 1 kW of power flowing one hour

LEU: low enrichment uranium, under 20% U-235 or 12% U-233

LFTR: liquid fluoride thorium reactor

LNT: linear no threshold, a model of radiation health risk

LWR: light water reactor, nuclear reactor cooled by ordinary water

M: mega, prefix meaning 1,000,000

Mt: megatonne

MSR: molten salt reactor

MW: megawatt, 1,000,000 watts

MW(e): MW of electric power

MW(t): MW of thermal power (heat flow rate)

NGCT: natural gas combustion turbine

OECD: Organization for Economic Cooperation and Development

ORNL: Oak Ridge National Laboratory

PB-AHTR: pebble bed advanced high-temperature reactor

PWR: pressurized water reactor, LWR, ~160 atmosphere pressure

quad: 1 quadrillion BTUs

RBMK: Russian high power channel type nuclear reactor

rem: Roentgen equivalent man, 0.01 Sv

Sv: Sievert, absorbed energy per kilogram of biomass

T: tera, prefix meaning 1,000,000,000,000

t: tonne 1000 kilograms, about 1.1 tons

TCF: trillion cubic feet; 1 TCF natural gas contains 1 quad

ton: 2000 pounds

transuranic: an element of atomic number 92 (uranium) or higher

W: watt, a unit of power, or energy flow rate

W: work, a product of kinetic energy

Foreword

THORIUM is the name of a heavy metal element that can release abundant energy, but thorium is only part of the story. The key technology is the _molten salt reactor_, enabling a fluid fuel form, reducing costs and enabling energy cheaper than from coal.

The subtitle, **energy cheaper than coal** presents the idea that economics and innovation are the means to displace coal burning for electric power. To check even more CO2 emissions, thorium energy must be cheaper than natural gas, as well. And for economically sustainable clean energy, thorium energy must be cheaper than wind, solar, or biofuel energy.

I studied mathematics and physics as a college undergraduate and graduate student, then spent my working career in information technology. After retiring back to Hanover NH I became interested in the continuing world energy and climate crises. I determined that nuclear power was much underutilized as a solution to these difficult problems and began making presentations about advanced nuclear power, including the pebble bed reactor.

With the rise of environmentalism and national focus on renewable energy I decided to learn more about these technologies. For four years I developed and taught a course for members of Dartmouth ILEAD, a program of continuing education at Dartmouth College. The course, _Energy Policy and Environmental Choices: Rethinking Nuclear Power_, reviewed fossil fuels, renewable energy sources, and nuclear power. It was followed by another course, _Energy Safari_, where we studied then visited many power plants: solar, wind, biomass, coal, hydro, nuclear, and natural gas.

In preparing the courses I came across several, advanced nuclear power technologies not known to the general public. Of these I concluded that molten salt reactors have the possibility of providing electricity cheaper than coal. This liquid fuel reactor can readily handle a wide variety of fuels, including thorium, uranium, plutonium, and waste from conventional nuclear reactors. I learned more about the liquid fluoride thorium reactor (LFTR) at the *Energy from Thorium* blog and forum, where I now occasionally write a short article.

It became clear to me that the world's nations will never adopt carbon taxes that economically disadvantage them individually, and that dissuading nations from burning coal will require an economically superior technology. If we can provide better energy technology, universally available, each nation's economic self-interest will lead it to retire coal-burning power plants.

I became an advocate for LFTR, presenting my *Aim High!* talk many times. *Aim High* exhorts us not just to build a better nuclear reactor, but one that can undersell coal power and be safely mass produced and used throughout the world, checking CO_2 emissions and ending energy poverty. Later Ralph Moir and I wrote an article, *Liquid Fluoride Thorium Reactors*, in the July/August 2010 issue of *American Scientist*, and *Liquid Fuel Nuclear Reactors* in an American Physical Society newsletter.

Since those articles a half dozen projects have been launched to develop the liquid fluoride thorium reactor; some are private and some are national, with hundreds of millions of dollars of funding. The United States government remains hardly interested.

Numbers in this book are rounded so you can remember them, make quick mental calculations, and gain insights into statements about energy. Cost analyses and models are simple approximations that illustrate relative energy costs in forms suitable for energy policymakers. For further study, references referring to page numbers are given after the final chapter and posted at:

http://www.thoriumenergycheaperthancoal.com.

1 Introduction

THE STORY OF FIRE

Imagine you're the cave man, and you go back to the cave, and you've got a stick, and it's burning.
> *One guy says, what's this?*

I call it fire.
> *What does it do?*

It can do a lot of things. It can keep us warm. It can cook our food, and scare the scary animals away.
> *Well, what about the waste?*

Well, as long as we keep the cave well vented and so forth, it should be all right, and don't put your finger in it, keep a safe distance from it, and fire should work out really good for us.
> *Ahh, I don't like the fire thing. I'm going to sleep out on the savannah tonight. I'm not going to be in the cave with the scary fire.*

And that night the saber-toothed tiger eats the other guy. The fire guy mates and has children; his progeny use fire, and so on. It wasn't long before the human race was really into fire, because everybody who wasn't was dead. Societies that use energy effectively succeed; societies that don't will diminish. Which one do we want to be?

Kirk Sorensen, The Good Reactor movie trailer

Astonishing benefits of nuclear energy

"... let's never forget the astonishing benefits that nuclear technology has brought to our lives. Nuclear technology helps make our food safe. It prevents disease in the developing world. It's the high-tech medicine that treats cancer and finds new cures. And, of course, it's the energy—the clean energy—that helps cut the carbon pollution that contributes to climate change."

US President Barack Obama, March 26, 2012

Energy and environment issues are harsh yet there are good solutions.

Global warming is harming us all.

CO2 in the earth's atmosphere is increasing from burning fossil fuels. The consensus of scientists is that this is causing the earth's temperature to rise, changing weather, altering climate, raising sea levels, acidifying oceans, stifling the algal birth of the ocean food chain, and melting glaciers that supply steady water for agriculture to feed us.

> Yet advanced nuclear power can provide plentiful power without CO2 emissions, checking global warming.

Increasing population stresses natural resources.

The world population is growing to an estimated 9 billion people, all competing for diminishing natural resources – fresh water, oil, agricultural land, and food. The largest population growth is in the most impoverished countries, where people die young from starvation, disease, and war; and bear more children.

> Yet affordable, reliable electricity is a key to economic prosperity in the developing nations, which suffer from energy poverty. Basic electric power allows modest economic prosperity, with time for women to learn, work, become independent, and make reproductive choices, leading to a sustainable population.

Cheap oil is ending.

World economies depend on oil for transportation fuels. As conventional petroleum resources dwindle, supplies are being extended by drilling deeper, in more hostile environments, refining heavy crude, and mining tar sands, at ever higher costs and ever higher CO2 emissions.

> Yet powering small vehicles with electricity from nuclear power plants will reduce oil dependency. And high temperature heat from advanced nuclear reactors can synthesize substitute liquid fuels.

Air pollution kills millions.

Soot from burning coal causes respiratory illness and annually kills tens of thousands of people in the US, hundreds of thousands in China, and a million worldwide.

> Yet nuclear electric power plants emit no soot.

Energy insecurity leads to conflict.

Nations lack energy security for stability and peace. Japan depends on imported liquefied natural gas for energy; the US on petroleum; France on uranium. Supply disruptions can wreck national economies.

> Yet domestic thorium energy resources are sufficient for every nation to attain energy security.

Carbon taxes increase contention between rich and poor.

Tens of thousands of people attended the United Nations Framework Convention on Climate Change meetings in Kyoto, Copenhagen, Tianjin, Cancun, Bangkok, Bonn, Panama, and Durban without agreements to impose carbon taxes to reduce CO_2 emissions.

> Yet advanced nuclear power can provide the world with energy without contentious carbon taxes or transfer payments that pit rich against poor and impair economic growth.

Less, more expensive food will not feed more, poor people.

Malnutrition is the largest cause of death in a world population growing from 7 to 9 billion people. Food prices are increasingly unaffordable as more land is dedicated to produce biofuels such as corn ethanol.

> Yet advanced high-temperature nuclear power can vastly improve land-area-to-biofuel productivity, by extracting the carbon from any biomass to synthesize hydrocarbon fuels similar to gasoline.

A MARKET-BASED ENVIRONMENTAL SOLUTION

We can solve our global energy and environmental crises straightforwardly – through technology innovation and free-market economics. We need a disruptive technology – energy cheaper than coal. If we offer to sell to all the world the capability to produce energy that cheaply, all the world will stop burning coal.

It's as simple as that. Rely on the economic self-interest of 7 billion people in 250 nations to choose cheaper, nonpolluting energy.

Energy is about 7% of the economy. We, and especially developing nations, can not afford to pay much more for energy. Many environmentalists advocate replacing fossil fuel energy with wind and solar energy sources, blind to the fact that these are 3-4 times more costly! Global economic prosperity requires lower energy costs, not higher costs from taxes or mandated costly wind and solar sources.

THORIUM energy cheaper than coal advocates lowering costs for clean energy – a market-based environmental solution.

2 Energy and Civilization

This chapter is about the relationship between energy, life, and human civilization. We will first introduce a bit of the science of energy, to better understand its integral role. Then we will see how life depends on energy flows, how humans learned to use energy as tools, how civilization's progress accelerated with the energy of the Industrial Revolution, and how the 21st century society is facing the crises of global warming and energy consumption.

ENERGY

Energy and mass are the substance of the universe – the sun, earth, animals, cells, proteins, molecules, and atoms. Energy can have many forms such as heat, light, kinetic energy, and potential energy.

Kinetic energy is mass in motion.

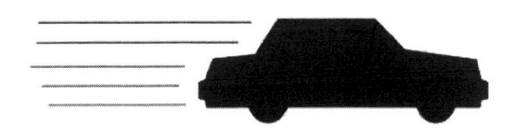

A moving car has energy proportionate to its mass, increasing with the square of its velocity. Liquid squirting from a water pistol has kinetic energy. A breath of air on a birthday candle has kinetic energy.

Flywheels are mass in rotational motion, storing kinetic energy. The flywheel in a gasoline engine stores energy between forceful piston strokes. Jaguar and Volvo have designed new hybrid cars that use flywheels to store energy less expensively than lithium-ion batteries.

There is kinetic energy in waves that break and dissipate on the beach. Wind energy, the kinetic energy of an air mass, can move sailboats or wind turbine blades. A big hurricane has kinetic energy stored in its hundreds-miles-wide rotating column of air; it dissipates energy faster than the entire world's electrical power plants can generate.

A roller coaster running down its track accelerates achieving its maximum speed at X. Its kinetic energy lets it coast uphill to Y, slowing down as its kinetic energy decreases.

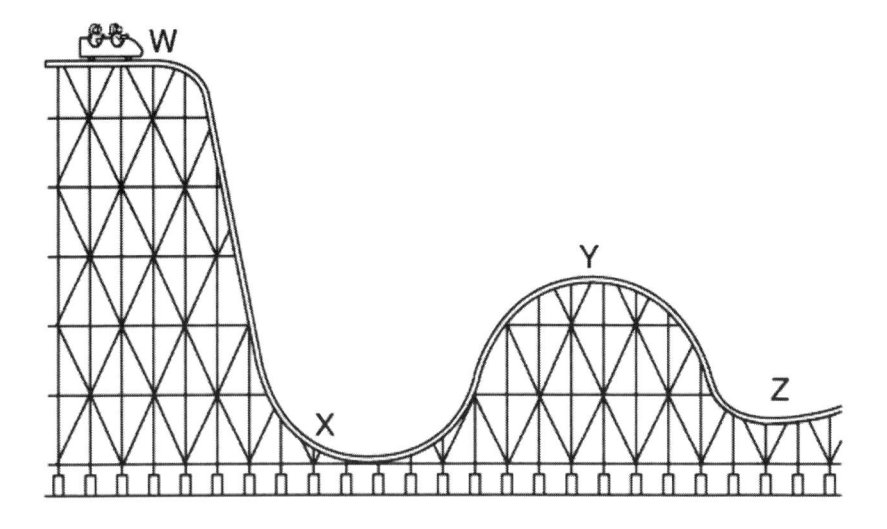

Kinetic energy from gravitational potential energy

Potential energy can be stored by the force of gravity.

In the example of the roller coaster, it has gravitational potential energy at W, which is transformed to kinetic energy by gravity as it falls to X. It then regains potential energy as it coasts uphill to Y, giving up kinetic energy. Except for friction losses, the sum of gravitational potential energy and kinetic energy is constant. The friction losses change some kinetic energy into heat.

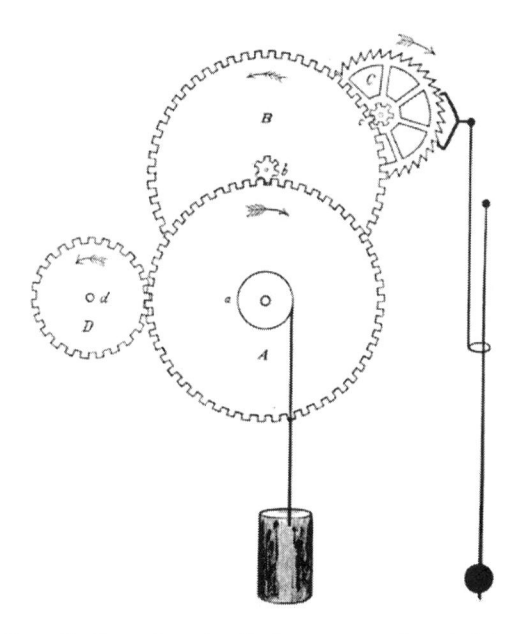

Gravitational potential energy of weight

The weight in a grandfather clock is slowly pulled down by gravity potential energy creating the bit of kinetic energy that moves the clock hands. The energy is slowly lost by friction to become heat.

The clock pendulum swinging back and forth stores some gravitational potential energy as it reaches the top of its swing. That energy becomes kinetic energy at the bottom as the pendulum rushes past vertical. Kinetic and potential energy are exchanged with every swing.

Gravitational potential energy powers a water wheel.

Water elevated behind a dam provides gravitational potential energy that is converted to kinetic energy when it flows down through a waterwheel or turbine. Near my home, the Moore hydro power dam generates 192 MW of power from water dropping 159 feet. Gravity can also be used to store energy; at Northfield, Massachusetts, water is pumped to a reservoir 800 feet above river level, then allowed to flow back down to generate power when needed.

Chemical bonds between atoms store energy.

Chemical potential energy is stored by electron interactions in chemical bonds between atoms of a molecule.

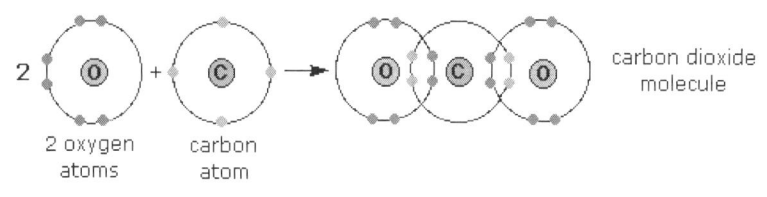

Chemical potential energy

The chemical bonds of CO2 are formed when the C and O atoms get close enough that their electrons are shared between them, by burning the carbon in coal. In that chemical reaction energy is lost to heat and electromagnetic radiation. The atoms are chemically bound; they can not be separated except by restoring the binding energy released as the bond was formed. Think

    energy {atoms} = energy {molecule} + binding energy

Chemical potential energy (binding energy) is that heat and radiation that might potentially be released when the chemical reaction binds the atoms into a molecule.

Chemical potential energy can also be created and stored. For example, charcoal (mostly carbon) can be made by heating wood, a carbohydrate made principally of carbon and hydrogen. The chemical bonds between carbon and hydrogen are broken by the added thermal energy, the hydrogen combines with oxygen, and the water vapor escapes. The resulting charcoal then contains more chemical potential energy than the original wood. It can be transported and burned later and hotter to release its stored chemical energy.

An internal combustion engine converts some of the chemical potential energy from burning gasoline into heat into kinetic energy that moves the car; most of that heat is lost to through the radiator and exhaust pipe.

Chemical potential energy can be stored and released in more ways. A lithium-hydride computer battery converts stored chemical energy into electrical energy, and can convert electricity into stored chemical potential energy.

Elastic potential energy is stored in a spring.

Elastic potential energy is related to chemical potential energy; the spring's elastic force arises from the chemical electron bonds in molecules where the atoms are displaced from their relaxed state. A bow and arrow changes elastic potential energy of the taut bow and string into kinetic energy of the flying arrow.

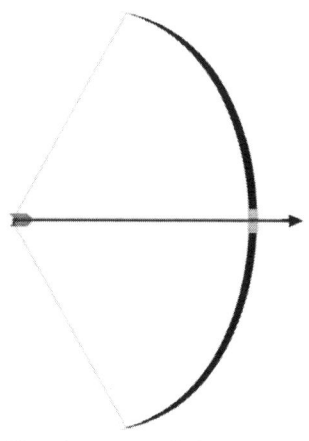

Elastic potential energy

Energy can change solid, liquid, and gas states of matter.

Ice absorbs heat when it melts to water, storing the heat energy. It's given up when water freezes. Winter snowstorms keep the temperature from falling, because freezing water into snowflakes gives up heat energy. The cooling rate of a "one ton" air conditioner is the heat absorbed by melting one ton of ice per day.

Similarly energy is stored and heat is absorbed when a liquid becomes a vapor, and released when it condenses. At the Alhambra in Spain, 14th century Moorish architects designed water flowing in fountains, beside walkways, and down banister troughs to evaporate water and cool the caliph.

ENERGY AND CIVILIZATION 37

Alhambra, cooled by liquid-to-gas state change

Electrostatic fields can store energy.

Two metal plates separated by an insulator create a capacitor that stores energy in the electric field between them. If electrons are moved from one plate to the other, creating equal and opposite charges, Q, an electric field is built up between the plates.

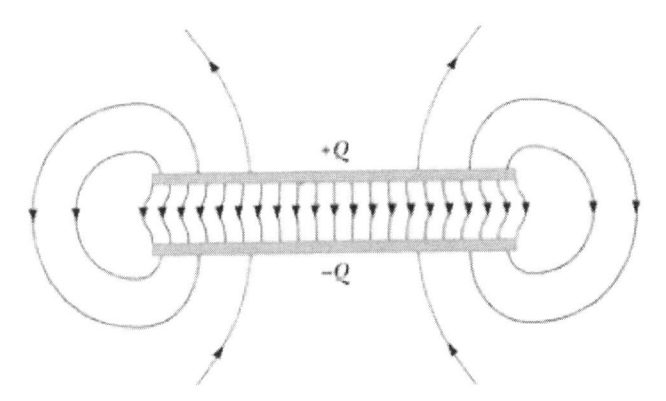

Electrostatic potential energy

The electric field is static because the electrons can not pass through the insulation between the plates. This static electric field contains energy that can be discharged rapidly. Capacitors can supplement batteries to boost electric car acceleration.

A magnetic field stores energy.

Electric currents in a coil of wire create a magnetic field that stores energy. That energy can be transferred to another coil of wire, or a spark plug, or the kinetic energy of a rotating electric motor. Magnetic fields store and transfer energy 120 times per second in the pair of coils of a typical power transformer.

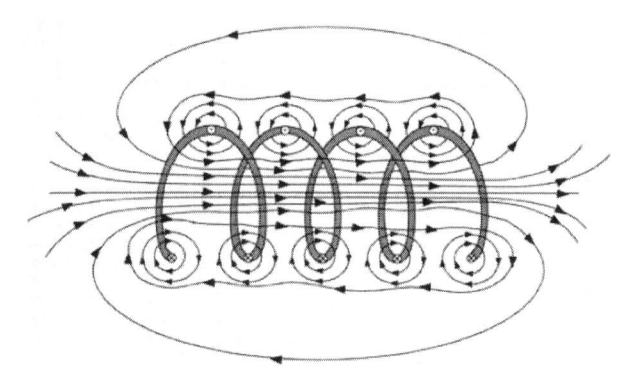

Magnetic field potential energy

Electromagnetic radiation is energy at the speed of light.

Photons are coupled, crossed electric and magnetic fields that oscillate as the photons travel through space at the speed of light.

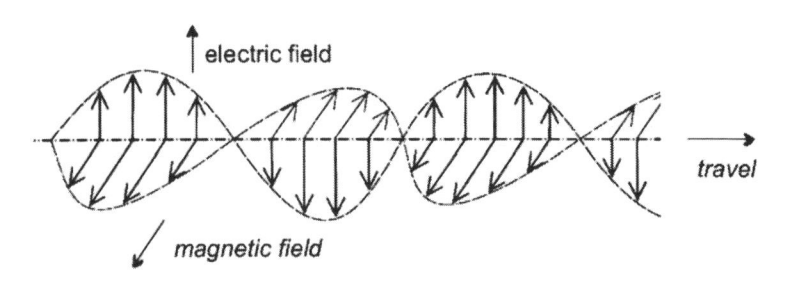

Photon energy

Visible light is composed of many photons that change electric and magnetic fields back and forth every half micron of travel. [A micron is one millionth of a meter.] The frequency is about 600 trillion times per second. Each photon is a small, discrete amount of energy proportional to its frequency. Ultraviolet light, X-rays, and gamma rays are photons with more energy than visible light photons. Infrared, microwave, and radio wave photons have less energy.

Photons are very small amounts of energy. A single 2.5 watt Christmas tree light bulb radiates about a million trillion visible light photons per second. Visible photons cause chemical changes in our retinas, so we can see. Accommodated to a dark room, a human eye can discern the flash of a single incident photon.

Tree leaves use photon energies to drive the chemical processes that use CO_2 from the air and hydrogen from water to manufacture hydrocarbons for the cells that form the growing tree.

Storing much energy in electromagnetic radiation is difficult, because it moves so quickly. Lasers reflect light back and forth internally, and then release the energy all in one pulse.

Mass is a dense form of energy.

Albert Einstein showed the equivalency of mass and energy in the famous equation: $E = mc^2$. Just as atoms are bound together to form molecules, neutrons and protons are particles bound together to make the nuclei of atoms. The binding energies for nuclear particles are about a million times stronger than the chemical binding energies that link atoms together in molecules.

Energy was thus stored in the nuclei of heavy metals when the earth's elements were created in a supernova 5 billion years ago. Today a nuclear power plant changes the bonds between neutrons and protons, transmuting the heavy metal elements into others, releasing the stored energy.

Thermal energy is the kinetic energy of many molecules.

Thermal energy is the microscopic, random, energetic motion of atoms and molecules in solids, liquids, or gases. Each molecule has velocity and kinetic energy that increases with temperature. It's easier to deal with the collective kinetic energy of a trillion trillion molecules than with them individually. The diagram below represents a close-up view of many molecules bouncing around in a constraining box. The more motion, the more thermal energy, the higher the temperature. The more collisions with the box, the greater the pressure.

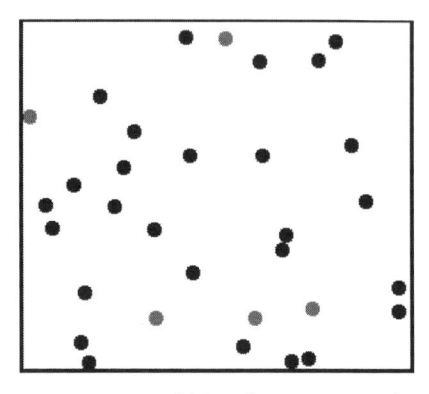

Thermal energy, sum of kinetic energy of molecules

Heat flows from higher temperature matter to lower temperature matter. The flow of heat from a hot to cold object can be partly harnessed to make more useful energy flows such as work or electric power.

Thermal energy radiates away.

A hot objects radiates electromagnetic energy as light, infrared, or microwave photons, depending on the object temperature. Hotter objects emit more, more energetic photons.

The atmosphere of the sun has a temperature of about 5,000°C and emits a range of electromagnetic radiation, with wavelength centered about the visible light spectrum that we see – 0.4 to 0.7 microns [thousandths of a millimeter]. Sight evolved to use this most common radiation – white light. Burning candles emit yellow light from the lower temperature flame, 1650°C. A blacksmith works red-hot iron at 700°C.

Hot objects emit much more radiation than cool objects because the emitted power is proportional to the fourth power of the object's temperature above absolute zero.

Even cooler bodies radiate measurable energy. An unclothed human's 37°C skin would radiate 1,000 watts of infrared light, but absorb 900 watts from the surrounding 23°C walls and ceilings. Indoors you absorb some infrared light from walls, but not much through windows transparent to the cold outdoors, so you feel cooler in a room with many windows, even though the indoor air temperature is normal.

Seen from space, the earth radiates energy as if it had an average temperature of -19°C. The cooling is balanced by heating from the sun, gravitational tides and from decay of thorium, uranium, and potassium in the earth's core.

ELECTRICITY

Power is measured in watts, energy in watt-hours.

Power is energy flow – the energy flowing past a point per second. Power can describe a rate of consumption or generation of energy. We are all familiar with one unit of measure of power – the watt, written W. A 100-watt electric light bulb consumes electric energy at the rate of 100 watts. A toaster may consume 1000 watts. A kilowatt is 1,000 watts, written 1 kW.

Energy meter

A measure of energy we are all familiar with is the kilowatt-hour, written kWh. We buy energy from the electric utility company, at a price such as $0.15/kWh – fifteen cents per kilowatt-hour. Although we often call the electric utility "the power company", consumers buy energy, not power. Power is the rate at which electricity is or can be supplied. A suburban home may have wires to the utility company that can supply power up to 48 kW.

Summarizing, power is measured in watts, not watts per second. Energy is measured in watt-hours, not watts. Many journalists get this wrong, so read carefully and critique accordingly.

Electricity is the flow of electrons.

Electricity is the flow of electrical current, typically electrons flowing in a metal wire. The amount of current (I) is measured in amperes. Power (W) is current times the voltage potential (V) through which the current flows. In analogy to the water wheel example, power is like the water flow (current) times the water height (voltage). For electricity:

Power, W = I x V (current x voltage)
Energy, E = W x t (power x time)

For electricity we usually use kilowatt-hour units, but for smaller quantities, watts and seconds are more convenient.

E (kilowatt-hours) = W (kilowatts) x t (hours)
E (watt-seconds) = W (watts)   x   t (seconds)

One kilowatt-hour = 1000 x 60 x 60 watt-seconds. Electric power is power transmitted by electricity, by an electric current across a voltage potential. Electric power is a rate of energy transfer.

Electric energy is fleeting.

Electric energy is electric power multiplied by time. Electric power is a transfer agent from one form of energy to another.

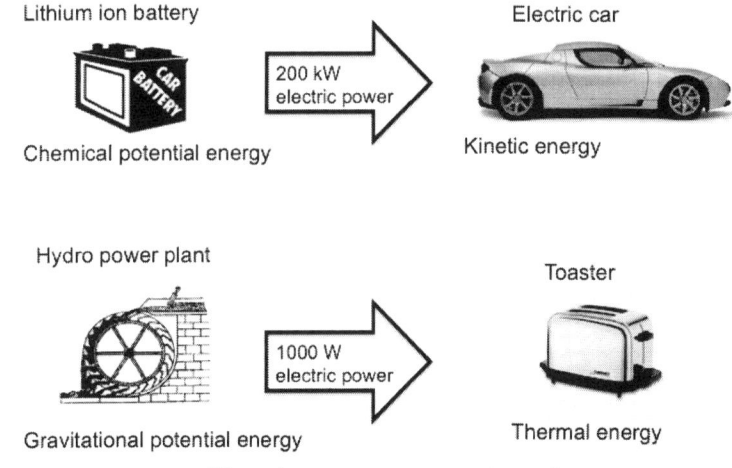

Lithium ion battery → 200 kW electric power → Electric car
Chemical potential energy → Kinetic energy

Hydro power plant → 1000 W electric power → Toaster
Gravitational potential energy → Thermal energy

Electric power energy transfer

In these examples, chemical potential energy of a battery creates electric power that becomes kinetic energy of the electric car. The gravitational potential energy of the water elevated behind a dam makes electric power that becomes thermal energy in the toaster.

Small amounts of energy can be stored as the electrostatic energy of an electric field or the electromagnetic energy of a magnetic field. In practice electric energy is rarely stored for long except by converting it to another such form of energy.

WORK AND HEAT

Work is force applied over a distance.

For example, lifting a weight of 550 pounds one foot up requires 550 foot-pounds of work. The standard rate of work for a horse hauling water up from a mine was 550 foot-pounds per second – defined as one horsepower. James Watt used this definition to account for royalties for his horse-substituting steam engine. One horsepower is 746 watts. The 2 kW electric motor on my small boat is equivalent to a 2.7 horsepower gasoline engine.

Work makes kinetic energy. The result of that kinetic energy is also energy, such as gravitational potential energy if the horse is lifting water from a mine, or thermal energy produced by friction if the horse is pulling a sledge.

A related unit of energy is the horsepower-hour – one horsepower of power applied for one hour. That's 0.746 kWh, about ten cents worth of electricity today – much cheaper than power from horses.

A bicyclist in good condition can exert about ¼ horsepower. If humans were paid competitively to electricity for their physical work they would receive 2.5 cents/hour.

Energy flows from the Big Bang to Heat Death.

The energy and mass of the universe were created by the Big Bang over 10 billion years ago. The universe of mass and energy expands, cools, mixes, and occasionally clusters to form stars and planets. Stars such as our sun burn hydrogen and dissipate the energy into space, with some radiated photon energy (eg light) absorbed by the earth. The effect on earth is weather, mixing, warming and cooling the atmosphere, oceans and land. That absorbed energy is subsequently radiated back into space, in all directions, but at the lower temperature of infrared, invisible light. Aside from radioactive decay and gravitational tides, as much energy arrives from the sun as is radiated away from the earth, or else the earth temperature changes.

At each stage of energy flow a kind of destruction occurs. Ordered, localized, hot energy is transformed to cooler, more random, dispersed energy. Energy is conserved, but its overall utility is partially destroyed. Energy flows exhibit this energy dispersion or diminishment of utility at every transition. As the universe continues to expand and cool, its energy becomes less useful and the universe approaches Heat Death – the other end of the timeline that started with the Big Bang.

It's easy to make thermal energy.

The most useful energy forms eventually become thermal energy – heat transferred to a system. Kinetic energy is diminished by friction making heat; rub your hands together for an example. Electric power flowing in wires heats them because of the inherent internal resistance of the wires. Potential energy (gravitational, chemical, elastic) can remain static and unused until their hosting structures decay.

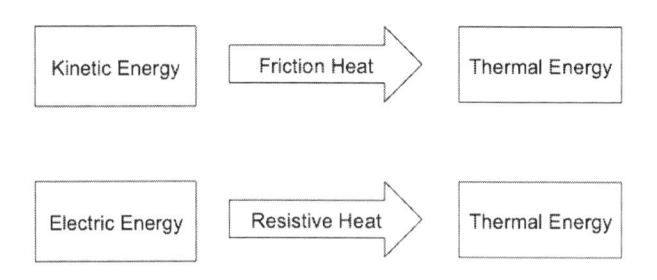

```
[Kinetic Energy]  →(Friction Heat)→  [Thermal Energy]

[Electric Energy] →(Resistive Heat)→ [Thermal Energy]
```

Thermal energy destiny

A car coming to a stop heats its brakes. The kinetic energy of the car is converted to thermal energy in the brake pads. Energy is always conserved; the process is 100% efficient.

An electric heater in your home similarly converts all the consumed electric energy to heat. A light bulb converts all its electric energy to heat, both as direct heating of the bulb and the absorption of the light by the walls of the room [except for the light that escapes through the window and continues on past Pluto]. The conversion is 100% efficient; you can heat your home as efficiently by opening the electric oven door or turning on an electric heater.

American-born, British Loyalist Count Rumford discovered the equivalence of work and thermal energy while boring cannons, providing evidence that led to the principle of conservation of energy. [He also invented the coffee percolator and thermal underwear.]

It's harder to use thermal energy.

The arrow of time points only one way. The processes that convert kinetic energy (work) to thermal energy (heat) are not 100% reversible. Physics will not let us convert all that thermal energy back to kinetic energy. However we can convert some of thermal energy flow between objects of different temperatures.

Heat is thermal energy. Heat flows from hot to cold; the molecular motion of hot thermal energy is normally dissipated into a larger, cooler system. If we do nothing, this heat flow is totally wasted. Alternatively, we can insert a heat engine into that heat flow and extract some (but not all) of that thermal energy into work (W) to make kinetic energy.

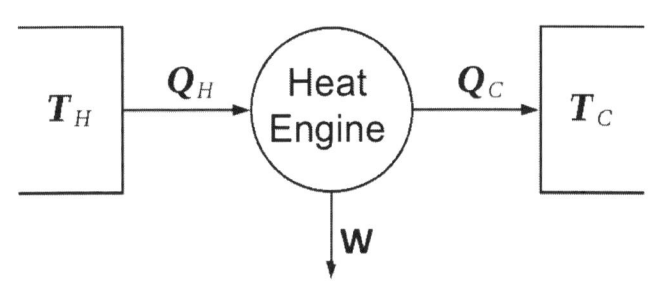

Heat source, heat engine, heat sink

We use W to symbolize work. T_H is temperature of a source of hot thermal energy, Q_H is the heat going into the heat engine, W is the useful work extracted by the heat engine, and Q_C is the rejected heat, the heat the engine was unable to convert, flowing into sink of temperature T_C cooler than T_H.

In an automobile engine, Q_H is the heat generated by burning gasoline, W is the work delivered by the rotating crankshaft, and Q_C is the heat lost to the atmosphere via the cooling radiator and the exhaust pipe. Other examples of heat engines are Watt's 18th century steam engine and an aircraft turbine jet engine. They convert some heat to work.

Thermal to kinetic energy conversion efficiency is always < 1.

Energy in equals energy out, so $Q_H = Q_C + W$. By Carnot's theorem, no matter what engine is devised, physics limits its thermal to kinetic energy conversion efficiency to be less than 1.

$$\text{Efficiency} = \frac{W}{Q_H} = \frac{T_H - T_C}{T_H} < 1$$

Temperatures (T) are in degrees Kelvin, K°, relative to absolute zero, -273°C. The higher the temperature difference between source and sink, the better the efficiency. Increasing the heat source temperature is one way to increase efficiency. Engineers raised new coal plant efficiencies from 32% to 44% by using pulverized coal burned at T_H of 1300°C. Decreasing the heat sink temperature also increases efficiency; cooling power plants with river or ocean water instead of air generally lowers T_C, increasing efficiency.

Rudolph Diesel's 1896 invention of a high-compression, high-temperature internal combustion engine had a theoretical maximum kinetic/thermal energy conversion efficiency of 75%, compared to 10% for the competitive steam engine, making him a millionaire.

In practice the typical efficiency of an automobile diesel engine is 40-50%. It burns fuel at a higher temperature than the gasoline engine with its 25-30% efficiency. The reciprocating steam engines of the 18th century had efficiencies near 1%; today's steam turbines heated by pulverized coal can reach over 40%. Large shipboard diesel engines can achieve over 50% efficiency.

Electric power is more valuable than thermal power.

An electric power plant uses a heat engine to convert thermal energy to kinetic energy of a rotating shaft, which is then converted to electric energy by the generator. Such generators can achieve electric/kinetic conversion efficiencies of 99%, so we will ignore losses of that step of the thermal-to-kinetic-to-electric energy conversion process.

Because power plants deal with both thermal power and electric power, a special notation can help prevent confusion. One GW of electric power can be written 1 GW(e). 1 GW of thermal power is written 1 GW(t). Typical average electric/thermal conversion efficiencies of US electric power plants are about 33%. Such a power plant would require 3 GW(t) of thermal power to produce 1 GW(e) of electric power.

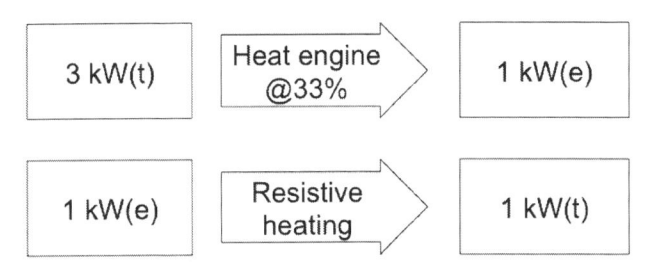

```
[3 kW(t)] → Heat engine @33% → [1 kW(e)]

[1 kW(e)] → Resistive heating → [1 kW(t)]
```

Asymmetric energy conversion

We can use the same suffix for energy units. Operating a typical 2,600 W electric stove burner for one hour consumes 2.6 kWh(e) of electricity, which is converted to 2.6 kWh(t) thermal energy (heat). That electricity cost at $0.15/kWh(e) is 39 cents.

We could get that same heat from burning natural gas; 2.6 kWh(t) of gas costs about 12 cents at retail – about a third as much as the electricity cost of 39 cents. Why such a difference? A 33%-efficient power plant needs 2.6 kWh(e)/0.33 = 7.8 kWh(t) of gas to generate the 2.6 kWh(e) of electricity. That much gas costs about 3 x 12 cents = 36 cents, roughly equal to the electricity cost of 39 cents. Cooking, drying clothes, and heating homes using electric energy is about three times as expensive as using thermal energy.

Heat pumps are heat engines run in reverse.

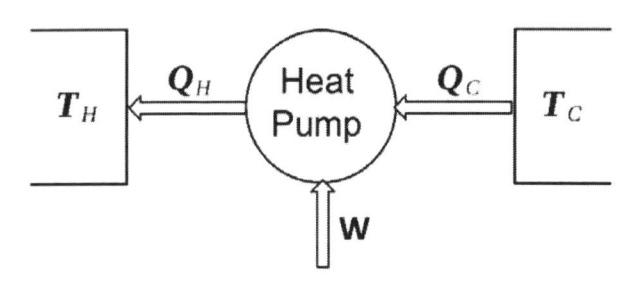

```
[T_H] ← Q_H — ( Heat Pump ) ← Q_C — [T_C]
                    ↑
                    W
```

Heat sink, heat pump, heat source

The heat pump is similar to the heat engine, except that the arrows of energy flow are reversed. Kinetic energy becomes work (W) used to pump heat Q_C from a cold source to a hot sink of temperature T_H. This is the reverse of the natural flow of heat from hot to cold, and it takes kinetic energy W to accomplish moving heat from cold to hot. Energy in equals energy out, so Q_C + W = Q_H.

An air conditioner removes heat from warm room air and transfers it to even warmer outside air. An air conditioner is judged by how much heat can be removed for the electric power it consumes. Its cooling coefficient of performance (COP) is the ratio Q_C/W. A typical window air conditioner has COP of 3; the thermal energy removed is three times the electric energy used.

Turning on a 100 W(e) light bulb in an air conditioned room will generate 100 W(t) of heat to be removed, requiring an additional 33 W(e) to power the air conditioner. Each person in a room also generates about 100 W(t).

The air source heat pump, like an air conditioner but used in reverse, extracts heat from cold outside air (chilling it more) and transfers its heat to the home interior. Its heating coefficient of performance, COP, is Q_H / W, the delivered heat transfer rate divided by the electric power used. For example, 9 kW(t) / 3 kW(e) = 3 for a typical home air source heat pump.

Geothermal heat pumps use chlorofluorocarbon liquids pumped through tubing buried in the earth as the heat source. A COP = 3 is typical for both geothermal and air source kinds. Such heat pumps can deliver 9 kW(t) of heat for only 3 kW(e) of electricity – three times better than electric space heaters. But generating that source of 3 kW(e) of electricity in a 33%-efficient power plant requires 9 kW(t) of heat to begin with. The home owner could have burned coal, oil, or gas in a home furnace consuming the same fossil fuel the power plant burned to power the heat pump. Consequently there is no CO_2 emissions benefit from heat pumps unless the electric power source is carbon-free, such as a nuclear power plant, hydro plant, wind turbine, or solar farm.

LIFE

Energy is the key to life.

As the energy of the universe flows and disperses, life temporarily borrows a stream of it for growth, reproduction, and motive force.

Life began on earth over four billion years ago, with energy bonding the essential elements -- hydrogen, oxygen, carbon, nitrogen, sulfur, and phosphorous -- creating amino acids, then proteins, and eventually prokaryotes (bacteria). Over 3 billion years ago new organisms developed, cyanobacteria. These use light energy to capture CO_2, use the carbon for building hydrocarbon structures, and expel the oxygen into the atmosphere.

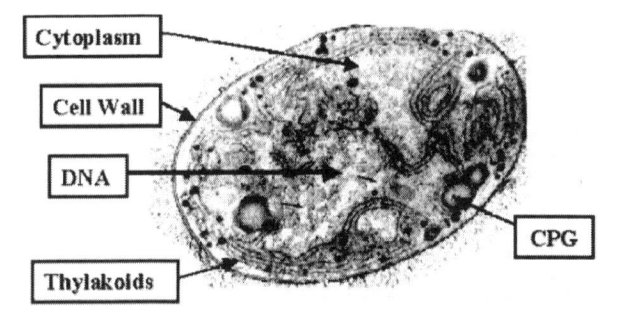

Cyanobacteria

Even today, about 20% of the world oxygen supply comes from these ancient aquatic cells. Within the cyanobacteria are thylakoids that accomplish the photosynthesis. Plants incorporate similar thylakoids in their structures to obtain the energy from sunlight.

In the evolution that led to modern animal cells, variants of these cyanobacteria evolved symbiotically to become mitochondria, energy generators within eukaryotic cells. Glucose food from the cell's environment crosses into the cytoplasm.

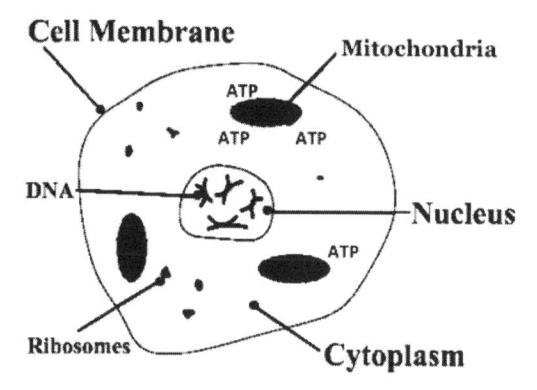

Eukaryotic cell with nucleus

The mitochondria use oxygen ions to break chemical bonds in the food and release energy to manufacture ATP molecules. This ATP (adenosine triphosphate) is the energy currency within the cell. Three phosphate molecules are popped off or onto the ADP molecule to release or store energy. The energized ATP is transported to provide the energy for other intracellular functions, such as causing muscle contraction.

In analogy, the ATP flow is like electricity flowing from generators (mitochondria) that create it from fuel (glucose) transported through cellular membranes into the cell cytoplasm and then and through the mitochondrial membranes.

Big fish eat little fish.

Animals eat and digest plants for energy. Some animals eat other animals for energy.

The digestive system obtains this food energy from the plant and animal tissue the animal eats. The ingested food carbohydrates are broken down in multiple steps to form sugars that carry the chemical potential energy distributed to the cells. The fluids circulating in the body transmit glucose food through cell membranes. Each cell distributes ATP throughout its cytoplasm for intra-cell energy.

HUMANS

Cooking with fire energy shaped human evolution.

Humans are composed of about a hundred trillion individual cells co-operating to make the single being. Humans, too, eat plants and animals for energy. Another energy source is the warmth of sunlight, reducing demand for energy from food metabolism; reptiles use this extensively. But the big, breakthrough energy technology for humans was fire. Fire provided alternative energy to metabolism of food.

Harnessing fire 1.8 million years ago made a singular difference for humans. Cooking food saved time and energy. Primates still spend half their day chewing raw food. By switching to cooked, softer, more energetically rich food homo erectus was able to devote time to more productive activities, making tools, farming, and interacting socially, as evidenced by records of their larger brains and smaller guts, jaws, and teeth. Reduced kinetic energy demands for metabolism permitted evolution of the human's large brain, which consumes a quarter of the body's energy.

Humans and animals consume energy to do work.

The human is also a source of work – energy directed to a motive task. On average a human uses 100 watts of chemical (food) energy flow. Underground coal miners expend energy at 300 W with peak power of 600 W. Human labor continued to be essential to US farming as late as 1918, when Quaker Oats promoted their food's high calorie content, needed for work, literally advertising 1810 calories per pound (2.1 kWh per pound).

Cattle and horses can supplement human labor to make work. Cattle feeding on grassland can provide 300 to 400 W of steady power. More powerful horses fed with higher protein grains can generate 500 to 1500 W for sustained periods. One horsepower is now defined as 745.7 watts. A well fed horse consumes grain that would feed six people, but provides ten times the energy. In 1910-1920 one fifth of US farmland was devoted to horse feed.

CIVILIZATION

With the invention of fire came the need for fuel.

In pre-industrial civilization tree branches, bark, and dead roots could be collected for fire fuel. With the invention of axes and saws heavy branches and tree trunks could be cut and dried for fuel. As cities grew in the temperate climates the demand for energy for cooking, heating, and industry was 20-30 W/m², which required a forest area 100 times larger than the city area. Approximately 1-2 tonnes of wood per person per year were needed.

Iron smelting required the high temperatures achieved from burning charcoal. Charcoal was made by using wood fires to heat piles of wood covered with turf or clay to keep oxygen away. This pyrolysis broke down the hydrocarbons and drove off water and other volatiles, leaving nearly pure carbon, used for smelting metals. Hundreds of thousands of people were employed making charcoal. As demand for wood for charcoal increased, Europe and England consumed much of their forests, causing an energy crisis in 17th century England. When wood became unavailable, coal mining became the energy supply. Indeed one of the first applications of the steam engine was pumping water from coal mines. Repeating history, making eucalyptus wood into charcoal for "green steel" is now practiced in Brazil.

With forests consumed, cattle dung is another fuel used even today in developing countries such as India, where it is collected, shaped, dried, and burned rurally or sold in cities for $0.14/kg, or about $0.03/kWh.

Dried cattle dung patties

Energy from farming surpassed hunting and gathering.

Agriculture was invented approximately 10,000 years ago as food from hunting and gathering became more difficult, possibly due to the end of an ice age, creating a dryer climate. Dry conditions favor annual plants, which store energy in seeds rather than woody, perennial growth. Their energy density made seeds an attractive food, but their shells limited digestibility by humans.

So another great invention of the time was grinding seeds into flour, which was made into bread. The grinding, the fermentation, and the cooking made an easily digestible, transportable, storable food energy supply that sustained people living in villages and cities rather than dispersed people for hunting and gathering.

Agriculture allowed accumulation of food, creating wealth. Increasing that wealth required human labor to tend more crops, and slave labor became an important source of energy for wealthy nations such as the Roman Empire. As Christianity spread and slaves were freed, this power source was lost, and with it the glory of Rome.

Water power provided energy to mill the grain.

Milling grain required human energy expenditure and time. One new invention was the use of water power to mill grain.

Early millstones were rotated horizontally, about a vertical shaft, which can be powered efficiently by a horizontal water wheel.

Friction losses are minimized because there are no gears. The farmer at the top feeds grain into a hopper. The top millstone rotates to grind seeds into flour.

More familiar vertical water wheels came into use after efficient gears were invented. In the first century the Romans built an aqueduct that supplied drinking water to Arles, France, and also powered 16 vertical water wheels that could produce 4.5 tons of flour per day, enough to feed 6,000 people.

Water power was key to populating New England, where I live. Settlers travelled up rivers and harnessed the streams with water wheels to power grain mills and lumber mills. The lumber came from trees quickly felled to make room for agriculture for food energy.

Wind kinetic energy was captured by horizontal windmills.

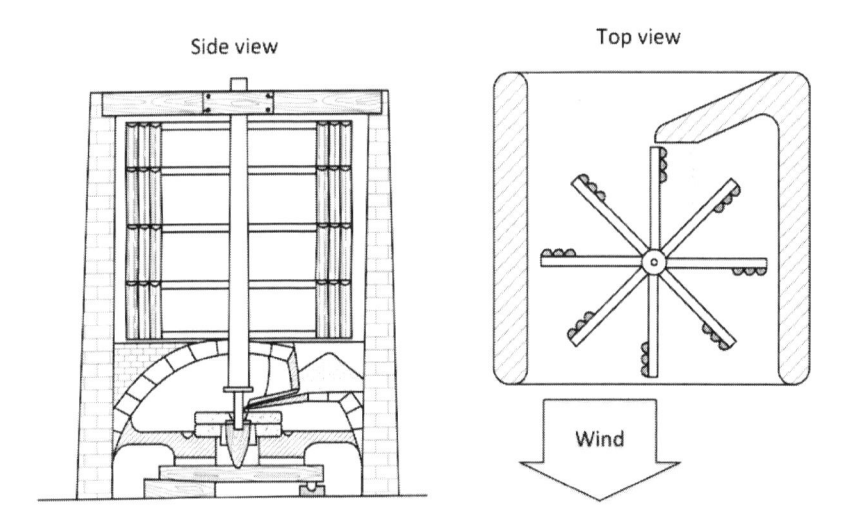

10th century Persian horizontal windmills

Early windmills were constructed to rotate about a vertical axis. This windmill design is from 10th century Persia. They were used for milling grain and pumping water. More familiar windmills with blades rotating in a vertical plane on a horizontal axis facing the wind were developed after low friction gear technology was able to transfer the kinetic energy.

Coal energy enabled the Industrial Revolution.

Up until the late 1700s, economies depended upon work from humans and draft animals. The industrial revolution, beginning in England in the latter part of the 18th century, was launched by energy from coal-fired steam engines, by expanded waterpower, and the expansion of trade over canals, highways, and railways. Innovations made use of more energy in productive ways. As textile manufacturing techniques were improved and patented, more automated cotton mills evolved, powered by horse power, then by water power, and then by steam power.

The steam engine propelled the industrial revolution, changing chemical energy of fossil fuels into kinetic energy. Newcomen's early, large steam engine had a thermal/kinetic energy conversion efficiency of < 1%, but coal was cheap and 3.7 kW of power was delivered. By 1800 nearly 500 of Watt's five times more efficient steam engines each provided up to 7.5 kW of power.

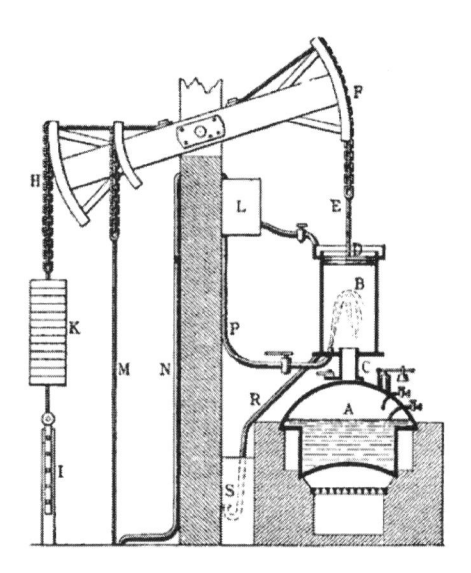

Newcomen's 1712 steam engine

Extensive coal mining was possible because of steam engines that pumped water from the mines and lifted coal to the surface. Steam engines enabled factories to be built where no water power was available. They also pumped water into canal locks to facilitate transportation for growing trade. Mined coal powered them, provided heating, and burned hot enough to smelt iron. Iron and steel, stronger than copper or bronze, enabled better machines to be built. Lathes and other metal working machine tools were fabricated.

Chemical energy from solid coal was transferred to a gas by heating coal and spraying it with steam. Gas street lighting was established in London by 1820 and it spread to factories and businesses, allowing them to stay open longer.

Heat from coal helped advance the chemical industry, enabling production of sulfuric acid and sodium carbonate used in the glass, textile, soap, and paper industries. Sintering ground limestone and clay at a high temperature of 1600°C created Portland cement for construction.

Powered paper mills provided plentiful, inexpensive paper for publication of books, helping spread knowledge. Canals, roads, and railways were built and used for commerce, including hauling coal.

Energy and the industrial revolution transformed the world.

The industrial revolution spread from the United Kingdom to Western Europe, North America, Japan, and the world. In two centuries the world average per capita income increased over tenfold. Since 1820 world population has increased five times and per capita income has increased eight times. Lifespans have more than doubled. The following graph of world GDP per capita comes from estimates by Angus Maddison, with the greatest growth rate coming at the time of the industrial revolution and coal energy.

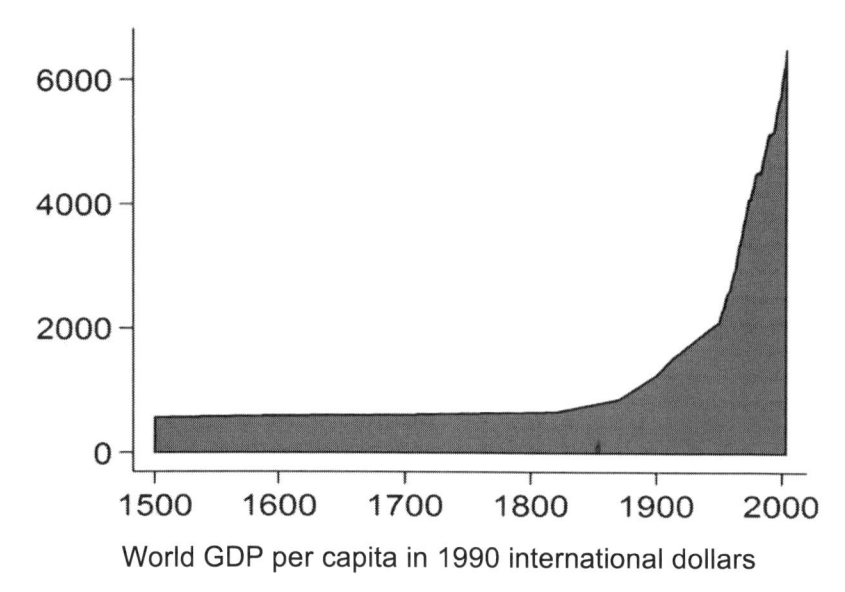

World GDP per capita in 1990 international dollars

Half of all historical world energy consumption occurred in the last two decades. Today the world consumes energy at an average rate of 16,000 GW, or 2,500 W per person, compared to a primitive sustenance rate of about 200 W per person. The US uses 3,000 GW of average power, or about 10,000 W per person.

Individual energy production rate	Watts
Modern man	100
Primitive sustenance man	200
Man at hard labor	300
Water buffalo	350
Horse	750
Human average energy use rate	
World citizen	2,500
US citizen	10,000

Developing countries will consume more energy.

The table above shows how the industrial revolution vastly increased human energy use. It also illustrates demand for energy outside the US may quadruple.

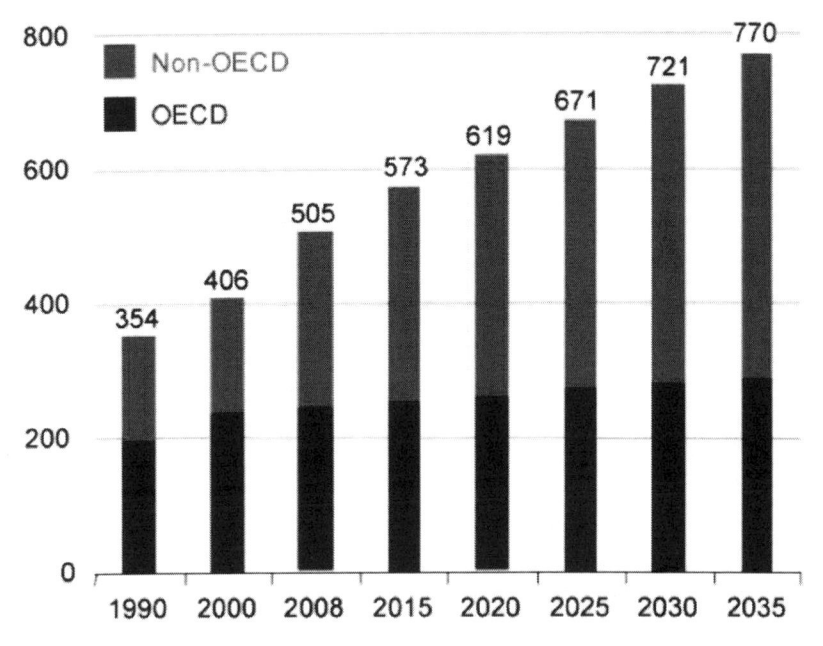

Projected world energy consumption in quads

The US Energy Information Agency (EIA) projects increasing energy consumption, particularly for the non-OECD, developing nations. The 34 member nations of the Organization for Economic Cooperation and Development are principally the world's leading economic democracies. The 2035 projection of 770 quads is an average rate of 25,000 GW, or about 3,000 W per person of a projected 8.3 billion people on earth.

Fossil energy use increased world atmospheric CO2.

In 1769 Watt patented his efficient coal-fired steam engine, which powered the industrial revolution and changed the world.

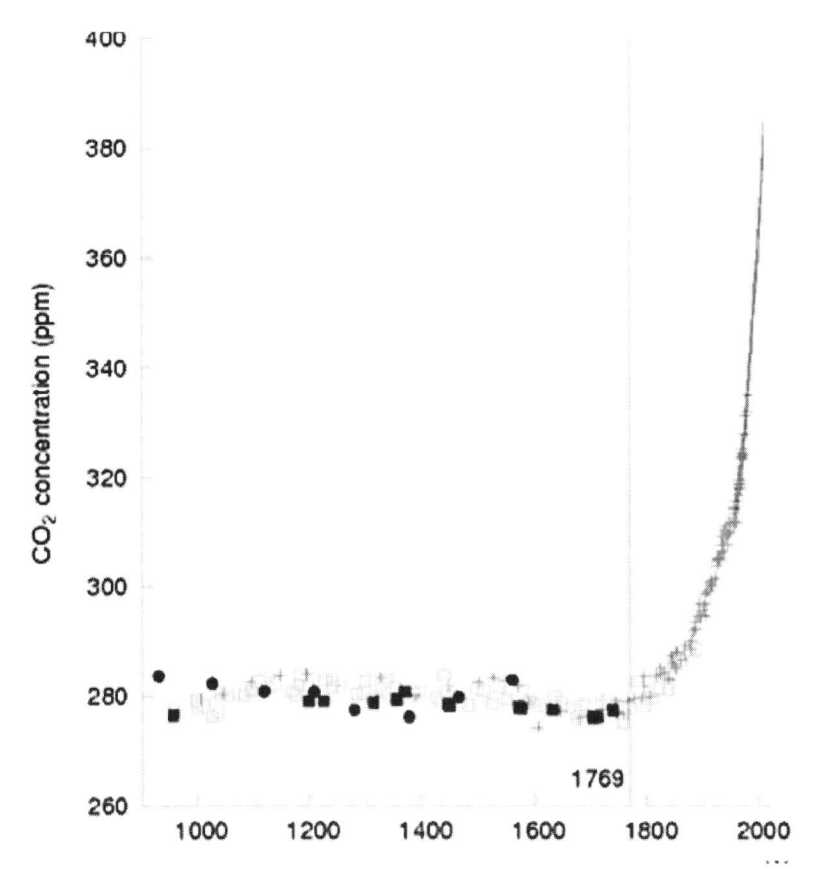

Atmospheric CO2 (parts per million) before and after 1769

Burning the coal emitted CO_2 into the air. The resulting increased atmospheric CO_2 traps infrared radiation much like a greenhouse, increasing earth's temperature. In 2012 CO_2 concentrations reached 400 ppm.

Energy related CO2 emissions will rise to over 30 Gt per year.

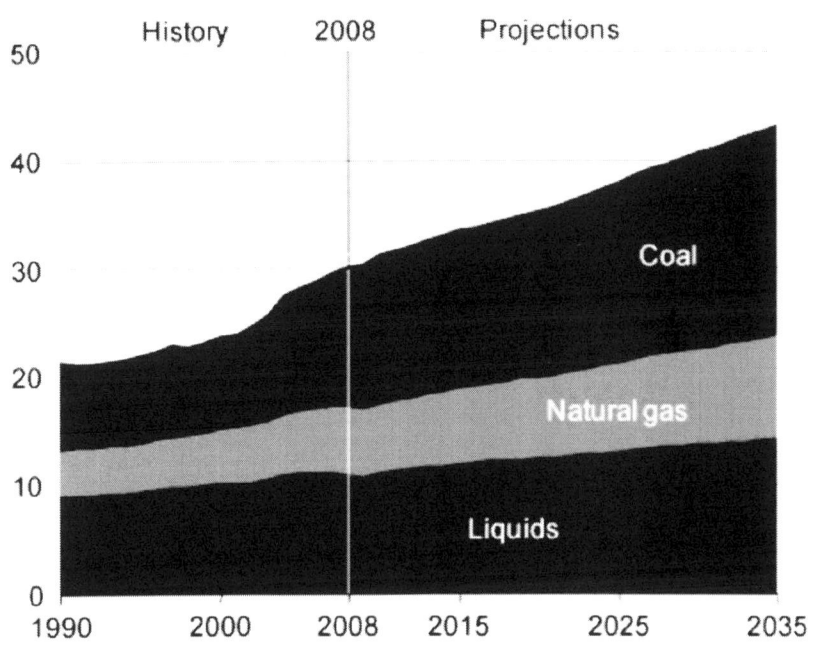

World CO2 emission in gigatonnes per year

The US Energy Information Agency projects continuing increases in the rate of CO2 emissions, worldwide. The EIA is an independent, professional organization that makes its best estimates based on data it collects and compiles, and based on current law and regulations. Unless some dramatic technical change occurs, CO2 emissions will continue to rise and accumulate ever faster in the atmosphere, adding over 30 gigatonnes (Gt) of CO2 each year. The mass of the atmosphere is 5,000,000 Gt, so this source annually contributes roughly 0.6 ppm (parts per million) to the current 400 ppm concentration.

ENERGY AND CIVILIZATION SUMMARY

Energy is the stuff of the universe, created at the Big Bang, continually expanding and cooling. Energy exists in several forms: mass, kinetic, electric, potential, and thermal. Although energy is conserved, it always degrades to thermal energy, the slow, unorganized vibrations of atoms. Here on earth, life borrows a stream of that energy for growth, self-replications, and motive action. The human life form first needed about 200 W of energy for primitive sustenance. Harnessing fire energy and farming food energy released time and effort for finding and eating food, allowing evolution of thought, social communication, and tools. Civilization evolved slowly until the Industrial Revolution, which harnessed the energy of burning coal. In today's advanced civilizations humans use energy at the rate of 10,000 watts.

3 An Unsustainable World

The earth's resources are finite.

Today the public is concerned with global warming and its terrifying implications for climate, water, agriculture, food, life, and civilization. But our global problems are much worse than simply climate change. We are running low on petroleum, which fuels our transportation. Fresh water sources are drying up as we pump out aquifers, irrigate deserts, and divert it to industrial processes such as natural gas drilling or extraction of oil from tar sands. Our coal plants spew particulates into the air, causing 34,000 respiratory deaths a year in the US alone. Worldwide, hunger for food results in 17,000 daily child deaths.

Limits to growth arise from finite resources.

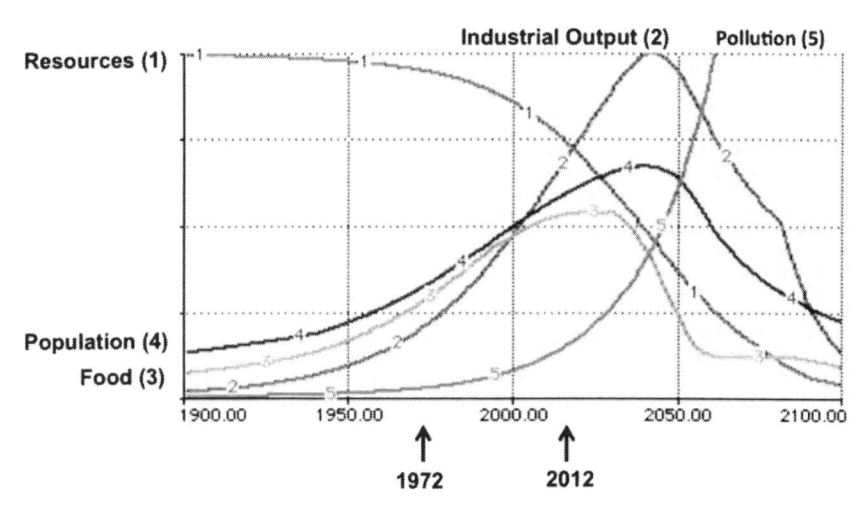

1972 model of the world economy

Forty years ago in 1972, Dennis Meadows' *Limits to Growth* modeled the effects of finite resources on the fate of the world. He projected that consumption of natural resources and rise of pollution from industry would diminish food and eventually population. [The spotty graph is from a 1972 teletype pin printer we both used at Dartmouth College's early computer systems.]

Meadows was rebuked by economists who pointed out that innovation and rising prices for resources have historically resulted in finding new resources and inventing new ways to increase economic productivity. New resources could be found at higher prices, but increased economic productivity would make them affordable. The world is finite, though. Since then the world has experienced the oil price shocks of the 1970s and now commodities price shocks as more energy is required for making iron, aluminum, corn, and other commodities more in demand from an increasingly demanding, expanding world population.

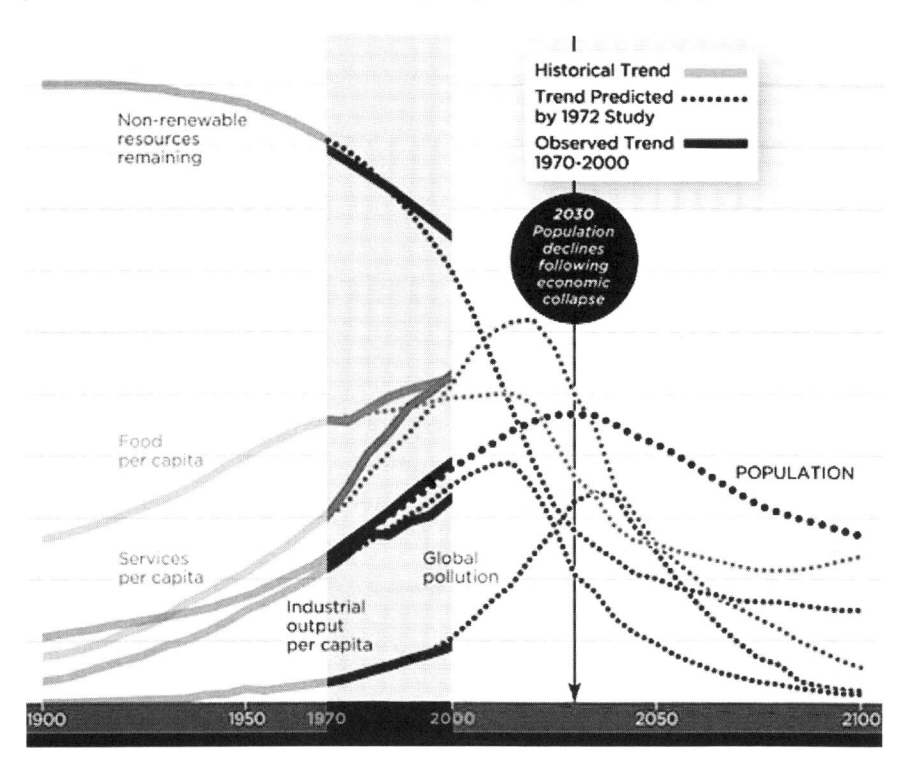

Comparison of limits-to-growth model to observations

Meadows' projections to date are consistent with observations, according to articles in *American Scientist* and *Smithsonian* magazine, whose graphic shows historical data in solid lines and projections in dotted lines, with a 30 year overlap.

But higher prices will not always secure new energy. Now economists are aware of EROI, energy return on energy invested. For example, obtaining energy from oil consumes energy for exploration, drilling, pumping, refining, transportation, distribution, and marketing. The ratio of the energy provided by the finished oil product to the energy used to obtain it is EROI.

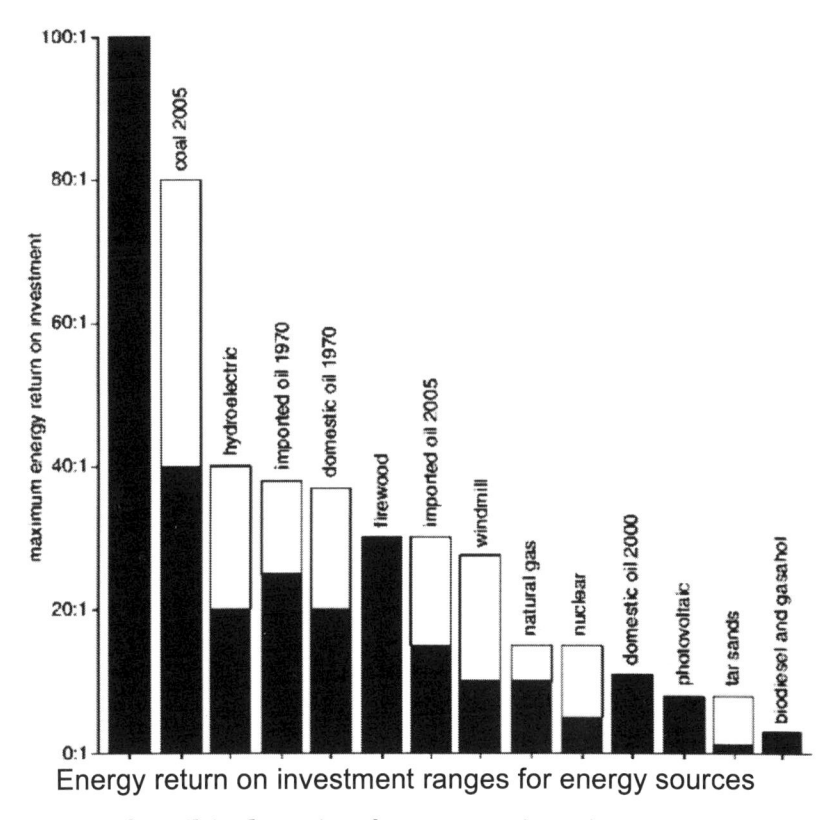

Energy return on investment ranges for energy sources

The EROI for oil is dropping from 100:1 (1930) to 40:1 (1970) to 14:1 (2000) to an estimated 5:1 (2009) for new exploration. Price is not the important limit, rather EROI is the hard stop limit, for when it drops to 1:1, we can get no more energy. The lighter parts of the bars represent the range of EROIs; for example coal EROI ranges from 40 to 80. The EROI for corn ethanol is already < 1 in many situations.

Resource depletion may be more severe than climate change.

Global warming is indeed a severe threat to our environment and human civilization. But resource depletion may be an even more immediate threat. Physicist Tom Murphy writes the blog, *Do the Math*, encouraging people to quantify the problems and envisioned solutions. In a 2012 interview with *OilPrice.com* he says:

> "I see climate change as a serious threat to natural services and species survival, perhaps ultimately having a very negative impact on humanity. But resource depletion trumps climate change for me, because I think this has the potential to effect far more people on a far shorter timescale with far greater certainty. Our economic model is based on growth, setting us on a collision course with nature. When it becomes clear that growth cannot continue, the ramifications can be sudden and severe. So my focus is more on averting the chaos of economic/resource/agriculture/distribution collapse, which stands to wipe out much of what we have accomplished in the fossil fuel age. To the extent that climate change and resource limits are both served by a deliberate and aggressive transition away from fossil fuels, I see a natural alliance."

Population is stable in developed nations.

World population is projected to grow from 7 billion to over 9 billion people. Most of this growth is in the developing nations. The US and other economically strong OECD nations have little population growth, attributable to immigration from the developing nations.

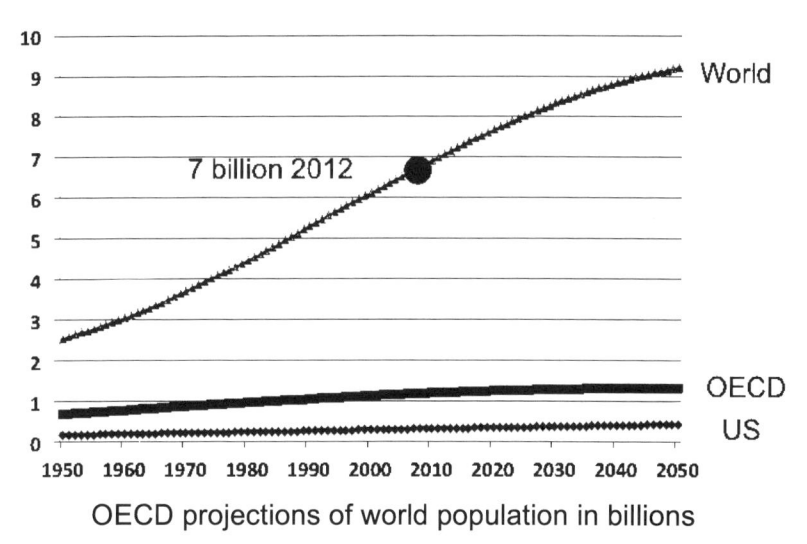

OCED projections of world population in billions

Increasing population will increase the demand for resources of food and energy. Increased demand leads to increased competition and possible conflict.

Impoverished countries birth the most children.

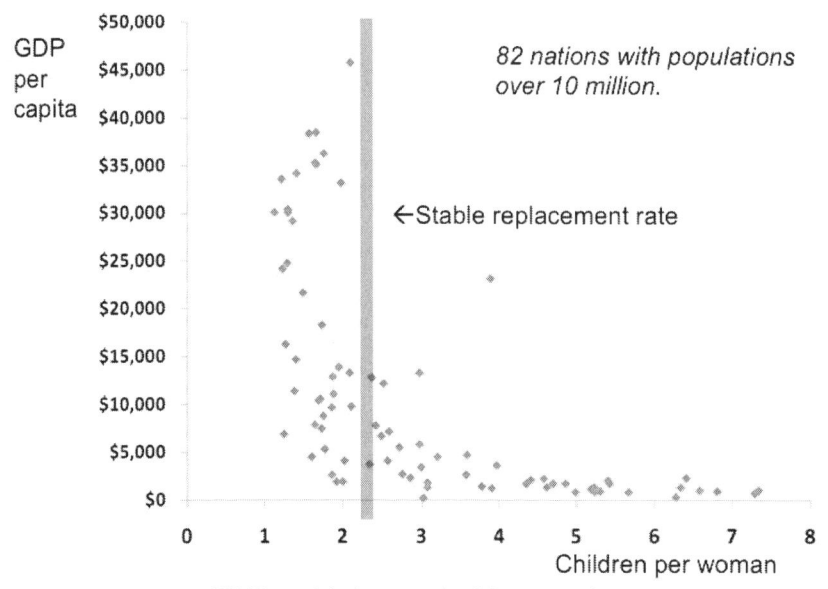

[Scatter plot: GDP per capita ($0–$50,000) vs Children per woman (0–8), showing 82 nations with populations over 10 million. A vertical bar marks the stable replacement rate at approximately 2.3 children per woman.]

GDP vs birthrates in 82 countries

This scatter plot uses data from the 2008 CIA world fact book. Each point corresponds to one nation, relating average number of children born to each woman and GDP per capita – closely related to income. It demonstrates that countries with high GDP per capita have birthrates that lead to a sustainable population. All the countries to the left of the vertical bar would have diminishing populations, except for immigration.

With increased income, there is less need to have children to work in agriculture, or to care for aging parents. There is less need to give birth to extra children to compensate for childhood deaths. With work saving technologies such as water pumps, efficient cook stoves, and washing machines, women are freed from constant labor. They are able to have time for education and to earn money. With more independence and access to contraceptives, women can choose to have fewer children, as evidenced above.

AN UNSTAINABLE WORLD

Prosperity stabilizes population.

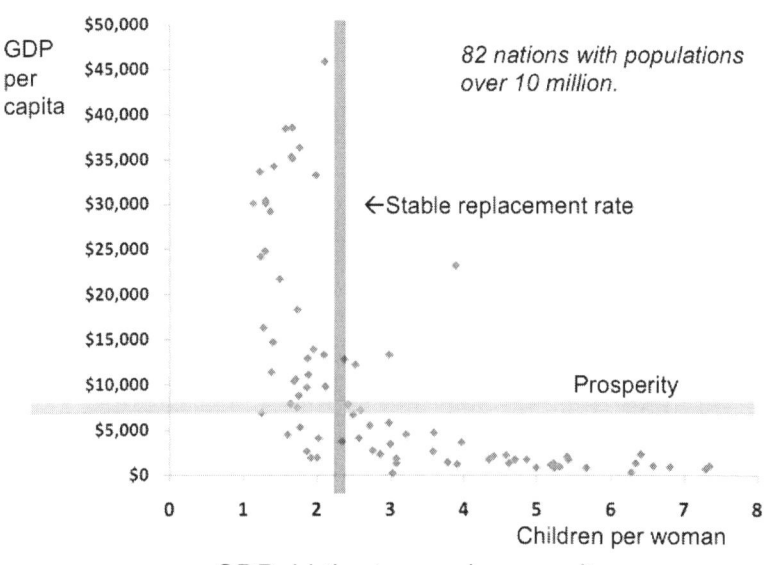

GDP, birthrates, and prosperity

In this same plot is added a horizontal bar at $7,500 GDP per capita, arbitrarily chosen and labeled "Prosperity". The poor nations, below $7,500, are those that have the highest birthrates. This strongly implies that improving the economic status of poor nations will lower birthrates, leading to a stable or shrinking world population. This plot cries out for a need to increase world prosperity to $7,500 GDP per capita, only 16% of the US number. With a stable or shrinking global population, world civilization can be sustainable.

At the *Wall Street Journal ECO:nomics* forum in March 2012 Microsoft founder and philanthropist Bill Gates remarked:

> "If you want to improve the situation of the poorest two billion on the planet, having the price of energy go down substantially is about the best thing you could do for them. ... Energy is the thing that allowed civilization over the last 220 years to dramatically change everything."

Prosperity depends on energy.

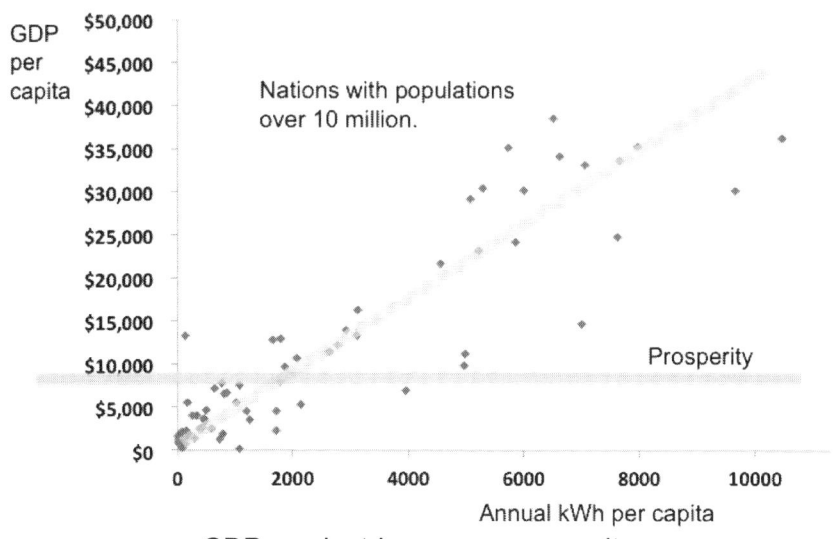

GDP vs electric energy, per capita

This plot, also with CIA data, shows the relationship between GDP and energy – specifically electric energy, measured in kilowatt-hours per capita per year. For our civilization, electric energy is the most valuable and useful form of energy. Unlike heat from fire, or power from falling water, electric power can be used for many purposes essential to economic development. Applications include water sanitizing and distribution, sewage processing, lighting, heating, refrigeration, air conditioning, cooking, communications, computing, transportation, food processing, medical care, manufacturing, industry, and commerce. These are all hallmarks of emerging prosperity.

Adequate electric power alone can not guarantee a prosperous economy and civilization without education, basic health care, rule of law, property rights, financial system, and good government. But electricity is essential for economic progress.

Over 1.3 billion people, 20% of the world population, have no access to electricity. Even rapidly developing nations such as India and South Africa can not provide full time electricity.

Electricity can power sewage processing systems, necessary to assure clean water. The World Bank says 2.6 billion people have no access to sanitation, leading to illness that reduces GDP by 6%. Diarrhea is responsible for more child deaths than AIDS, TB, and malaria combined. UNESCO reports that 8% of worldwide electric power is used for water pumping, purification, and wastewater treatment.

Clean water distribution is one example of how affordable, reliable power can free women from hauling water, helping to lead to a standard of living with time for education, gainful work, women's independence, and choices about reproduction.

The previous plot suggests an annual 2,000 kWh per capita supply leads to the $7,500 GDP per capita level that leads to sustainable birthrates and population. This minimum electric energy supply rate is 230 watts per person, about 16% of the US rate.

In summary, an economy with minimum electric power availability of 230 W per person is needed to achieve the modest prosperity level of $7,500 per person leading to a sustainable population.

In India today, average electric power consumption per capita is 85 W; 40% of the people have no access to electricity, and another 40% have access only a few hours per day. The long term goal of India's government ministers is 570 W per capita, compared to 1400 W in the US.

Energy use is growing rapidly in developing nations.

The developing nations understand the need for more electric power to increase the economic prosperity of their citizens. They have limited money to spend and must build affordable power plants with low fuel costs – coal-fired power plants. Energy annual demands on this chart are denominated in quads – quadrillions of BTUs per year. For comparison, the US uses about 100 quads of energy annually.

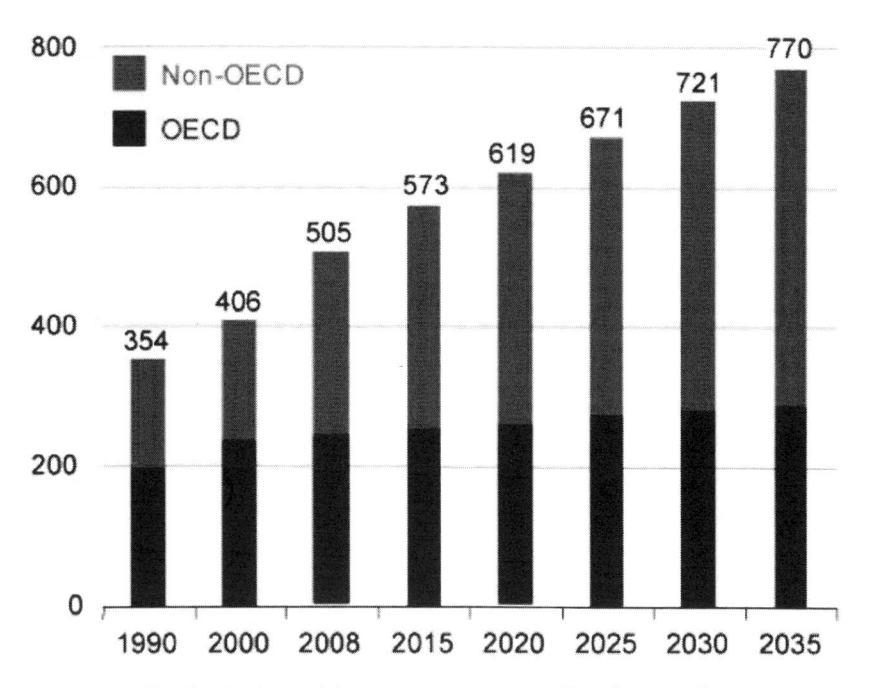

Projected world energy consumption in quads

The 34 OECD nations have a population of 1.2 billion people with an average GDP per capita of $34,000, adjusted for purchasing power parity. OECD is the organization of the world's wealthy nations; the non-OECD nations are developing nations. The OECD outlook is that "World energy demand in 2050 will be 80% higher ... and still 85% reliant on fossil fuel-based energy."

Coal burning is increasing sharply in developing nations.

China and India, with large populations, are driving the projected increases in coal use.

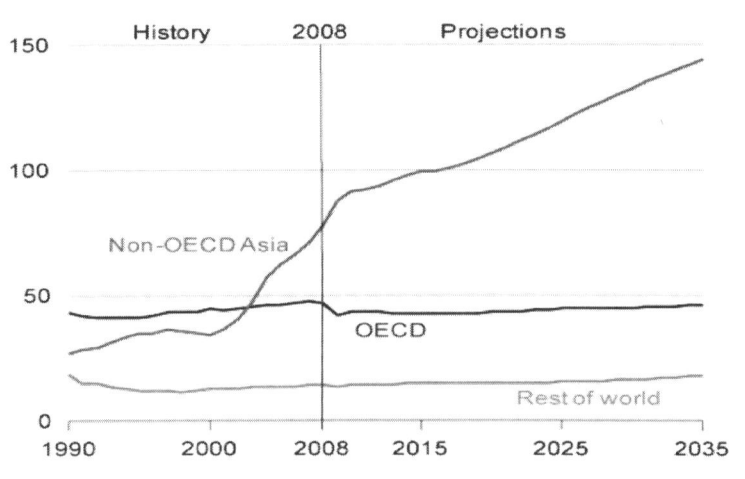

DOE projections of annual energy consumption, quads

Global carbon dioxide emissions are rising.

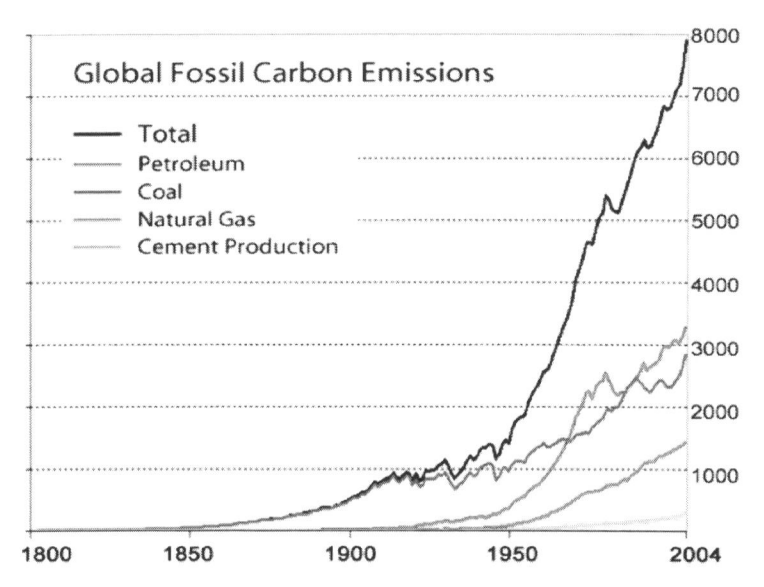

World CO2 emissions, reported as millions of tonnes of carbon

The chart shows 2004 total emissions as 8,000 million tonnes (8 Gt) of carbon, equivalent to 29 Gt of CO2. The bottom, cement production line includes the coal, heavy crude, and natural gas used to fire the kilns that make the world annual production of 3.3 Gt of cement used for making concrete for construction, mostly in China. After a recessionary dip, annual CO2 emissions continue to rise, up 5% in 2010 to 30.6 Gt.

GLOBAL WARMING

Global temperatures are rising.

The US National Oceanic and Atmospheric Administration has tracked mean monthly temperatures for over a century. Their graph illustrates the change in temperature relative to the last century average. The vertical scale is temperature in °C. Temperatures have risen about one degree in the last century.

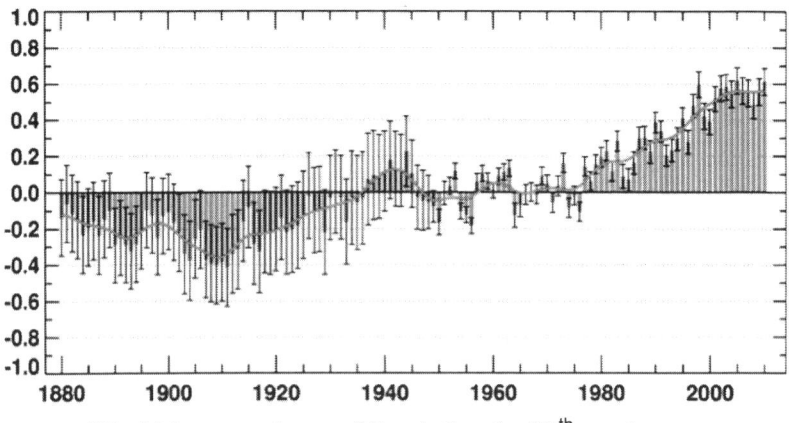

World temperatures, °C relative to 20th century average

Carbon dioxide emissions increase global warming.

The following chart from climate scientist James Hansen shows the history of CO2 emissions, methane emissions, and temperature. The horizontal scale is thousands of years before

1850, at "0". The scale of time from 1850 to 2000 is expanded, by 400:1, to show in detail civilization's rapid effects since the industrial revolution. The units for CO2 are parts per million, for CH4 are parts per billion, and for T are degrees °C relative to last century's average.

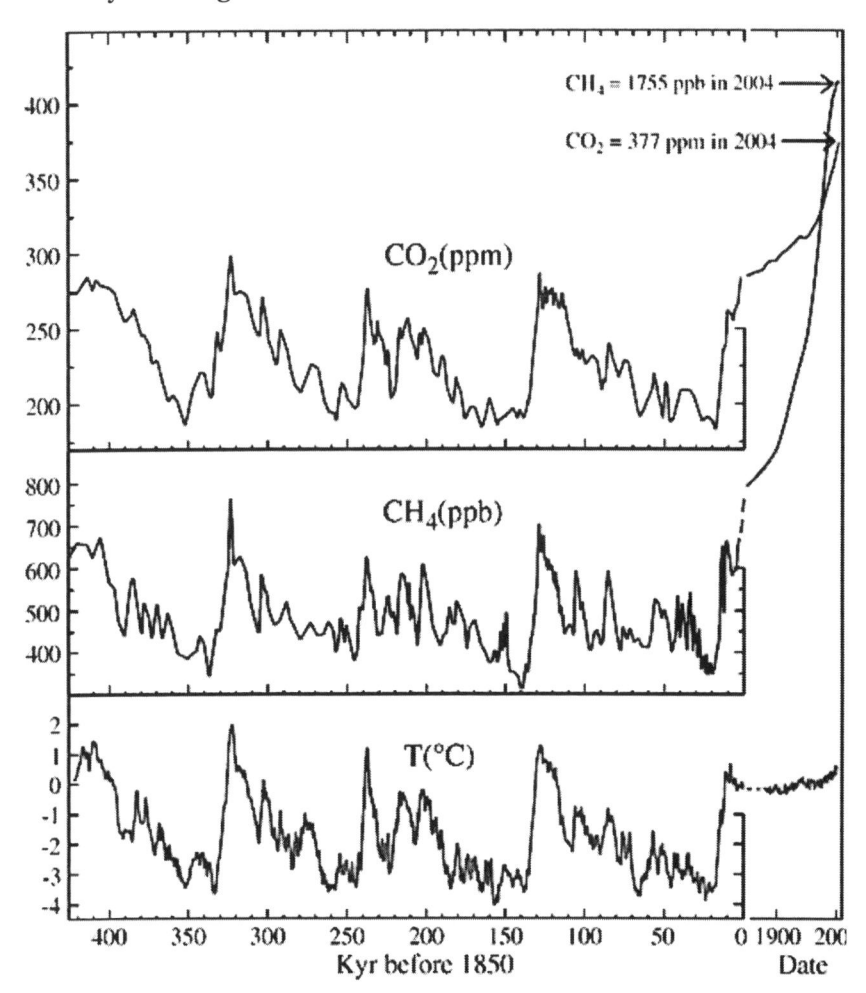

Atmospheric CO2 and CH4 concentrations, before and after 1850

CO2 and T are strongly correlated, implying that the recent rapid increase in atmospheric CO2 will force a rise in temperature T. Correlation is not causation, so climate models were used to compute how the CO2 greenhouse effect changes the world climate.

Much of the sun's energy reaches earth as visible light, which passes through the transparent atmosphere and heats the earth. Though much cooler than the sun, the earth does re-radiate its heat as less energetic, infrared radiation (IR). The atmosphere is not so transparent to IR, so the atmosphere absorbs IR and heats up the earth. The amount of absorption depends on the amounts of H2O, CH4, and CO2, each of which absorbs energy differently.

Computer models were developed to simulate the earth's climate. These are complex, taking into account many factors that affect the earth temperature. This chart, from climate scientist James Hansen, illustrates some factors that change the normal balance of energy radiated, absorbed or reflected by the earth.

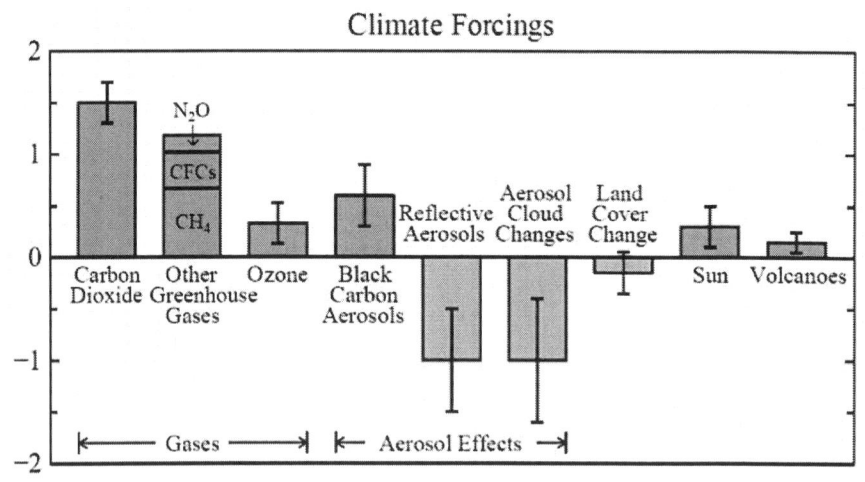

Climate Forcings

Factors included in computer models of climate

- Greenhouse gases such as CO2, chlorofluorocarbons, methane, and ozone increase infrared absorption, warming the earth.
- Black carbon soot from inefficient stoves in developing nations also increases absorption and warming.
- Aerosols from SO2 and NO2 reflect incident light, cooling the planet.

Contrasting to criticisms, this chart shows that the climate models do in fact take into account many factors that affect global temperatures. There is consensus by scientists that the computer models are sufficiently accurate to illustrate (1) the planet is warming, and (2) human civilization's atmospheric emissions are a big cause.

IPCC climate modeling projects a warming earth.

IPCC is the Intergovernmental Panel on Climate Change.

"For the next two decades a warming of about 0.2°C per decade is projected for a range of SRES emissions scenarios. Even if the concentrations of all GHGs and aerosols had been kept constant at year 2000 levels, a further warming of about 0.1°C per decade would be expected."

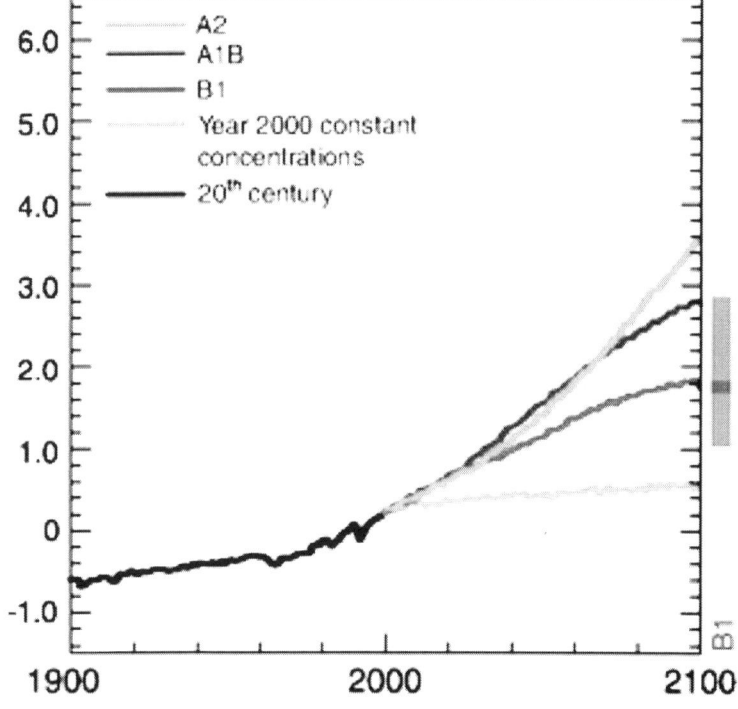

IPCC projections of world average temperature change, °C

The IPCC made several projections of world average temperature, relative to year 2000, based on differing civilization scenarios. For example, the B1 scenario is a world of very rapid economic growth, a global population that peaks in mid-century, with rapid introduction of new and more efficient technologies and changes in economic structures toward a service and information economy. The B1 line, ending at the error bar, forecasts a 1.8°C increase in temperature in 2100. The bottom line forecasts temperatures if CO2 in the atmosphere does not rise any more at all. The IPCC makes no prediction of which scenario is most likely, but they all project global warming, varying with CO2 emissions.

Unchecked global warming will end life as we know it.

One of the biggest effects of rising temperatures will be the melting of sea ice and glaciers. Sea levels may increase by as much as 1 meter by 2100. The loss of habitat symbolized by the iconic polar bear will affect many other arctic animals, such as seals and walruses. Agriculture in India and other places depend on water from rivers sourced by summer glacier melting.

Rongbuk glacier change

The Rongbuk glacier in the Himalayas all but disappeared between 1968 and 2007. As such glaciers vanish they will not provide seasonal melt-water to rivers used for irrigation for growing food in the dry season, possibly causing famine for hundreds of millions of people.

Animals change locales and habits to adapt to the climate; for example, spruce bark beetles have thrived in now-warmer Alaska, destroying 4 million acres of spruce trees. Weather will become more extreme; hurricanes and storms will be more common; floods and droughts will become more common.

Changes to life in the ocean will also be dire. Ocean life thrives in cold water; Caribbean water is blue and clear because it has less life than temperate and polar oceans. Algae, the start of the ocean food chain, require cold, polar water to grow, and this cold area that is the source of ocean life is shrinking. Depletion of dissolved oxygen is causing more dead zones. Warming water causes corals to expel symbiotic algae, then bleach and die.

Bleached Indian Ocean coral

Temperature is not the only problem. Carbon dioxide from the atmosphere slowly dissolves in the ocean, making it more acidic. Scientists report that today's 29% increase in dissolved hydrogen ions since the Industrial Revolution will rise even more. Dissolved CO2 depletes the carbonate ions that corals, mollusks, and some plankton need for reef and shell building. By mid-century this will threaten survival of shellfish and the marine food chain.

Fossil fuel burning kills 34,000 US citizens per year.

Air pollution is a more immediate problem than global warming. Many US coal plants have installed scrubbers and other equipment to reduce the toxic waste spewed into the atmosphere. The Clean Air Task Force has lobbied for legislation and regulation to reduce the death rate from breathing atmospheric pollutants to 13,000 per year in the US.

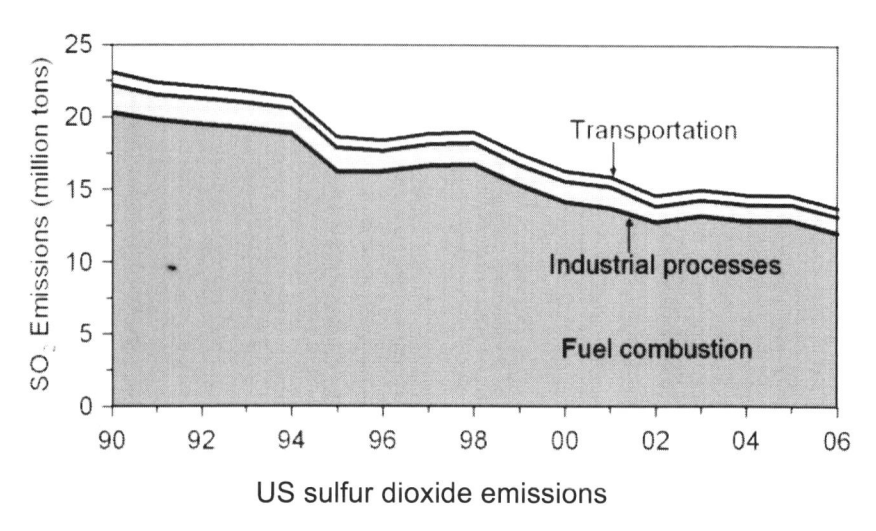

US sulfur dioxide emissions

The US Environmental Protection Agency has tracked the slow, steady reduction in one pollutant, sulfur dioxide. In an abrupt policy change on July 7, 2011 the US EPA issued a very restrictive, contested rule to reduce emissions to a much safer level.

> "The Cross-State Air Pollution Rule will protect communities that are home to 240 million Americans from smog and soot pollution, preventing up to 34,000 premature deaths, 15,000 nonfatal heart attacks, 19,000 cases of acute bronchitis, 400,000 cases of aggravated asthma, and 1.8 million sick

days a year beginning in 2014 – achieving up to $280 billion in annual health benefits."

The EPA estimate is 13,000 to 34,000 deaths per year. Most of these deaths arise from sulfur dioxide from coal plant flue gases that nucleate fine particles (< 4% the diameter of a hair) that are inhaled. Nitrogen oxides and mercury are two other fatal pollution contributors. EPA estimates the annual economic benefit from reducing emissions to be 120 to 280 billion dollars.

Air pollution is even worse in China, where hundreds of thousands of people annually die prematurely from respiratory disease from coal burning. Worldwide, the UN estimates a death rate over 1 million per year from carbon particulates from all sources.

The March 2012 OECD outlook projects that with business-as-usual policies

"Urban air pollution is set to become the top environmental cause of mortality worldwide by 2050, ahead of dirty water and lack of sanitation. The number of premature deaths from exposure to particulate air pollutants leading to respiratory failure could double from current levels to 3.6 million every year globally, with most occurring in China and India."

Shipping emits more air pollution than all the world's cars.

Just the 15 largest container ships emit as much air pollution as the world's 760 million cars. Large ship diesel engines are

powered by refineries' residual oil, essentially asphalt that contains 2000 times the sulfur of automobile diesel fuel. The 2300 ton engines generate up to 90 MW of power while burning 16 tons of fuel per hour. Shipping has increased as China has become the largest manufacturing country. The industry consumes 7 million barrels of fuel per day. The entire ocean fleet annually emits 20 million tons of SO2. Shipping is responsible for 18-30% of the world's NOx (nitrogen oxides) pollution, 9% of SOx pollution, and 4% of all climate change emissions.

The US EPA is working to reduce costal ship emissions, causing 12,000 to 31,000 premature deaths, 1.4 million work days lost, and from $110 to $270 billion dollars of health care costs.

The US is addicted to imported oil.

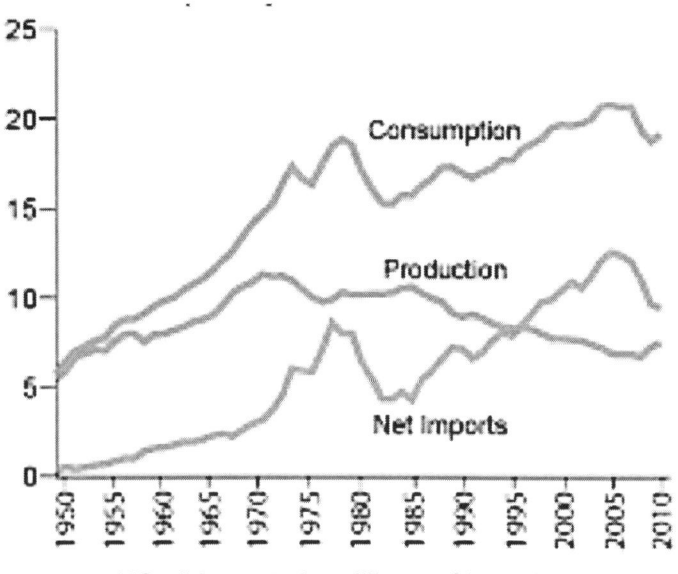

US oil imports in millions of barrels per day

This graph illustrates US oil imports of approximately 10 million barrels of oil per day. With oil at $100 per barrel, this cost of $1 billion per day affects our balance of trade payments, increasing our trade deficit by roughly $365 billion per year, a dominating fraction of the total US trade deficit of $500 billion. Cumulatively, that trade deficit amounts to $10 trillion -- money borrowed to import foreign oil and other goods. In the future the US must export a net of $10 trillion of goods and services to pay the money back.

The US imports half of the petroleum it consumes. The rankings of major suppliers of imported oil change with market conditions:

Canada	25%
Saudi Arabia	12%
Nigeria	11%
Venezuela	10%
Mexico	9%

Other oil suppliers are Colombia, Iraq, Ecuador, Angola, Russia, Brazil, Kuwait, Algeria, Chad, and Oman. There is a robust, efficient, international market for petroleum, so the US is not specifically dependent upon any one foreign source. However short-term demand is inelastic, and worldwide production capacity only slightly exceeds demand, so disruptions in oil supply can cause shortages and price spikes.

Reduced resources and increased population spark conflict.

The invasion of Kuwait by Iraq in 1990 was an attempt to seize possession of one of the world's largest energy sources. The Kuwait oil fields constitute 8% of the entire world's oil reserves. As the defeated Iraqis withdrew, they set fire to 700 oil wells, burning 6 million barrels of oil per day over ten months, causing widespread pollution.

Pentagon studies conclude the greatest danger posed by climate change is not the degradation of ecosystems per se, but rather the disintegration of entire human societies, producing wholesale starvation, mass migrations, and recurring conflict over resources.

Current world energy flows will not sustain civilization.

The current pattern of sources and uses of energy can not be sustained in a civilized world. In summary:

Population	World population is rising, especially in poor nations.
Energy poverty	Over 20% of the world population has no access to electricity -- critical to achieving even modest prosperity. Liquid fuels are needed for transportation for commerce and industry.
Energy growth	Developing nations are increasing energy demands to enable economic growth and individual prosperity.
Coal burning	Burning coal for electric power is the least expensive way for developing nations to generate electric power. Coal plant construction continues even in OECD nations.
CO2 emissions	World coal burning dumps 31 Gt per year of CO2 into the atmosphere, more than from petroleum burning.
Temperature	World temperatures are rising from excess man-made atmospheric CO2 causing planetary changes in terrestrial and ocean life, fresh water supplies, and the ability to produce food.
Pollution	Coal plant emissions spew particulates into the air, responsible for 34,000 deaths per year in the US and over 1 million deaths globally.
Oil	Petroleum is critical to transportation, and worldwide demand is increasing. The US is the biggest importer of this diminishing resource, increasing its trade deficit by a third of a trillion dollars annually.
Conflict	A growing world population, increasing demand for shrinking resources, stresses from pollution, and resulting social unrest lead to war.

Carbon taxes are not a global solution.

An oft-proposed political solution to stop global warming is to impose a tax on CO_2 emissions by all emitters. In concept, the price of electricity from a coal-fired power plant does not include externalities – the costs of the damage to the environment by dumping combustion emissions into the atmosphere. Economists try to compute the damage and propose CO_2 taxes in the range of $40-100 per tonne, increasing the cost of such energy. The cap-and-trade variant of the carbon tax similarly increases energy costs affecting economic productivity. Similar attempts to displace CO_2 generating fuels are feed-in-tariffs and renewable portfolio standards mandates, which require utilities to buy some renewable fuels at high prices.

The US and other nations have unsuccessfully attempted to tax emissions. Although Europe has experimented with cap-and-trade forms of taxation, CO_2 emissions are rising there. Prior to the Kyoto Treaty, the US Senate voted unanimously not to approve any emissions-limiting treaty that exempted developing nations.

The United Nations Framework Convention on Climate change sponsored international climate change meetings in Kyoto, Copenhagen, Tianjin, Cancun, Bangkok, Bonn, Panama, and Durban without consensus on how to impose carbon taxes or reduce CO_2 emissions, which continue to rise. Tens of thousands of people attended each of these meetings. It's hard to conceive how all the world's nations would agree to a treaty against their individual self-interests.

Developing nations argue that the prosperity of the wealthy OECD nations has come about through inexpensive energy from burning fossil fuels, raising current CO_2 concentrations to 400 ppm. They argue that they should have the same opportunity to increase CO_2 for energy that would propel their prosperity to an OECD-like GDP, *per capita*.

This bar chart illustrates China's argument.

- United States
- China

China is four times as populous as the United States'
- United States: 307 million
- China: 1,339 million

China's GDP per-person is around one eighth of the United States'
- United States: $45,592
- China: $5,383

China's per-person GHG emissions are around one quarter of the United States'
- United States: 19.2 tons
- China: 4.9 tons

China's historical contribution to climate change (i.e. cumulative historical emissions) is around one third of the United States'
- United States: 29.00%
- China: 8.62%

China's per-person historical contribution to climate change is less than seven percent of the United States'
- United States: 1,125.6 tons per person
- China: 76.0 tons per person

China Daily News, October 7, 2010

CO2 emissions growth accelerated in 2012.

NASA photos show the Greenland surface ice sheet melted on July 12, 2012, going from 60% to 3% ice cover in 4 days.

In 2012 the Organization for Economic Cooperation and Development published dire warnings about the continuing rise is world CO2 emissions. The largest growth is expected in the power generation sector, where emissions are expected to rise from 10 Gt in 2012 to 18 Gt in 2050 – an 80% increase. Graphs from this OECD report follow.

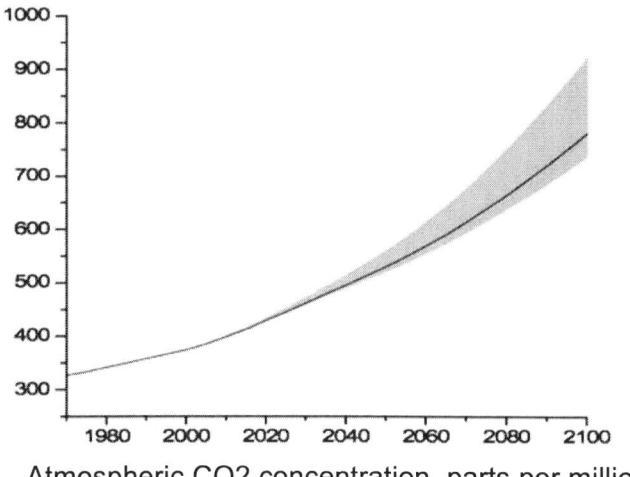

Atmospheric CO2 concentration, parts per million

At this rate, atmospheric concentrations will reach 685 ppm by 2100, compared to 450 ppm which most climate scientists say is the maximum climate-stable concentration. OECD predicts average temperatures will rise from 3 to 6°C at 685 ppm – a catastrophe.

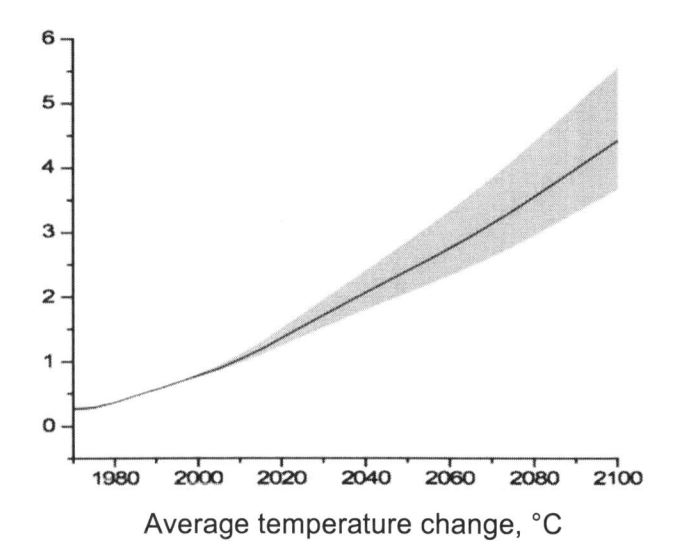

Average temperature change, °C

NEW ENERGY TECHNOLOGY

New energy technology can solve our environmental issues.

Prof. Jeffrey Sachs, Director of the Columbia Earth Institute and advisor to the UN Secretary-General, is an economist who advocates new energy technologies over carbon taxes. He writes in *Scientific American,*

> "Technology policy lies at the core of the climate change challenge... If we try to restrain emissions without a fundamentally new set of technologies, we will end up stifling economic growth, including the development prospects for billions of people... We will need much more than a price on carbon ... technologies developed in the rich world will need to be adopted rapidly in poorer countries."

Nations resist carbon taxes that would increase the cost of energy from burning coal, because the taxes will impede economic development. Much of the contention in attempted climate treaty negotiations is from proposals for OECD nations to pay billions of dollars to developing nations to help them reduce their current and future net CO_2 emissions.

There is a better solution – energy cheaper than coal. If new technologies such as the liquid fluoride thorium reactor undercut coal economics, nations will forego coal power plants in their own economic self-interest. There is a clear economic tipping point here, set by the cost of coal electricity. Success is a new energy technology that provides power below this coal price point. Contentious international treaty negotiations and economically burdensome taxes will not be needed

The liquid fluoride thorium reactor is potentially that new energy source, cheaper than coal, that dissuades all nations from burning coal, without carbon taxes, and simultaneously improves economic productivity.

New technology makes clean energy, cheaper than coal.

New energy technology solves more problems than just global warming. Some people are still skeptical that man-made CO_2 emissions are responsible for global warming. They are concerned that increasing energy costs will harm the US economy. Moreover they are concerned that international treaties might disadvantage the US and other OECD nations, by exempting developing nations from emissions constraints and by paying them to avoid CO_2 emissions.

There are multiple reasons to develop an energy source cheaper than coal. Any one of these reasons can justify the investment in developing a solution such as the liquid fluoride thorium reactor.

- Stopping particulate air pollution will save million of lives.
- Lowering energy costs will increase economic productivity.
- Ending energy poverty leads to a sustainable population.
- Reducing CO_2 emissions will check global warming.

Even climate skeptics should support advanced energy technology for improved economic productivity, population sustainability, and improved human health.

In the US conservative Republicans and liberal Democrats bicker over impairing economic growth by imposing taxes to address global warming. Both sides should agree to an energy technology that improves both the environment and productivity.

Stopping particulate air pollution will save million of lives.

The US Environmental Protection Agency estimates that 34,000 lives could be saved annually by stopping particulate air pollution. For making environmental analyses, EPA assigns a value of a human life at $7.9 million. Multiplying these EPA numbers gives an annual savings of $267 billion in the US alone.

The emissions of coal plants include SO_2 and NO_2, which interact with water in the atmosphere to form aerosol particulates, many less than 2.5 microns in diameter, which are the respiratory health hazard.

Hundreds of thousands of lives would be saved annually by cleaning China's dirty air. Enough prosperity to replace primitive cook stoves would stop their carbon soot emissions, bringing annual lives saved to over one million.

Lowering energy costs will increase economic productivity.

Electric energy costs are a component of all goods and services. Reducing that cost improves economic productivity. 500 GW-years at today's approximate cost of 5 cents per kWh amounts to $200 billion out of the US GDP of $15 trillion. Dropping that production cost by 2 cents would free up ½% of GDP for other uses. Compare this improvement to the economic damage that would be caused by *raising* the cost of electric energy with taxes or with expensive wind and solar power.

Ending energy poverty leads to a sustainable population.

Over a billion people have no access to electricity, a key to economic development and a lifestyle in which women have chore-free time, are educated, work, gain independence, and make their own decisions about reproduction. Providing developing nations with affordable energy can help them reach such goals. Even rapidly developing nations such as India and South Africa can not provide full time electricity. The world's prosperous nations generally have a sustainable or diminishing populations. A sustainable population reduces natural resources competition and causes for war.

Reducing CO2 emissions will check global warming.

Carbon dioxide emissions are increasing global warming and destroying the environment. Electric power generation from burning coal is the largest source of CO2 emissions, worldwide. The most effective way to start reducing CO2 is to stop burning coal in power plants.

LIQUID FLUORIDE THORIUM REACTOR

This book is about the potential for the liquid fluoride thorium reactor (LFTR) to provide energy to address humanity's crises of global warming, energy poverty, prosperity, and resource conflict. This disruptive technology for energy cheaper than coal is described in more detail in Chapter 5.

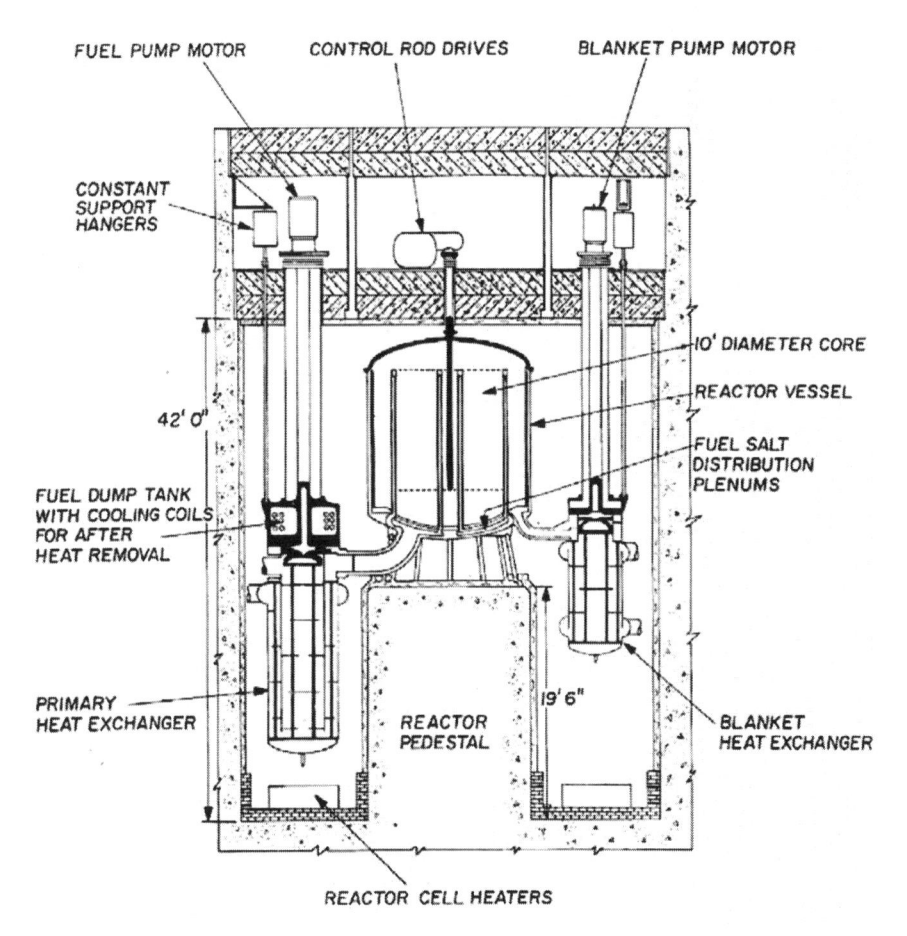

ORNL liquid fluoride thorium reactor, ~1975

Oak Ridge National Laboratories developed the liquid fluoride thorium reactor concept in the 1970s. New interest in LFTR has developed because of these characteristics.

Liquid: The fuel in this reactor is dissolved in molten salt. This liquid fuel form can be continuously circulated through the reactor, allowing complete fuel burn-up with continuous processing of the fuel and continuous addition of new fuel.

Fluoride: Fluoride salts are the most chemically stable elements on earth. They don't change under high temperature or high radiation. They lock up dangerous radioactive materials chemically to prevent them from being released to the environment, even in a severe accident. The fluoride salts stay liquid at high temperature, at normal atmospheric pressure.

Thorium: Thorium is an abundant natural nuclear fuel, found in literally every country on earth. It is so energy dense that every nation can be energy-independent.

Reactor: This new, high-temperature nuclear reactor is safe; fuel can't melt down because it is already molten. Any leaking radioactive salt would solidify in place. It costs less to build because it is efficient and compact. It costs less to run because thorium is relatively cheap and plentiful.

LFTR is a type of molten salt reactor (MSR). Different MSR designs may keep thorium and uranium separate or together, or use chloride salts, or burn just uranium or plutonium fuel, or burn spent fuel from conventional water-cooled nuclear reactors.

LFTR energy technology is cheaper and better.

LFTR is a new energy technology, far better than we have today, that

- produces electricity *cheaper* than from coal,
- is *inexhaustible*,
- provides energy *security* to all,
- reduces *waste*,
- is *affordable* to developing nations,
- synthesizes vehicle *fuel*,
- is walk-away *safe*.

Advanced nuclear power sources such as the liquid fluoride thorium reactor can satisfy these requirements. These benefits will be validated in Chapter 5. Following is a summary.

LFTR produces electricity cheaper than from coal

Small, modular LFTRs can be factory-produced. Capital costs for LFTR electric power plants can be about $2/watt. Recovering capital expenses will cost about 2 cents per kWh for a plant operating 90% of the time with money borrowed at 8%. Thorium fuel cost is insignificant compared to coal costs. LFTR can produce power at about $0.03/kWh, cheaper than coal.

Energy from LFTR is virtually inexhaustible.

Thorium is energy-dense and as plentiful as lead. All US electric power could be generated with just 500 tons per year. Just one single mining claim in Lemhi Pass has enough to power the US for 500 years.

Thorium fuel provides energy security to all nations.

Thorium is found all over the world. Every nation has enough to power its own needs, providing energy security for all.

LFTR produces little waste.

The amount of long-lived radiotoxic waste generated by LFTR is < 1% of that from today's nuclear power plants. LFTR can even consume long-lived radioactive transuranic elements in LWR spent fuel.

LFTR is affordable to developing nations.

Because the LFTR can be produced in small modular units for as little as $200 million, LFTRs can be purchased by developing nations that can not afford $5 billion investments for the advanced nuclear power plants such as are now being installed in China and the US.

LFTR can synthesize vehicle fuels.

The 700°C high-temperature heat produced by the reactor enables process heat technologies that can dissociate water, producing hydrogen, which can be a feedstock for producing synthetic fuels to replace gasoline and diesel.

LFTR is walk-away safe.

LFTR requires no external electric power to provide passive cooling. Radiotoxic fission products such as cesium and strontium are nonvolatile fluorides kept within LFTR salt.

LFTR can zero world coal power plant emissions.

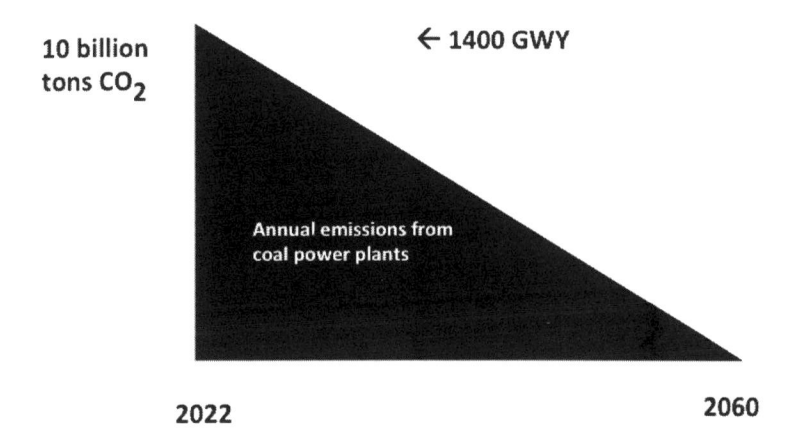

LFTR-reduced CO2 emissions coal power plants

Daily production of 100 MW LFTR power plants can replace all the world's coal power plants by 2060. World coal plant electric energy production is about 1400 GW-years annually. Each year's production adds approximately 10 billion tons (10 Gt) of CO2 to the atmosphere. This can eliminate the single largest global source of this gas that drives global warming.

The benefits of LFTR will be well worth the cost.

LFTR development cost is estimated to be near $ 1 billion, to develop the design and a working prototype. These R&D investments might be made by a government with results available to capable industry. The conversion of the prototype to a complete design for mass production would be considerably more – perhaps $5 billion invested by nuclear industry participants.

The 100 MW LFTR units, costing as little as $200 million each in mass production, might be manufactured and first sold in the US, with later potential for export. Daily sales of $200 million would amount to a $70 billion export-oriented industry, potentially improving the US balance of trade deficit. China may well compete in this market.

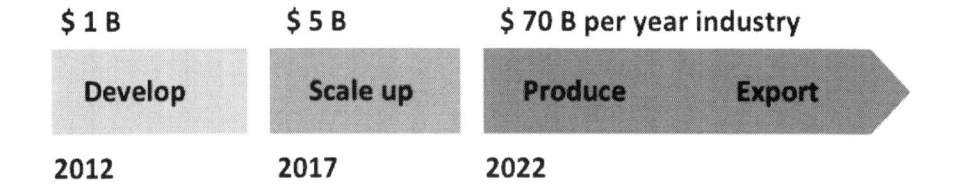

$1 B	$5 B	$70 B per year industry
Develop	Scale up	Produce — Export
2012	2017	2022

- *Cut 10 billion tons/year CO2 emissions to zero by 2060.*
- *Avoid carbon taxes.*
- *Stop deadly air pollution.*
- *Improve developing world prosperity, and check growth.*
- *Use inexhaustible thorium fuel, available in all nations.*
- *Walk away safe.*

In the next chapter we review the various sources of energy the world uses, in preparation for a more detailed discussion of the liquid fluoride thorium reactor solution in Chapter 5.

4 Energy Sources

The title of this book, *THORIUM: energy cheaper than coal*, is an incomplete description of the full objective. To fully meet the objective of checking CO2 pollution, thorium energy needs to be cheaper than all fossil fuel energies, including natural gas. Thorium energy, less expensive than energy from wind and solar sources, can avoid the economic blow from substituting renewable sources that produce electric energy at four-times higher cost. Of the renewable sources, only hydro is economically competitive, but there are few suitable hydro sites remaining. This book analyzes many such energy sources used to generate electric power.

We generate electricity from a diversity of energy sources today – fossil fuels, nuclear power, and renewable energy. This section explores the character of each and discusses how each could be used in a future, sustainable world.

I recently led a Dartmouth ILEAD (Institute for Lifelong Education at Dartmouth) class to visit many different electric power plants. All in the class said it was worthwhile to be able to understand the physical scale of the different sources. I encourage the reader to make similar visits, or possibly take a virtual tour using the class website.

People sometimes describe various power plants as dirty, or clean, or greedy, or polluting, or unsafe, or renewable. In our visits we were impressed with the professionalism and pride of the managers of all these plants. All operators strive to run their plants economically and efficiently within the bounds of law and regulation under which they operate. Labeling power sources with emotion-arising words is not helpful to solving our energy and climate crises. Understanding and analyzing the kinds of power sources and their costs is key to the solution. This is the purpose of this section of the book.

ENERGY DEMAND

To analyze energy demands we consider two kinds of energy: electric energy and thermal energy, which the EIA often terms primary energy. Electric energy is expressed as power times time. In one year a big 1 GW(e) power plant generates 1 GW(e)-yr of electric energy.

Thermal energy is often stated in quads (quadrillion BTU); we'll convert to GW(t)-yrs to simplify comparison to electric energy. Thermal energy is used for heating, fueling vehicles, making cement, running refineries, and also heating the boilers that make steam to generate electricity. The efficiency of conversion of thermal energy to electric energy varies with the power plant; 33% is typical. Each 1 GW(e) accounts for about 3 GW(t) included in the thermal energy supply. The table below expresses energy at its average annual consumption rate, GW-yr per year, or GW.

Thermal energy and electric energy consumption rates				
		2015	2035	*increase*
Thermal energy GW(t)	US	3,300	3,800	15%
	World	19,000	26,000	37%
Electric energy GW(e)	US	500	600	20%
	World	2,600	4,000	**54%**
Thermal, GW(t), net of elec. gen.	US	1,800	2000	11%
	World	11,200	14,000	**25%**

EIA projections of US and world energy demand

The last two rows subtract out the thermal energy that is converted to electric energy, leaving just thermal energy demand for other uses such as heating, internal combustion engines for transportation, and industrial process heat. World growth in electricity energy demand (54%) is more than double the growth of other energy uses (25%).

Growth of civilization in both the US and world especially requires the valuable energy of electricity, more so than the energy of heat. For example, computers run on electricity, and the growth of the internet services has become possible with warehouses of computer servers. Worldwide, data centers use 1.3% of world electric power.

The ongoing third industrial revolution demands more electric energy. New digital technologies are making manufacturing more efficient. *The Economist* (April 21, 2012) describes how some automobile firms have doubled car production per employee. New 3-D printers create parts in layers, making low-volume, on-demand production simple and economical Industrial robots are becoming more functional and more flexible.. Labor costs for an iPad are just 7% of the sales price. Manufacturing is returning to the US.

2010 US energy demand was 98 quad BTU.

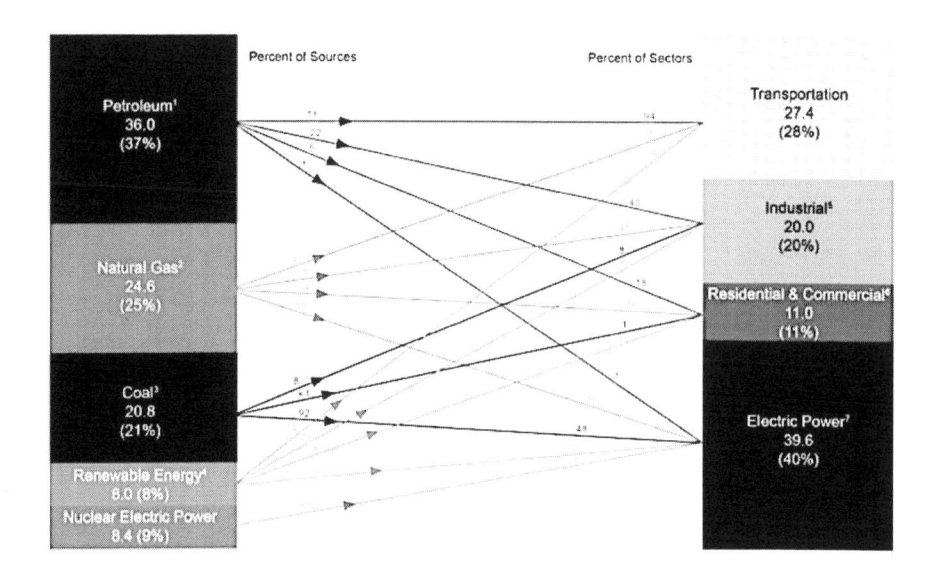

The sources for US energy represented above in 2010 were:

2010 US energy sources		
	Quads	GW(t)-years
Petroleum	36	1202
Natural gas	25	835
Coal	21	701
Renewable energy	8	267
Nuclear energy	8	280
Total	98	3272

GENERATION CAPACITY

Average US electric power is 45% of 1,100 GW total capacity.

2010 US electric power generation		
Energy source	max capacity GW	average GW
Coal	319	211
Oil	60	4
Natural Gas	439	113
Nuclear	103	92
Hydro	78	30
Wind	39	11
Solar	1	0.14
Biomass	11	6
Other	29	20
Total	1079	471

Data from EIA in the second column above shows maximum power generation capacity of each energy source. Generators do not operate continuously. The last column is the average power generation rate for the year. The total electric generation capacity is about 1,100 GW, supplying 500 GW on average.

EIA calculates electric power generation capacity factors.

The capacity factor is the ratio of average power to maximum, nameplate, nominal power generation capacity. For all US electric power generation facilities, the overall capacity factor is 45%. EIA's observed 2010 capacity factors of several sources are presented in this table below.

2010 US electric power capacity factors	
Energy source	average / nominal
Coal	64%
Petroleum	8%
Natural gas CCGT	42%
Other natural gas	10%
Nuclear	90%
Hydroelectric	40%
Other renewables	34%
All sources	45%

EIA estimates capital costs of electric power generation.

The US DOE Energy Information Agency makes annual, detailed analyses of the costs of electric power generation. The estimates presented in the table are in 2010 dollars. The "watt" in this table is the generation capacity when the unit is operating at full power.

Power generation technology	Capital cost $/watt
Advanced pulverized coal	2.84
Integrated coal Gasification CC	3.22
Natural Gas Combined Cycle	1.00
Natural gas turbine	0.67
Fuel cell	6.80
Nuclear	5.33
Biomass	3.86
Hydro	3.08
Wind	2.44
Wind, offshore	5.97
Solar, thermal	4.69
Solar, photovoltaic	4.75

These capital costs are so-called "overnight" costs. They exclude interest on the money borrowed during construction. This is the cost if the plant were paid for and built overnight. For example, for a three-year construction period with steady payments of money borrowed at 8% this would be approximately $0.12 per $1.00 of capital expenditure.

Coal

Coal is a very important source of energy in the US and worldwide, ever since it powered the Industrial Revolution. Petroleum energy consumption exceeds that of coal because gasoline and diesel are so well suited for transportation. Coal is the largest source of energy for electric power, and also the largest source of CO2 emissions world-wide.

Coal-provided energy is now about 21 quads per year or 701 GW(t). Coal is plentiful in the world, and especially in the US. At current consumption rates the US has enough for 222 years and the world as a whole has a 126-year supply.

Coal burning is the world's largest CO2 source.

The US Energy Information Agency projects that world CO2 emissions from burning coal will continue to exceed those from other fuel sources. This graph illustrates atmospheric CO2 emissions approaching 20 Gt/year, worldwide, by 2035.

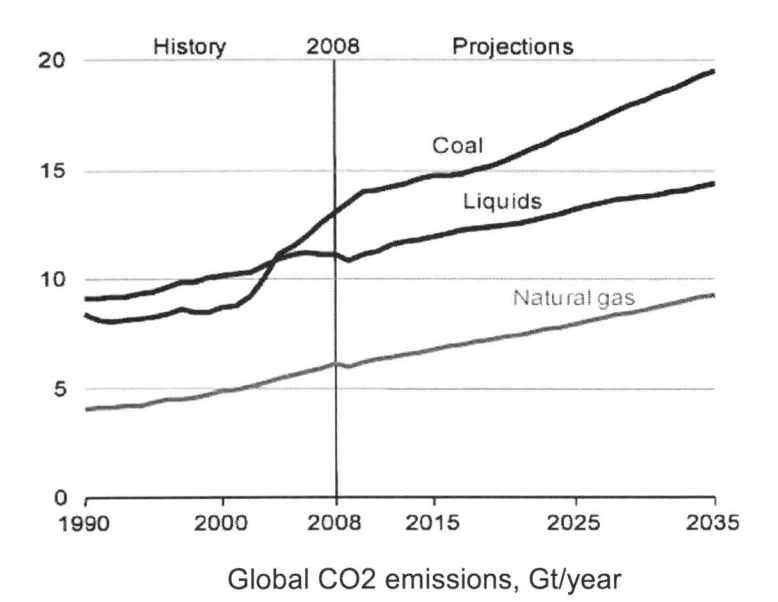

Global CO2 emissions, Gt/year

US coal production and use declined in 2008 recession. The EIA projects it will not return to 2008 levels until 2025, due to increased competition from natural gas, nuclear, and renewables. Coal use may then grow 1% per year for increased electricity generation and conversion to synthetic fuels.

China already mines three times more coal than the US, and China is the world's biggest coal importer. In spite of China's famous investments in hydro power and nuclear power, it derives 80% of its electric power from coal, compared to 30% in the US.

China is adding coal power plants at the rate of about 1 GW per week. The US added over 6 GW of capacity in 2010. Fewer plants will be added in the future, because of EPA restrictions and repayment concerns of potential lenders.

In the US, 34,000 annual deaths are attributed to coal plant emissions by the EPA, which has issued new regulations to further restrict pollutants emissions, such as mercury and sulfur dioxide. The cost of installing new pollution control equipment will persuade many old coal plant owners to shut them down.

More efficient coal plants could use less coal.

CO_2 emissions could be reduced in new plants by using more efficient, high-temperature technologies such as supercritical pulverized coal. However, such investments are more costly. A new ultra supercritical pulverized coal plant can achieve a 44% electrical/thermal conversion efficiency, compared to 33% typical of older plants. So a new plant would only use 33/44 of the coal, emitting 25% less CO_2. This would require 25% less coal mining.

Worldwide there are approximately 1,000 GW of traditional technology coal plants that could be replaced with more efficient ones. This would avoid 1.5 gigatonnes of annual CO_2 emissions according to Peabody Coal. China has retired 71 GW of capacity of old, inefficient coal power plants as they install more efficient units.

In the US since 2010 only 9 of these more efficient supercritical pulverized coal and IGCC (integrated gas combined cycle) plants came into operation, compared to 43 less efficient plants. In 2012 there are 12 of the high efficiency plants under construction, and 9 of the less efficient ones. Only half the 100 announced and proposed new US coal-burning power plants plan to use the more expensive, more efficient technologies.

Perfection being the enemy of practicality, the potential for cleaner burning coal plants is being obscured by "clean coal" – carbon capture and sequestration.

"Clean Coal" is a marketing achievement.

The coal industry, mindful of the possible profit impact of carbon taxes, has supported research into CCS, carbon capture and sequestration. If CO_2 could be buried forever, we could continue to burn coal for electricity without emitting CO_2. By changing the public issue from current climate damage to future possible clean coal solutions, the coal industry has convinced the public to defer taking action to reduce coal plant emissions or to develop effective alternative energy sources.

The sheer scale of coal mining and transport for power generation is vast. Most of the coal comes by diesel-fueled rail transport, adding 20-59% to the mined coal cost. A large coal plant requires a one-hundred-car coal train daily to keep it supplied. The resulting CO_2 is $(12+2*16)/12$ times more massive and would require a daily train of 367 refrigerated, pressurized tank cars to remove it for burial, somewhere. The US uses 227 GW of coal power, so the resulting daily tank car trains would be a total of 833 miles long. Putting the CO_2 back into mines won't work, because the volume exceeds that of the coal by a factor of 3.

A side-effect of annual coal plant power production is 130 million tons of solid wastes that remain near the power plant sites, nearly as much as the total of US municipal solid waste sent to landfills.

Carbon capture and sequestration projects are tiny.

US advanced technology coal plant demonstrations have gone through cycles of on-again/off-again funding. The Mattoon IL demonstration project has recently changed into two proposals: to separate CO_2 at one power plant and to pipe it 175 miles to an underground storage site.

No large scale power plant CCS projects are in operation. However in natural gas extraction, CO_2 flows from the well along with the valuable methane. The CO_2 is removed and vented to the atmosphere, except in a few pilot projects where the CO_2 is re-injected into the earth to help force out more methane. A few CCS projects are motivated by enhanced oil extraction. None sequester CO_2 from coal-burning power plants.

The most advanced coal power plants such as IGCC (integrated gasification combined cycle) first gasify the coal, so the resulting carbon monoxide and hydrogen gases can be burned in an efficient gas turbine, much like a natural gas plant uses. The IGCC plants separate oxygen from air to facilitate the coal gasification. Burning coal gas in oxygen, rather than air, means there is little nitrogen in the turbine exhaust, which is mainly CO_2 and H_2O. This facilitates capture of CO_2. Clean coal power plants with carbon capture and sequestration (CCS) are promoted as having zero carbon emissions.

In the US at Edwardsport IN the Duke Energy $3 billion, 630 MW IGCC coal plant provides for an optional add-on CCS facility. If the $390 million unit were added it would capture only 23% of the emitted CO_2. The capture plant would also use power, so this would reduce the net power production by 10%, leading to CCS overall CO_2 emissions reductions of just 13% compared to ultra supercritical pulverized coal. IGCC power plants operate at 31-40% efficiencies, less than ultra supercritical pulverized coal plants at 44%.

The Congressional Budget Office reported in 2012 that CCS equipped coal plants would be 35-75% more expensive than regular coal plants, and that more than 200 GW of capacity would need to be constructed to meet DOE's goal. Congress has appropriated $6.9 billion for CCS with little result.

Sleipner natural gas field with CO2 sequestration

CCS advocates exemplify Norway's Sleipner natural gas field project, which sequesters about 1 megatonne of CO_2 per year. World coal power plant CO_2 emissions are 10,000 times this.

China is progressing with its CCS project GreenGen with Peabody Coal and other partners including MIT. This experimental project has received $3.5 billion in funding to construct a 400 MW, 55-60% efficient coal power plant with 80% of the CO_2 sequestered, by 2020.

Injection of CO_2 into brittle rocks of continents can trigger small earthquakes that threaten the seal integrity of repositories, releasing the CO_2.

Electricity from coal is inexpensive.

We estimate the power cost as the sum of three costs: upfront capital investment cost recovery, fuel costs, and operational costs.

Capital	The cost of the plant plus a return on investment to be recovered over the lifetime of generation,
Fuel	The cost of the coal burned to produce power.
Operations	Costs of labor, services, and supplies to operate and maintain the plant.

This simple model ignores taxes and fees imposed by governments because our objective is to model comparable economic costs for multiple energy sources to guide policy makers, not to reflect complex laws and regulations that form existing energy policy.

Capital plus return on the amount invested must be recovered through kilowatt-hours of energy produced and sold. We can compute capital cost recovery on a financial calculator or Excel spreadsheet. For example a $1 per watt investment with money borrowed at 8%, recovered by selling electricity 90% of the time, over 40 years, 365 days/year, 24 hours/day, requires a repayment of $0.00001 per watt-hour or $0.01/kWh. We use an optimistic 90% capacity factor in this and other examples.

The revised MIT study of the future of nuclear power estimates new coal plant construction technology at $2.30/watt of generating capacity. Costs can be higher: Duke Energy's IGCC plant will cost $4.76/W without CCS. The EIA projects a 2010 cost of $2.84 for advanced pulverized coal. A $2.84/watt investment costs about $0.028/kWh.

The fuel cost of the coal delivered to US power plants runs about $45 per ton in the US. Each ton releases 16-26 million BTU of thermal energy when burned, depending on coal quality. The average cost is about $0.00785/kWh(t). Assuming the coal is burned in a modern, new supercritical pulverized coal or IGCC plant achieving a high electrical/thermal efficiency of 44%, the fuel cost will be $0.00785/0.44 or about 1.8 cents per kWh(e).

EIA data shows 2012 Appalachia coal costs at the low end of the historical range, partly depressed by a warm winter and cheap natural gas. Our $45/ton estimate is at the low end of the range.

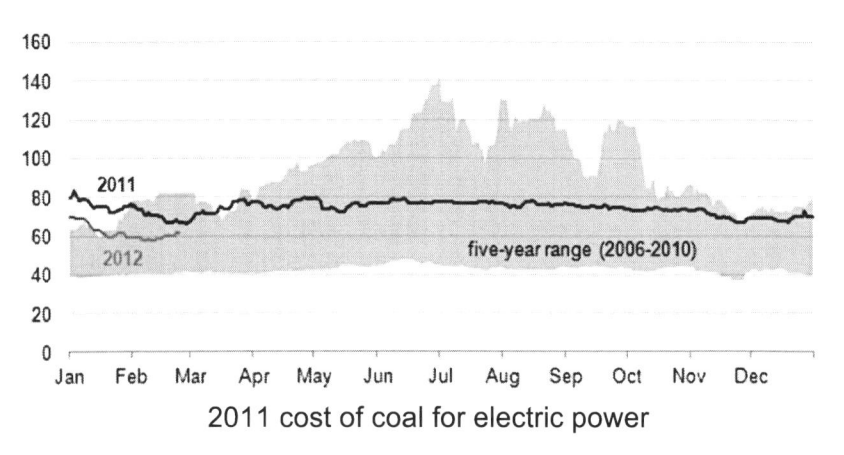

2011 cost of coal for electric power

Operational costs cover labor, maintenance, waste disposal, services, compliance, etc are estimated at 1 cent per kWh.

Advanced coal electricity cost	
Capital cost recovery	2.8 cents/kWh
Fuel	1.8
Operations	1.0
Total	5.6 cents/kWh

This 5.6 cents/kWh is the cost for electricity leaving the power plant, to be sent over transmission lines and power distribution systems to consumers. A US business or homeowner might pay 15 cents/kWh for delivered energy, including the costs for transmission and distribution, grid management and maintenance, billing and collection, regulatory fees and taxes. These added costs should be similar for all form of electric energy sources, so we will compare just the costs of electricity produced by the power plants, not the total costs paid by consumers.

Indirect pollution damage adds to the cost of coal power.

The National Academy of Sciences studied the hidden cost of environmental damage from burning coal and estimated it at 3.2 cents/kWh. This excluded damage from mining or disposing of the chemicals scrubbed from coal plant smokestacks. It did not count impacts of CO_2-caused global warming. A Harvard Medical School report estimates the total additional damage costs five times higher, at 18 cents/kWh.

A political approach to solving our climate and energy crises is to impose taxes to raise revenues to offset the damage, and to raise the coal power costs to levels where other, cleaner energy technologies can compete. This book instead endorses innovative technologies that produce energy cheaper than coal, counting only direct costs. Our analyses exclude these costs.

Energy cheaper than coal must be < 5.6 cents/kWh.

We conclude that our objective for LFTR energy "cheaper than coal" must be less than 5.6 cents kWh.

Natural Gas

Natural gas will dominate the near-term energy scene.

- Natural gas is our fastest growing source of energy. In the US it represents 25% of all energy supplies, exceeding coal and nuclear sources.

- Natural gas supplies have grown in the US with the advent of new hydraulic fracturing, drilling, and extraction techniques

- Natural gas was recently one of the costliest sources of energy for electric power generation. Today it competes with low-cost energy sources such as coal, hydro, and nuclear power.

- Per unit of thermal energy released, burning natural gas releases half the CO2 of coal. Per kWh of electricity generated, natural gas can emit less than a third the CO2 of coal.

Natural gas burns twice as cleanly as coal.

Burning natural gas emits half the CO2 of burning coal. Coal is largely carbon; natural gas is methane. Compare the chemical reactions for burning coal and methane.

Coal $\quad$ C + O2 → CO2

Methane $\quad$ CH4 + 2 O2 → CO2 + 2 H2O

For each molecule of CO2 produced, the natural gas (methane) derives additional energy from also oxidizing the 4 hydrogen atoms associated with each methane molecule.

From chemistry, the heat of combustion of burning 1 mole (6 x 10**23) molecules of methane is 800 kJ (kilo joules) of thermal energy. Burning 1 mole of carbon releases only 394 kJ. Having the same number of carbon atoms, both release the same amount of CO2, but the carbon only provides 394/800 the thermal energy –

about half. Burning natural gas emits half the CO2 of burning coal to generate the same thermal energy.

Burning natural gas does not release the sulfur dioxides emitted from burning coal. Natural gas impurities are largely removed in the natural gas production stage before being sent into pipelines.

For equal amounts of heat produced, burning natural gas emits half the CO2 of burning coal. Most news-reported comparisons of CO2 emissions from coal and natural gas correctly report this factor-of-two improvement in thermal energy per ton-CO2, but fail to emphasize the electric/thermal efficiency advantage of some natural gas turbines, discussed below.

Natural gas can burn more efficiently than coal.

Earlier we noted that the efficiency of conversion of BTUs of thermal energy depends strongly on temperature.

$$\text{Efficiency} \leq \frac{T_H - T_C}{T_H}$$

Coal plant efficiencies range from 33 to 44%. Natural gas combustion turbine (NGCT) efficiencies can be higher or lower.

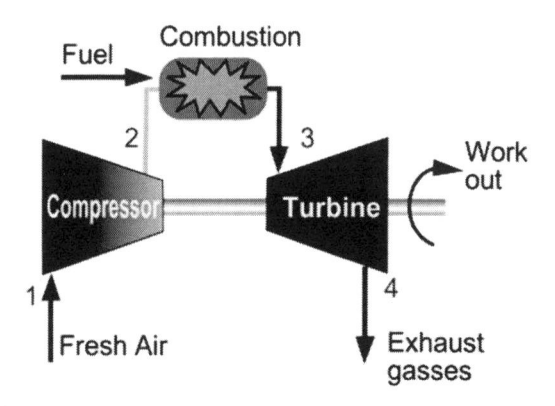

Natural gas combustion turbine

Unlike coal-fired boilers, the gas turbine is an internal combustion engine, which achieves a high temperature by burning the fuel in air at high pressure created by the compressor blades. The methane gas burns, and the resulting high-temperature gas (CO_2 and H_2O) enables high kinetic/thermal conversion efficiency, thus high electric/thermal efficiency.

Natural gas combustion turbines supply peak power.

Until 2009 natural-gas-generated electricity was more expensive than electric power from coal, nuclear, or hydro, because natural gas was expensive. Electric utility companies bought and dispatched to customers the least expensive power available. As more power was demanded by customers at peak demand times during the day, the utilities bought supplemental, more expensive power from natural gas generation. Utility companies could only recoup capital investments during the fraction of the time that power was generated, so it was important to keep invested capital minimal. The NGCTs operate at a capacity factor less than 11%, which is economically feasible because their capital cost is the lowest for any energy source -- $0.67/W. They operate at an electric/thermal efficiency of only about 29%.

Combined cycle gas turbines are most efficient.

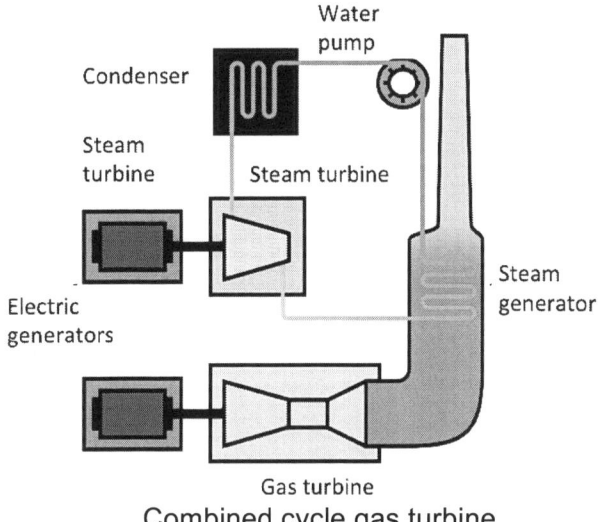

Combined cycle gas turbine

In the combined cycle gas turbine (CCGT), natural gas is first burned in the gas turbine, powering an electric generator. The still-hot exhaust gases flow through a steam generator. That steam turns a steam turbine to make additional electric power. The steam is then condensed back to water and pumped through the steam generator again. The CCGT is called "combined" because it is a combination of the gas turbine power conversion cycle and the steam power conversion cycle.

Such more expensive CCGT generators operate at a higher efficiency than simpler NGCT generators. CCGT efficiency averages 45% in the US. New CCGTs from GE and Siemens can operate at conversion efficiencies of up to 60%. These are more expensive investments, costing about $1.00/W. There are few CCGT turbines compared to more common NGCTs.

CCGT CO2 emissions are much less than those of coal.

Not only do CCGT plants use cleaner fuel, they burn it more efficiently. From the same thermal energy, a CCGT with an efficiency of 60% generates 60/33 the electric power of an older

33% efficient coal plant. Thermal energy from natural gas (methane) emits half the CO_2 emissions of coal fuel (carbon). So the CCGT CO_2 emissions per kWh are lower by a factor of (1/2) x (33/60) = 0.28, or 72% less.

Even the older 29% efficient NGCT plants emit less CO_2 than typical coal plants, by a factor of (1/2) x (33/29) = 0.44, or 56% less.

Even new 44% efficient ultra supercritical pulverized coal plants can't compare to CCGT plants, which release less CO_2 by a factor of (1/2) x (44/60) = 0.37, or 63% less.

CCGTs emit less CO2 than proposed clean coal with CCS.

CCS (carbon capture and sequestration) projects do not propose to capture all the CO_2 generated by a coal plant. The CCS facility at the Duke Edwardsport IGCC coal plant would capture 23% of the CO_2, reducing CO_2 emissions/kWh by 13% in comparison to the best coal burning technology -- ultra supercritical pulverized coal.

In summary, compared to the best available advanced coal burning technology, the Edwardsport plant with CCS would reduce CO_2 emissions/kWh by 13%; CCGT natural gas plants would reduce CO_2 by much more, 63%, with no need for unproven costly CCS.

CCGT generation depletes natural gas reserves at half speed.

Natural gas turbine electric/thermal efficiencies are respectively 60% (CCGT) and 29% (NGCT). So for the same amount of generated electric power, the CCGT uses only 29/60 the fuel, lowering fuel costs and more than halving the rate of depletion of the natural gas reserve. The capital cost difference, $1.00/W (CCGT) vs $0.67/W (NGCT), can be made up in fuel savings.

Hydraulic fracturing taps natural gas trapped in shale.

The recent increases in natural gas reserves have come about through the development and refinement of hydraulic fracturing to release natural gas from shale. This is enabled by the new technology of horizontal, directed drilling. More than half of US drill rigs are boring horizontally. In 2009 76% of the increase in proved reserves was from shale gas drilling.

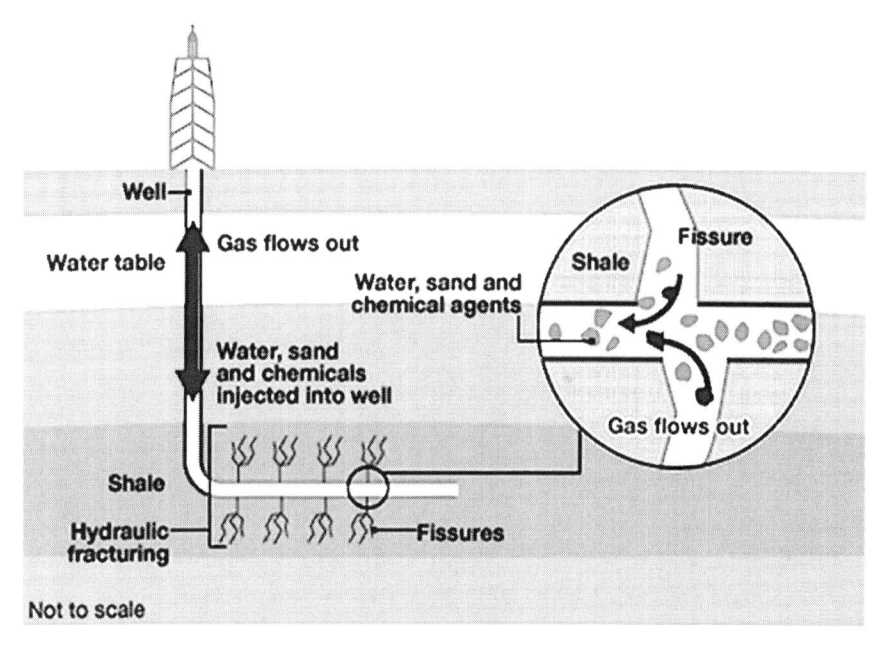

Natural gas hydraulic fracturing technology

Environmental concerns with fracking will be addressed.

Methane is a greenhouse gas more potent than CO_2, so methane leaks should be minimized. After hydraulic fracturing of the rock, the liquids are pumped up, bringing along methane gas. About 2% of the lifetime methane production of the well is vented to the atmosphere during this flow-back process at the start of production. An additional 4% is lost from leaks, pressure relief valves, and distribution common to the natural gas industry.

Sunlight and atmospheric chemical processes decompose methane to CO2 in about a decade, subsequently reducing this warming effect. Nevertheless, over a century, methane contributes 25 times as much to global warming as does an equal mass of carbon dioxide. Some authors conclude that substituting fracked natural gas for coal actually increases global warming. These emissions of methane natural gas can be reduced by 90% by improved techniques of liquids transfer, emissions control technology , and building pipelines in advance.

The chemicals in fracturing fluid may be hazardous if accidentally released in spills, leaks, or faults. The fracturing fluid (99.9% sand and water) is injected into shale a mile below fresh water aquifers used for public water supplies. The lubricants, anti-microbials, hydrochloric acid, and scale inhibitors are benign in the dilutions that might reach groundwater. Large amounts of water are used to drill and fracture the wells, but not during gas production. The retrieved wastewater must be treated before disposal or reuse. Although hydraulic fracturing can cause small tremors, too small to be a safety concern, the wastewater injection into deep wells may cause larger small earthquakes.

To summarize the environmental concerns, the benefit and need for cleaner energy is so strong, and the cost of clean, abundant natural gas is so low, that environmental concerns will be addressed. There is enough potential profit to ensure this.

In April 2012 the EPA issued a rule requiring the industry to limit emissions of global-warming methane and toxic benzene and hexane from 13,000 new wells drilled annually. The industry estimates the annual costs to be hundreds of millions of dollars, but the EPA says the net effect will be a savings of $11 to $19 million per year from selling methane otherwise lost.

The Marcellus shale alone contains 55% of US natural gas.

The Marcellus shale formed at the bottom of an ancient lake bed. The black shale covers an area of 50 million acres in the northeast US, 50-200 feet thick, at a depth of 5,000-8,000 feet.

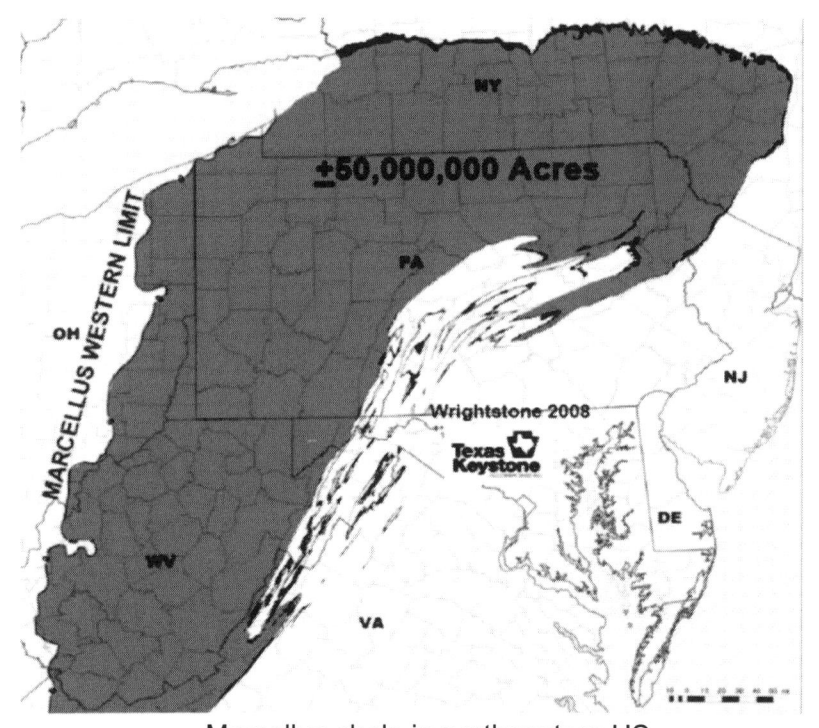

Marcellus shale in northeastern US

The shale contains methane in impermeable shale. Wells are drilled down to the 1 mile depth of the shale, then drilled horizontally through the 50-200 foot-thick shale, which is then fractured with high pressure injected water along with sand to hold the resulting fissures open.

The US natural gas pipeline network is diverse and growing.

The EIA reports that the US natural gas pipeline network added 2400 miles of new capacity in 2011.

US natural gas pipelines

The US has abundant natural gas reserves.

The recent discoveries of potential shale gas reserves has led to a bewildering range of estimates of reserves, proven reserves, technically recoverable reserves, inferred reserves, and undiscovered reserves. Reserves also vary with the price to be paid to extract the natural gas. There is too little history of fracked wells to be confident of their lifetime output, leading to a wide range of estimates of reserves.

Natural gas measurement units are commonly cubic feet, at standard temperature and pressure. One cubic foot of natural gas contains approximately 1,000 BTU, so 1 trillion cubic feet (TCF) is 1 quadrillion BTU, or one quad, or 33 GW(t)-years.

Proved reserves of all natural gas (including shale gas) are 273 TCF, reports EIA. Adding unproved shale gas brings the total to 755 TCF. The Intek report estimated 750 TCF (trillion cubic feet).

EIA reports additional, more speculative unproved reserves of 1,460 TCF in Alaska and off-shore, bringing total possible reserves to 2,215 TCF. The Potential Gas Committee of the Colorado School of Mines estimates 2,074 TCF, while IHS-CERA estimates 2,000-3,000 TCF. Less optimistically in 2011, the USGS reports 84 TCF in the Marcellus shale.

US natural gas reserves will not be depleted for decades.

How long can these reserves last? Let's review three cases: low (273 TCF), mid (750 TCF) and high (2000 TCF) of US natural gas reserves.

US annual consumption of natural gas is now just 25 TCF. Suppose the US rapidly retires its coal power plants and replaces them with combined cycle natural gas plants. This would require an additional 21 quads (701 GW-year) of energy, which could be supplied by 21 TCF of natural gas, increasing total gas demand to 46 TCF. So let's define two cases: now, and with coal replacement.

Years of natural gas reserves			
Natural gas consumption rate	Natural gas reserves estimates		
	Low, 273 TCF	Mid, 750 TCF	High, 2000 TCF
2012 25 TCF/yr	11	30	80
replace coal 46 TCF/yr	6	16	43

After inspecting the table, we would never replace coal plants if this action would make natural gas reserves run out in 6 or 16 years. If gas supplies were ample, we might well retire coal plants in favor of cleaner burning natural gas. On the other hand the US has 200 years of coal reserves. So natural gas supplies will probably persist for much of this century. These reserves could go up if more shale gas is discovered, or go down if shale-gas-well production lifetimes are less than expected, a concern of many. Exporting LNG (liquefied natural gas) from the US will use reserves faster.

The US will not run out of natural gas soon. In 2012 there is a temporary glut from overexploitation and a warm 2012 winter. What will happen to the price?

Natural gas arbitrage opportunities beckon.

The Federal Energy Regulatory Commission estimated the landed costs of LNG for February 2012. The lowest cost is $2.83 per million BTU at the Lake Charles LA import terminal. The cost is, for the moment, low because the US supplies of domestic natural gas have increased with shale gas extraction.

LNG prices, $/million-BTU

The values exceeding $14/MBTU in India, China, and Japan create market opportunities for natural gas exporters to build LNG refrigeration plants and fleets of LNG tankers, in order to export LNG to the profitable markets in the Far East.

Post-Fukushima LNG imports ended Japan's trade surplus.

After Fukushima, Japan shut down 52 nuclear reactors. To make up for lost electric power Japan increased imports of liquefied natural gas (LNG) to power natural gas turbine generators and so raised the cost of electricity. LNG demand also increased natural gas prices. In 2011 fossil fuel imports increased over $200 billion. Japan's balance of trade swung from positive to negative. A report by Japan's Energy and Environmental Council estimates the GDP will fall over 7% if nuclear energy is phased out.

CNG vehicles may also increase natural gas demand.

Compressed natural gas (CNG) can fuel vehicles. The US is increasing R&D support for CNG-fuelled vehicles. Honda already sells CNG passenger cars. In 2012 Chrysler announced plans to sell CNG vehicles. GM will also sell pick-up trucks that can run on either natural gas or gasoline. Compared to gasoline, the vehicles will emit 25% less CO_2 with CNG fuel, costing 33% less. Gasoline prices have risen, increasing the impetus to seek cheaper fuels such as CNG. In 2012 there are only 1000 US CNG fueling stations in the US. Another possible use is converting methane (natural gas) to methanol for vehicle fuels as a direct gasoline replacement.

US natural gas prices ($/MBTU) are historically volatile.

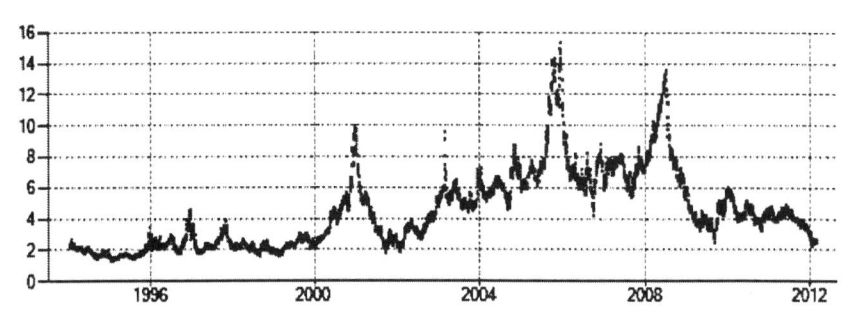

Natural gas prices in 2012 were low. The 2012 winter season was warmer than normal, decreasing home heating demand. Hydraulic fracturing technology attracted producers to a natural gas rush that has increased capacity. Reportedly many wells are unprofitable at the current low prices, so the rush to drill more wells may not continue until prices rise.

EIA projects natural gas prices will rise.

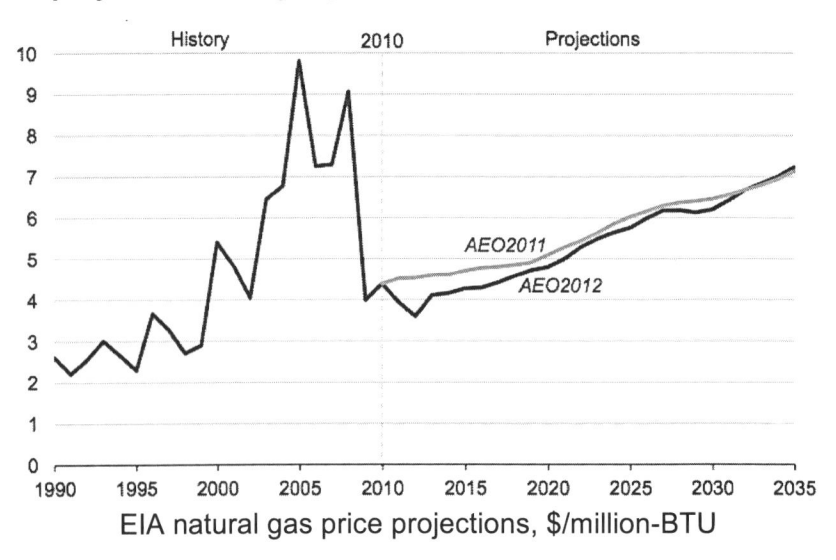

EIA natural gas price projections, $/million-BTU

Natural gas prices are projected by the EIA in its Annual Energy Outlook report for 2012.

Economics will move oil and natural gas prices closer together. At $100/bbl oil costs $17 per million BTU, so demand for natural gas will increase wherever it can be substituted for more expensive oil. An analysis by Lynn Pittinger on the *Oil Drum* suggests a future price of at least $8/MBTU. We will use $5 per million BTU in future comparisons.

Electricity from natural gas is inexpensive.

We consider only modern combined cycle gas turbine (CCGT) generators running in competition with other low cost energy sources such as coal, nuclear, and hydro. We again estimate the power cost as the sum of three items: upfront capital investment cost recovery, fuel costs, and operational costs.

Capital	The cost of the plant plus a return on investment to be recovered over the lifetime of generation,
Fuel	The cost of the natural gas burned to produce power.
Operations	Labor and supplies to operate and maintain the combined cycle plant.

Capital plus return on the amount invested must be recovered through kilowatt-hours of energy produced and sold. The capital cost of a CCGT plant is estimated by EIA and plant operators to be about $1.00/watt. As in the coal example, we assume an 8% cost of capital. Operating at 90% capacity factor for 40 years this adds 1 cent/kWh for capital cost recovery.

We estimate 2020 fuel costs at $5/MBTU, higher than the 2012 low price. For a modern, 60% efficient CCGT power plant this works out to $5 x 0.003412 / 0.60 = 2.8 cents/kWh. Operational costs cover labor, maintenance, waste disposal, management, compliance, etc estimated at 1 cent per kWh. So our estimate for electric power produced from a new, high technology, combined cycle gas turbine power plant is 4.8 cents per kWh.

Natural gas electricity cost	
Capital cost recovery	1.0 cents/kWh
Fuel	2.8
Operations	1.0
Total	4.8 cents/kWh

Electricity cheaper than natural gas must be < 4.8 cents/kWh.

Natural gas electricity will be even cheaper than coal electricity. We have set our objective for LFTR energy "cheaper than coal" to be less than 5.6 cents/kWh. To be cheaper than either coal or gas, LFTR energy must sell for less than 4.8 cents/kWh.

If, however unlikely, natural gas prices persist at $3/MBTU for decades, it would drop the required rate for LFTR to undersell gas to only 3.5 cents/kWh.

Natural gas electricity is the strongest competitor to potential LFTR produced electricity. CCGT technology already exists and is available commercially from GE and Siemens. Ample supplies of natural gas are available. The fuel distribution system is mature and robust. The capital cost is lower than coal, nuclear, and most other energy generation technologies. The cost per produced kWh is less even than that of coal.

Natural gas will displace much coal-fired electric power generation because of economics, health hazards of the emissions, opposition to coal mining, and concerns for CO_2 emissions causing global warming. In April 2012 the natural gas share of US electric power generation grew to 32%, matching that of coal.

Cheap natural gas electricity has its drawbacks.

Natural gas generated electricity has substantial weaknesses relative to LFTR energy. Using natural gas to generate electricity

- emits substantial CO_2, while LFTR emits none,

- drives up the price of natural gas and its generated electricity,

- depletes a fossil fuel resource within a century, while thorium is inexhaustible, and

- may increase global warming unless fugitive methane emissions are much reduced.

- makes utility companies reluctant to commit to a single source for all power generation.

Nevertheless, the goal for LFTR energy to be price competitive with natural gas generated energy is at 4.8 cents/kWh.

Wind

Wind provided 3% of US electricity in 2011.

Brazos wind farm, Fluvanna TX, 160 1 MW wind turbines

Wind turbines in the US generated electricity at an average rate of 14 GW during 2011, supplying 2.9% of total electric generation. The installed, nameplate capacity of all wind generators is 47 GW. On average these wind farms operated at 29% of installed capacity, largely because winds are not always blowing. In analyzing statements about the growth and market share of the wind industry, it is important to distinguish generated power from installed capacity.

generated power = installed capacity x capacity factor

Wind strength is suitable for electric power generation on mountain ridges, the plains, and offshore along the coastlines. The darker areas on this map correspond to stronger average wind intensity.

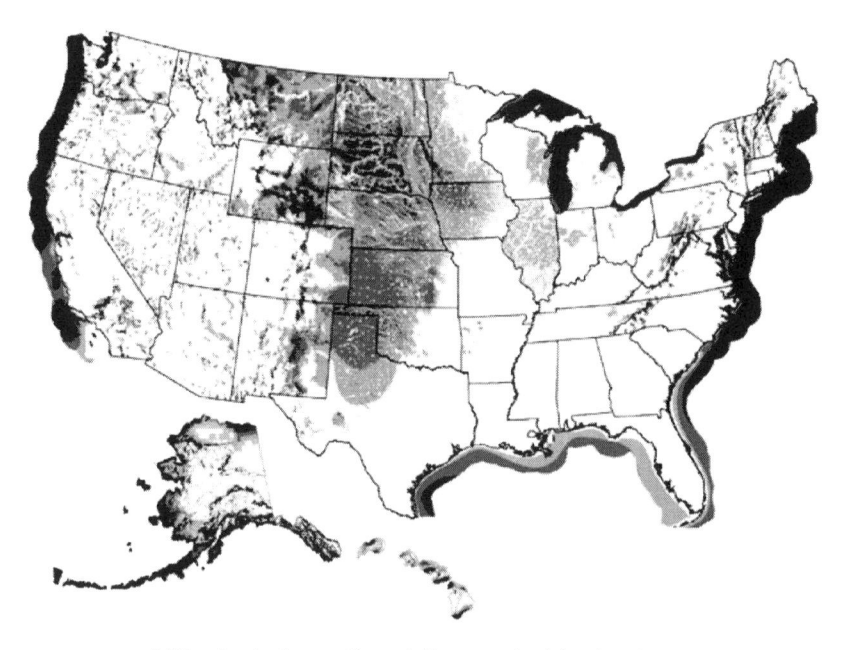

US wind strength: white weak, black strong

Wind strengths are slower near the earth's surface due to friction with the ground, trees, and structures, so efficient wind turbines are built on towers about 100 m high.

Wind turbines slow down the wind, so they must be spaced not to interfere with one another. Considering the possible density of windmills and the variability in wind speed, we can expect the average power produced to be 2 W/m^2. A wind farm capable of delivering 1 GW of average power must occupy a land area of 500,000,000 m^2, or about 200 square miles. In the windy plains of the US this land can also be used for agriculture.

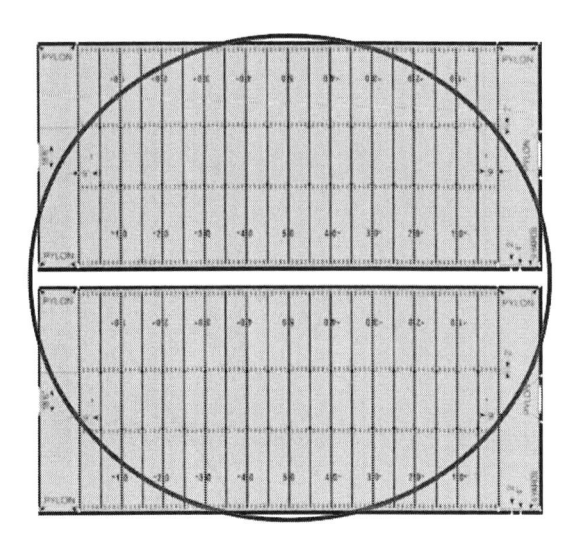

Wind turbine swept blade area laid out on 2 football fields

Utility-scale wind turbines are huge. Here is a typical circular area swept by the blades, laid out on two US football fields to illustrate. The supporting tower is nearly 100 m high.

Offshore wind power has a higher capacity factor.

Because winds are steadier over the sea, the capacity factor for offshore wind can be as high as 40%. The US has no offshore wind power.

Cape Wind, off the coast of Massachusetts, may become the first US offshore wind farm. The plan calls for 130 turbines on 87 m towers, each capable of generating 3.6 MW of power, for a total capacity of 454 MW, expected to produce 170 MW of electricity, for a capacity factor of 37%.

The expected capital cost of the investment is $2.62 billion, or $5.80/W of nameplate capacity, or $16/W of average power delivered. Capital costs of this power generation investment can only be recovered when the plant is generating power, 37% of the time.

Assuming the units operate for 40 years, recovering capital invested at 8% would add 14 cents/kWh to the price of wholesale electricity sold to the utility company. Since the current cost for electricity bought from nuclear, hydro, and natural gas generators is about 5-6 cents/kWh, the Cape Wind project would clearly be uncompetitive or uneconomical based on selling unsubsidized electricity.

The state of Massachusetts has forced public utility company National Grid to buy one half of all the power that might be generated by Cape Wind, at 18.7 cents/kWh, escalating annually by 3.5% for 15 years, reaching 31 cents/kWh. The utility will spread this cost premium over all other sources of electric power paid for by consumers. These prices will increase another 2.2 cents/kWh with the 2012 expiration of production tax credits, unless Congress acts to continue the subsidy.

Nstar in 2012 also agreed to buy 27.5% of the power at the same above-market prices, in order to gain state approval to merge with Northeast Utilities. So with the prospect of being able to sell 77.5% of its electricity at a price four times the free market rate, Cape Wind may yet be built.

In Rhode Island, Deepwater Wind plans to build a 30 MW offshore wind farm for $200 million, or $7/watt of nameplate capacity; National Grid has agreed to buy its power at 24.4 cents/kWh. Deepwater Wind also plans a future 100 turbine, $1.5 billion, 385 MW project off Block Island.

Intermittent wind power requires backup power sources.

Wind power is intermittent and variable in intensity. A wind turbine generates about 30% its nameplate capacity on average. To evaluate wind power it is important to distinguish installed capacity from generation. A 2.5 MW wind turbine installed in a favorable land location provides power of 0.75 MW on average.

When wind power decreases, some other power source must increase to make up for the loss of wind power. There is no energy storage in the electric power grid. Coal plants and nuclear power plants are expensive and designed to run at full power 90% of the time. The two prime candidates for backup are hydro power and natural gas power. Hydro power can be dispatched on demand by raising or lowering gates in sluices that convey water to the water wheel or turbine. A natural gas turbine can power up and down much like a jet aircraft turbine.

Hydroelectric power can back up wind power.

Hydro power is also usually intermittent. Power can only be generated if there is water available in the reservoir above the dam. The US capacity factor for all hydro power is about 38%, principally limited by available water. The hydro plants normally generate electricity just at times of peak daily demand, reserving water for times when the power is most valuable. The water reservoir contains stored gravitational potential energy.

Working together, wind turbines and hydroelectric facilities can postpone drawing down water levels when the wind is blowing, leaving potential energy in the reservoir water. The capacity factor of the combined hydro/wind sources will be about 31%. The reliability of the combined wind/hydro sources will be good as long as there is potential energy in the hydroelectric water reservoir. This hydro/wind feature is exploited by Denmark's wind turbines, which work in combination with hydro power from nearby Sweden and Norway.

With coal backup power, wind does not reduce CO2.

In many jurisdictions, rules require utilities to accept power from wind generator operators whenever it is available no matter the cost or effect. In Ontario, for example, operating coal power plants must reduce power output when wind power becomes available. To curtail electric power generation and still be able to provide power when the wind lulls, the coal plants continue to burn coal and vent 300 kg/sec of 250°C high pressure steam instead of directing it to the steam turbine generators. The coal plant continues to burn coal and emit CO_2, otherwise it could not increase its power rapidly enough during a wind lull. So no CO_2 reductions are realized.

Natural gas turbines commonly back up wind turbines.

Peaking natural gas turbines are natural gas combustion turbines (NGCT) installed to provide power at times of peak electricity demand, when continuously operating base-load power plants such as coal and nuclear are not sufficient. If winds are blowing at the same time as peak demand occurs, the NGCT units need not be turned on, cutting their CO_2 emissions.

If wind power were to become a major source of US electric power and reduce CO_2 emissions by closing some coal power plants, as much backup power would be required as wind power would be installed. Since there are few opportunities to expand hydroelectric power, because all the best sites have already been exploited, natural gas turbines are the only realistic backup choice for wind turbines.

Alternatively, think of this gas and wind pairing as principally a natural gas generator providing steady power, with occasional episodes of wind power allowing the gas power plant to curtail power generation and temporarily burn less fuel and emit less CO_2.

Introducing wind turbines can increase CO_2 emissions.

This seemingly illogical statement depends upon the choice of natural gas backup turbine types. Every 1,000 MW of wind turbine generation capacity requires 1,000 MW of natural gas generation backup capacity. Compare two choices a utility would make to build a 1,000 MW power plant system:

   1   wind turbines with natural gas backup
   2   natural gas turbine only, with no wind turbine.

In choice (1) the NGCT is used because it can start quickly when the wind lulls. With the wind turbine operating 30% of the time, the NGCT 1,000 MW(e) plant operates 70% of the time. The

NGCT electric/thermal efficiency is just 29%, so it consumes 70% x 1000 MW(t) / 0.29 = 2410 MW(t) of natural gas.

1,000 MW power plant alternatives					
	Power source	Cost $/W	Capacity factor	Efficiency	Gas burned
(1)	Wind turbine +	2.44	30%	-	-
	backup NGCT	0.67	70%	29%	2410 MW(t)
(2)	CCGT only	1.00	100%	60%	1670 MW(t)

In choice (2) the CCGT 1,000 MW(e) power plant operates steadily at 60% electric/thermal efficiency, consuming 1000/0.60 = 1670 MW(t) of natural gas.

The wind+NGCT power plant uses 44% more natural gas than the CCGT, venting 44% more CO_2 into the atmosphere. The CCGT plant costs $1 billion, while the less CO_2-efficient wind/NGCT plant costs $ 3.11 billion.

EPA proposes 454 g/kWh CO2 emissions limits.

What will be the effect of EPA's proposed rule? We can compute CO_2 emissions for various power alternatives.

- To compute grams of CO_2 per kWh, start with the heats of combustion per gram of coal and natural gas, kJ/g of fuel.
- Figure the fraction of C in CH_4 is the ratio of atomic weights of C to CH_4, 12/16, and approximate C in coal as 1.
- The fraction of C in CO_2 is the fraction of atomic weights of C to CO_2, 12/44. Multiply by these to get kJ/g of CO_2.
- The electric/thermal efficiency of power conversion varies with technology. One joule (J) = 1 watt-second, so 1 kWh = 3,600 kJ.

Grams of CO2 emissions per kWh					
Energy source	Heat, kJ/g of fuel	Heat, kJ/g of CO2	Effici-ency	Electricity, kJ/g of CO2	g of CO2 per kWh
Coal, conventional	33	9	33%	3	1200
Coal, advanced	33	9	44%	4	900
Natural gas, CCGT	50	18	60%	11	333
Natural gas, NGCT	50	18	29%	5	700
Wind + 70% NGCT backup					490
Wind + 70% CCGT backup					233

The last two rows in the table are just weighted averages for CO2 emissions for wind turbines at 30% capacity factor and gas turbines filling in the remaining 70%. Wind + CCGT backup is not common because CCGT power plants are not designed to stop and start. They can not achieve their maximum 60% efficiency during their ramp up times as they fill in for lulls in wind. If the wind remains steady or changes slowly enough, then the low 233 g/kWh CO2 emissions of CCGT might be achieved.

In March 2012 the US EPA proposed limits of 454 g/kWh for CO2 emissions from new fossil fuel power plants. One implication is that coal and NGCT power plants would be prohibited. Wind power backup might be possible from CCGT natural gas power plants.

In the future, wind with CCGT backup may emit less CO2.

Why not back up wind turbines with combined cycle gas turbines to achieve the least CO2 emissions? Increased cost is one reason. The main problem is that the more efficient CCGT power plants take hours to start up both their primary gas turbines and the secondary steam boiler and steam turbine. The situation may improve in the future. GE in Europe is introducing its FlexEfficiency 50 CCGT designed for use with intermittent solar power and wind turbines. It will go from zero to full power in 30-60 minutes, reaching 60% efficiency by the time it produces 87% of full power. None have yet been built, though EDF and GE have agreed to build one in France in 2015.

Gas turbines running at less than full power emit more CO2.

Gas turbines operate most efficiently at full power. Operating at half power reduces the electric/thermal efficiency by a factor of 0.85, increasing CO2 emissions per kWh by about 18%. They must continue to run at full speed to synchronize with the power grid. Gas turbines continue to consume fuel as they run at full speed when in spinning-reserve status, ready to provide power to the grid in an instant if some line or generator failure occurs.

Ramping, the process of changing power levels, also diminishes efficiency. Over a ten minute period, a gas turbine changing power levels from 60% to 40% back to 60% uses more fuel and emits more CO2 than running at 100% the whole time. An analogy is automobile mileage, which might be 22 mpg at a steady 55 mph speed, but only 15 mpg starting and stopping in city traffic.

The CO2 emissions impact of curtailed power, spinning reserve status, and ramping power is not included in the previous CO2 estimates.

Diversified wind turbines may reduce gas ramping penalties.

Wind power advocates claim that combining the outputs of many intermittent-power wind turbines will average out the lulls and make the total wind power less variable and more reliable. This is partially true.

Consider the example of wind farms in southeast Australia interconnected by a 1000-mile grid. This real-time data from 24 wind farms with a total capacity of 2 GW is published at http://windfarmperformance.info. For February 8, 2011, the capacity factors and totaled power are shown below.

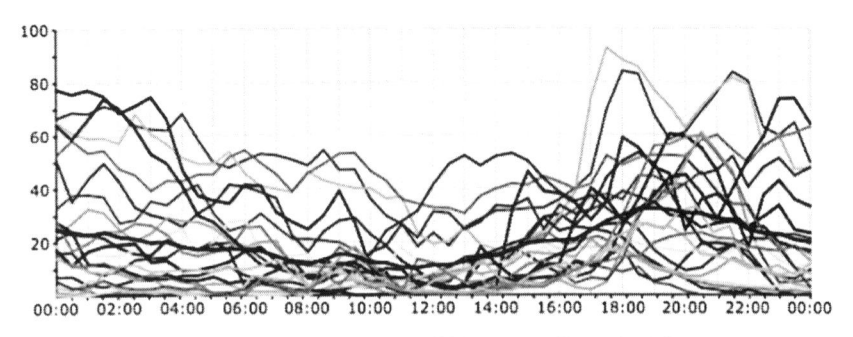

Capacity factors of 24 SE Australia wind farms

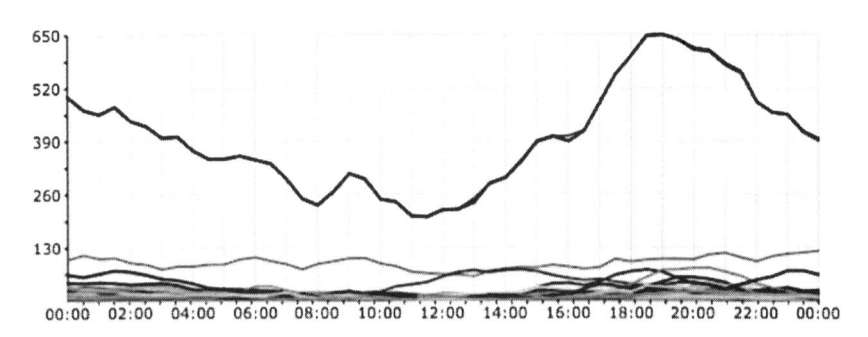

Total MW of power from all 24 SE Australia wind farms

In the graph above, the top line is the total of all the wind-generated power, ranging from about 250 MW to 650 MW; the 24 individual wind farm contribution lines are hard to distinguish. Total power generation ranges from 18 to 26 GW that day, for a capacity factor of about 20%. By inspection total wind power is clearly steadier than that of individual wind farms. Power ramp rates appear to be about 100 MW/hour, which should be within the capabilities of combined cycle gas turbines. So in this case the 80% of nameplate capacity not provided by wind turbines might be provided by the efficient combined cycle natural gas turbines, provided that the energy could be transported efficiently throughout the 1000 mile grid.

Field studies show minor CO2 reductions from wind power.

Field studies in Ireland, Colorado, and Texas measured the effects of introducing intermittent wind power and curtailing coal and natural gas power.

The Ireland electric grid is powered 65% by natural gas and 10% by wind power. The operator published detailed data of CO2 emissions, total power demand, and wind power supplied at 15 minute intervals. The data for April 2010 show that if wind supplies 12% of the power, CO2 is reduced by only 3% compared to no wind. This is because the fossil-powered plants became less efficient as their power output was ramped up and down as wind power contributions changed.

Public Service of Colorado must curtail output from coal power plants to accept wind power reductions. They operate less efficiently at lower power and while ramping power up or down. The Denver area carbon dioxide and nitrogen oxide emissions increased about 5%, while SO2 emissions rose by 18 to 172%.

The Texas grid is 58% powered by natural gas, important to allow large power swings as summer air conditioning demands change. Wind power was more easily accepted because of this, but some coal power plant cycling was also necessary. The Bentek study found that SOx and NOx emissions were higher than if the coal plants had continued to run at full capacity, and that the CO2 emissions savings were minimal at best.

Electricity cheaper than from wind must be < 18 cents/kWh.

Determining the cost of wind power is difficult. Most news articles report costs that are net of state and federal government subsidies, tax preferences, tax credits, renewable energy credits, production tax credits, etc. To make the proper energy decisions for society, full costs must be considered. However, prices are often confidential. Capital cost examples above are: $2.44/W (EIA), $5.80/W (Cape Wind), and $7.00/W (Deepwater Wind). We don't even count the necessary capital costs for backup generators adding at least $0.67/W for natural gas NGCT. We'll take Cape Wind as our capital cost example, $5.80/W. We assume a 40-year lifetime, 30% capacity factor, 8% interest rate, leading to 17.4 cents/kWh for capital cost recovery.

In our cost computation we can set the cost of fuel to zero for wind.

Wind electricity cost	
Capital cost recovery	17.4 cents/kWh
Fuel	0.0
Operations	1.0
Total	18.4 cents/kWh

Reported electricity cost examples are 16-24 cents/kWh (Cape Wind) and 24 cents/kWh (Deepwater Wind). So our cost computation of 18.4 cents/kWh is in the ballpark for reported prices. LFTR energy must cost less than 18 cents/kWh to be competitive with unsubsidized wind power.

Solar

Harnessing energy directly from the sun can be accomplished in several ways.

1. Passive solar building heating.
2. Solar hot water heating.
3. Photovoltaic solar electricity generation.
4. Concentrated thermal solar electricity generation.

We'll treat growing plants separately in the biofuels section.

Passive solar heating absorbs sunlight within a building.

Especially in winter when the sun is low in the sky, sunlight radiates through large, transparent, insulated glass windows into the building.

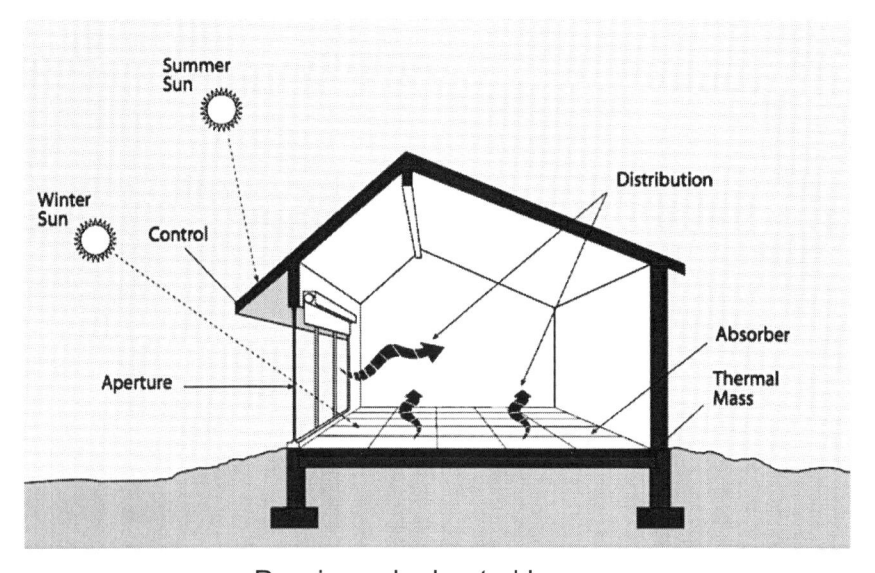

Passive solar heated house

The light is absorbed and so converted to thermal energy by all solid objects inside, including a thermal mass that heats up storing energy for later use. The thermal mass radiates invisible infrared light that does not pass through glass and is absorbed by other objects in the room, much like a greenhouse works.

Along with super insulation and low air infiltration, solar heating is a key component of Passivhaus design, which seeks to have an average heat demand < 2 watts per square meter (15 kWh/m² per year) or a maximum peak heat demand < 10 W/m². Such a 2000 square foot (200 m²) home could be heated with less than 3,000 kWh per year, costing $450 if heated with just electricity costing 15 cents/kWh, or about $150 with a heat pump instead of simple resistance heating.

Solar hot water readily displaces fossil fuel CO2.

Many domestic hot water heaters in the US use natural gas or electricity to heat water. A typical US electric hot water heater draws 4,500 W(e). If the electricity is generated by a power plant with the typical 33% electric/thermal conversion efficiency, then 13,500 W(t) of heat flow from burning CO2-producing natural gas or coal is required. Water heated directly from natural gas would require only 4,500 W(t), plus any heat lost out the flue. This reduces costs for the homeowner and reduces CO2 emissions.

Chinese apartment building with solar hot water heatser

Worldwide, China has the most solar hot water heating by far, over 100 GW(t) in 2009, exemplified in this photo of an apartment building. Water simply passes through the solar collector into interior plumbing. Where freezing might occur, the exposed collector may heat glycol, which indirectly heats hot water in a tank. Supplemental electric or natural gas heaters can increase the temperature if demand requires or the sun is not shining.

Photovoltaic cells convert sunlight directly to electricity.

Thermal energy powered generators use a heat engine to extract kinetic energy from heat flowing from hot to cold. Examples are coal fired plants, nuclear power plants, and concentrated thermal solar plants. All require cooling towers or water to absorb the heat than can not be used. PV solar cells are quite different. PV cells convert photons of light energy directly into electric energy at efficiencies up to 10%.

AllEarth Renewables solar farm in Vermont

The solar panels in this photograph of the AllEarth Renewables solar farm in Vermont rotate and tilt to face the sun squarely and maximize the power produced. At the equator, incident radiation from the sun at noon is 1000 W/m^2. At latitudes near 45° a solar farm like this generates an average of about 5 W/m², accounting for nightfall, clouds, and conversion efficiency. A solar farm in the southwest US could double this productivity to 10 W/m². Such a solar farm with an average capacity of 1 GW would require 100,000,000 m² of land area – nearly 50 square miles.

Scientific American published the *Solar Grand Plan*, generating 69% of US electricity in the southwest with compressed-air gas-reheated energy storage, requiring a $420 billion subsidy. It proposes 46,000 square miles of solar arrays with continental high voltage transmission lines supplying 69% of US electricity.

Concentrated solar power creates electricity from heat.

In the following photo of a concentrated solar power plant, sunlight falling on the parabolic reflector is concentrated on the thin tube along the focus of the parabolic trough.

Andasol concentrated solar power plant in Spain

Oil flowing through the tube is heated and used to make steam that drives a steam turbine turning an electric power generator. Another design is a field of mirrors that direct sunlight to a power tower where very high heat can be obtained. A concentrated solar power plant uses thermal to electric power conversion, as do coal, nuclear, and natural gas plants. Achieving high temperatures, the conversion efficiency can be 41%. Such plants similarly require heat rejection systems such as cooling towers that evaporate water, which may not be readily available in the hot, arid regions with clear skies where the sun shines intensely.

Concentrated solar power plants can store energy as heat.

The Andasol 50 MW(e) concentrated solar power plant in Spain also uses tanks of molten salt (60% $NaNO_3$ + 40% KNO_3) to store thermal energy. The 30,000 tons of molten salt can store 1 GWh(t) of energy and can accept or discharge the heat at over 100 MW(t). Thermal energy storage has also been proposed for molten salt reactors such as LFTR.

Electricity cheaper than from solar must be < 24 cents/kWh.

The costs of photovoltaic (PV) solar cells are decreasing. China has become the low cost producer, putting US solar cell companies out of business. Costs are expected to drop to $1/watt of capacity and such numbers are widely reported in news articles, but that's just for the PV solar cells, not for the entire solar farm. In a utility scale solar farm there are many other cost elements besides the PV solar cells. Examples are weather resistant frames and panels supporting the cells, tracking motors that point the panels at the sun, DC/AC power conversion equipment, control and monitoring systems, and interconnecting cables.

It can be difficult to find true total costs in media reports; often the information is confidential. Here are four examples of published total cost and total capacity.

Solar cells cost about $1.75/W at the AllEarth Renewables solar farm we visited outside of Burlington VT, reportedly representing about 35% of the construction cost. This 2130 kW capacity solar farm project cost $12 million, or $5.63/W of capital investment. Even if solar cells had zero cost, this solar farm would still cost about $4/W. The measured capacity factor over 7 months was 18%. This Vermont solar farm sells electricity to the public utility company at 30 cents/kWh by means of a feed-in-tariff that the legislature requires of the utility.

In 2009 Spanish company Albiasa announced a $1 billion, 200 MW concentrating solar power project to be installed in sunny Kingman, Arizona, implying a cost of $5/W. The project was abandoned in 2011. Albiasa is constructing a 50 MW solar plant in Caceres, Spain, with a contract to sell its energy at 27 eurocents/kWh, about 35 cents/kWh.

Abengoa is building a 280 MW solar thermal power plant outside Phoenix, Arizona, at a cost of $1.6 billion, or $5.71/W of capital investment.

Also in the US, Brightsource is building a concentrating solar plant on 3600 acres of the Mojave desert in California. Moving mirrors will direct sunlight to a 459 foot tower to collect the thermal energy. The reported cost is $2.2 billion for 370 MW, or $5.60/W.

The AllEarth and Albiasa electricity prices were 30 and 35 cents/kWh. Vermont recently reduced its feed-in-tariff from 30 to 24 cents/kWh.

Solar farm investments and electricity prices		
Builder	Capital cost $/W	Electricity price cents/kWh
AllEarth Renewables	5.63	30
Albiasa	5.00	35
Abengoa	5.71	-
Brightsource	5.90	-
US DOE EIA est.	4.70	-

The US EIA estimates capital costs for PV or solar thermal at about $4.70/W. Comparing this to the above examples, we'll accept that the investment in solar is about $5.00/W. The capacity factor for solar varies with latitude and weather patterns; we'll model costs using 20%. Capital investments are recovered when the sun is shining and power is being generated, so at 20% capacity factor and an 8% cost of capital and 40 year life (our standard assumptions) we have the following cost model.

Solar generated electricity cost	
Capital cost recovery	22.5 cents/kWh
Fuel	0.0
Operations	1.0
Total	23.5 cents/kWh

Electricity cheaper than from solar must be < 24 cents/kWh.

INTERMITTENT WIND AND SOLAR POWER

Intermittent power can raise emissions and costs.

In 2011 MIT published *Managing Large-Scale Penetration of Intermittent Renewables*, which delves more deeply into the difficulties of supplying steady electricity with intermittent sources.

> "In addition, fuel efficiencies will decrease when thermal generation plants are operated at partial load. Lower fuel efficiencies increase emissions rates and total costs, potentially diminishing the benefits of renewable generation. Continuously altering plant output also increases the need for operation outside of normal, steady-state procedures and the likelihood of operator error."

Solid Biofuels

Biofuel typically means a liquid fuel extracted from vegetation and often refined for use as a vehicle fuel. However the most energy that could be extracted from vegetation such as wood is the energy released simply by burning it. Making ethanol from wood reduces the potential chemical energy that could be obtained from burning wood. Many electric power plants do burn wood, so we start our analysis with burning wood.

About 20% of forest mass is carbon.

In the 15th century Van Helmont discovered that a growing tree gained weight from matter in the air – water vapor and "wood gas" which was later determined to be carbon dioxide. Lavoisier identified carbon and the plant respiration process before being guillotined in the French Revolution.

Carbon is about 50% of the mass of the carbohydrates that constitute dry wood. Trees growing in a forest can be 60% water and the other 40% dry wood. Thus carbon, taken from carbon dioxide in the air, represents about 20% of the mass of the wood forest.

Forests absorb carbon dioxide from air until maturity.

As trees grow and absorb carbon dioxide from the air, they incorporate the carbon in their hydrocarbon structures. The rate depends on variety and climate. For our model we'll use 3 tonnes of carbon per hectare per year (3 t-carbon/ha/yr). A hectare is a square 100 meters on each side – 2.47 acres. Green trees contain 20% carbon; they add mass at five times this rate, 15 t-wood/ha/yr.

After a century or so, forests achieve maturity as trees die and rot at the same rate they are regrown. This carbon sequestered in the mature forest is 100 to 600 tonnes/hectare, depending on geography, climate, variety of trees, and fires. The EPA estimates 250 t/ha in the southern US. IPCC default value is 200 t/ha.

Carbon is removed from the atmosphere as forests grow from start to maturity, but not thereafter. Wood-fired electric power plants are said to be carbon neutral because the wood will create atmospheric CO_2 whether it rots on the ground or burns in a furnace. Burning a forest puts CO_2 into the atmosphere until the forest regrows and absorbs it.

Wood-fired electric power plants are not sustainable.

Wood fired electric plants burn wood chips from trees that have been harvested but not dried. How much would it take to fuel our standard 1 GW electric power plant? The USDA says burning one tonne of such green wood chips produces almost 2 MWh(t) of heat. Assuming the usual 33% electric/thermal efficiency, for 1 GW(e) of electricity we would need

$$1 \text{ GW(e)} = \frac{24 \times 365 \text{ h}}{\text{year}} \times \frac{3 \text{ kWh(t)}}{1 \text{ kWh(e)}} \times \frac{\text{tonne-wood}}{2 \text{ MWh(t)}}$$

= 13 million tonnes of green wood chips each year.

At a tree growth rate of 15 t/ha/yr this requires almost a million hectares, nearly the land area of the state of Connecticut. Connecticut consumes more than 3 GW of electricity on average, but could only grow enough wood for 1 GW(e).

Wood-fired electricity generation is insufficient and clearly unsustainable. This is observable in history; England was largely deforested by burning wood before turning to coal for energy in the Industrial Revolution. Biomass-fueled electricity generation in the US is about 8 GW(e).

Energy cheaper than burning wood must be < 10 cents/kWh.

The reported replacement cost of the Springfield NH 19 MW wood-chip burning power plant is about $90 million, or $4.74/watt. The proposed Berlin NH 75 MW plant investment cost was $275 million, or $3.67/watt. EIA estimated $3.86/watt. The

Southern Company is building a $500 million, 100 MW plant in Nacogdoches TX, costing $5/watt. Let's model capital costs at $4/watt. We also assume our standard 8% cost of capital, 40 year life, and 90% capacity factor, leading to 4 cents/kWh for capital cost recovery.

Whole tree green wood chip prices vary with location, market conditions, and transportation costs from forests to power plants. The Burlington VT power plant pays $15-30/ton; Ryegate NH paid $12-20/ton; Worcester MA plans $22-34/ton; Springfield NH averages $28/ton. For our model we estimate $28/ton, or $31/tonne, and an optimistic 33% electric/thermal efficiency.

$$\frac{\$31}{tonne} \times \frac{tonne\text{-}wood}{2\ MWh(t)} \times \frac{3\ kWh(t)}{1\ kWh(e)} = \frac{\$47}{MWh(e)}$$

or about 4.7 cents/kWh for the fuel. Using our standard assumptions on capital cost recovery and operations, the cost of wood-fuel generated electricity is about

Wood fuel generated electricity cost	
Capital cost recovery	4.0 cents/kWh
Fuel	4.7
Operations	1.0
Total	9.7 cents/kWh

To undersell the cost of electricity from burning wood, the cost of LFTR electricity must be less than 10 cents/kWh.

Energy from renewables costs more than from fossil fuels.

Below is a summary of the cost analyses above. Note that the costs for wind, solar, and biomass electricity are much more than the costs for CO2-emitting coal and natural gas. Moreover the costs for intermittent wind and solar do not include any costs for backup power needed for intermittency.

Electricity costs from alternative sources, cents/kWh					
	Coal	Gas	Wind	Solar	Biomass
Capital cost recovery	2.8	1.0	17.4	22.5	4.0
Fuel	1.8	2.8	0	0	4.7
Operations	1.0	1.0	1.0	1.0	1.0
Total	5.6	4.8	18.4	23.5	9.7

Why are the relative costs for renewables so high? And how are they paid for? One factor is energy density. Wind turbines and solar collectors are spread over large land areas. Distant forests supplying wood-fueled generators require long-distance trucking.

Subsidies pay most of the costs for renewables. Federal and state governments pass many different laws and regulations that serve to make it profitable for developers to invest in wind and solar power. State subsidies have included tax credits of 30% of capital costs. Federal incomes tax subsidies are an additional 30%. Production tax credits pay developers 2.2 cents/kWh of generated electricity. Feed-in tariffs require utilities to pay high prices (20-30 cents/kWh) to buy wind and solar power, whenever it is available. State renewable portfolio standards require utilities to buy minimum amounts of wind and solar power, whatever the cost. Utilities recoup these above-market costs by raising prices to consumers for all power.

Liquid Biofuels

It makes no economic sense to burn biofuels in a power station, because less expensive biomass such as wood can be burned directly and more efficiently, as described above.

The principal objective for liquid biofuel manufacturing is to replace gasoline and diesel petroleum-derived liquid fuels in vehicles. These carbon-based fuels are exceptionally valuable for transportation. They have high energy density, which allows the energy source to be carried onboard a car, truck, train, or airplane. Trains can be electrified, small cars can carry the weight of batteries, but large trucks and airplanes need an energy-dense fuel such as diesel. Biofuels are but one replacement for petrofuels.

Substituting ethanol for gasoline *may* reduce CO2 emissions.

Burning petroleum fuels emits almost as much CO2 as burning coal, worldwide. Using carbon-neutral fuels derived from biomass is an attractive way to check global warming. In concept, emissions from biofuels are reabsorbed from the air as the next biomass crop is grown. Unlike forests that take a century to reach maturity, crops such as corn or sugar beets can be grown annually or more frequently. However, if a 100-year-old forest is cut, burned, and replaced with a 1-year crop, the net 99% of the carbon becomes CO2 that stays in the atmosphere; this is a land use change.

Ethanol biofuel in the US is derived from fermenting the starches and sugar in kernels of corn grown for this purpose. Today most gasoline in the US is diluted 10% with corn ethanol, with the objective of reducing CO2 emissions somewhat by substituting some renewable ethanol energy for petroleum-based energy. Another objective is gaining some energy security by reducing the demand for imported oil.

It is not clear that these objectives are met. Burning the ethanol in a gasoline engine yields only 2/3 the energy of gasoline, so the gasoline savings are only about 7%, not 10%. Much of the energy needed to save that gasoline comes from fossil fuels.

US corn ethanol return on invested energy is poor.

Making ethanol by farming and refining corn requires fertilizer, irrigation, transportation, and refinery operation, all of which require energy; approximately 74% is derived from fossil fuels. Energy Return on Investment (EROI) is the ratio of energy produced by burning ethanol to the sum of the energies consumed from all sources except sunlight to make the ethanol. Studies of EROI have been controversial, with some showing a net energy loss from manufacturing ethanol biofuel. A recent analysis of many studies concluded that EROI varied with many locations and other factors; the average value is EROI = 1.07 ± 0.2 within a 95% confidence interval – close to no gain at all. In comparison, EROI for new discoveries of unconventional oil is at least 5, so ethanol energy is much more energy-expensive than petroleum-derived energy.

The climate of Brazil allows farming of sugar cane, which has more starch and sugar content than corn, so the Brazil ethanol EROI is near 8. Ethanol use in the US has been encouraged by subsidies, by tariffs against imports, and by mandates.

Cellulosic ethanol is not yet economically feasible.

About half the mass of farmed corn is the grain; the remainder is stover – the residue of stalks, leafs, husks, and cobs. About 77% of that corn grain is the starch and sugars that can be commercially fermented and refined into ethanol. The remaining stover is principally cellulose. Cellulosic ethanol would be the result of also processing the more plentiful parts of vegetation such as stover in corn and cellulose in wood.

Commercializing the process for producing cellulosic ethanol admits a wide range of vegetative feedstocks, including trees and switchgrass. The production process breaks down the long chain molecules of cellulose and hemicellulose into sugars, which can be fermented and refined, in laboratory conditions.

As of 2012 there are no commercial-scale cellulosic ethanol refineries in the US. POET-DSM advanced biofuels is building an ethanol production factory in Emmetsburg IA at a cost of $250 million and capable of producing 20 million gallons per year from 250,000 tons of biomass. The company's pilot plant in Scotland, South Dakota, produces 80 gallons of ethanol from one ton of biomass, daily, at an estimated production cost of $2.50 to $3.00 per gallon, more expensive than corn ethanol.

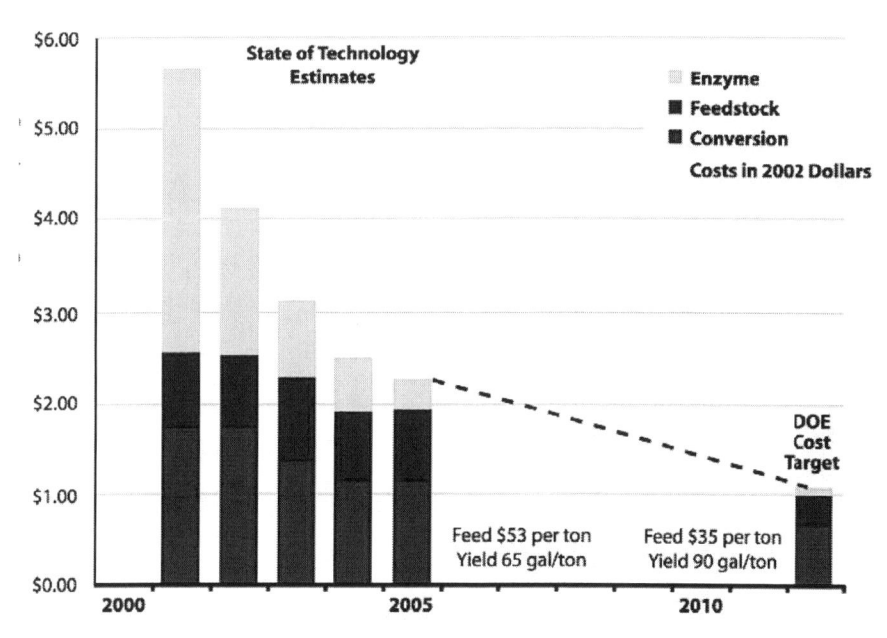

DOE 2005 projections for cellulosic ethanol cost

US National Renewable Energy Laboratory predicted cellulosic ethanol prices could drop to $1.35 per gallon in 2012 dollars, at a yield of 90 gallons per ton of biomass. Subsequently Congress mandated that refiners purchase 250 million gallons of cellulosic ethanol in 2011, but only 7 million were available.

Farming biomass for fuel raises food prices.

A side effect of farming biofuels is the increase in prices for food. Doubling the price of corn caused riots about tortilla prices in Mexico. In the US nearly 40% of corn is raised for biofuels, not food for people or beef cattle. The US mandated biofuel consumption of 36 billion gallons per year by 2022, or which 16 billion gallons must be cellulosic ethanol. In contrast, food shortages led China to ban the conversion of grains into biofuels. China now imports cassava for biofuels from Thailand, doubling its price there and encouraging land use for this crop. Europe's 10% by 2020 biofuel mandate led to forcible displacement of 3,200 farm people in Guatemala to instead grow sugar cane for ethanol for European cars and trucks.

Biomass to ethanol energy conversion efficiency is < 32%.

The US Department of Energy calculates the theoretical yield of ethanol from dry biomass for several crops; 100 gallons per dry ton is typical, but an economic yield will be about 60 gallons per dry ton of biomass. Burning a ton of dry wood or other biomass liberates about 15,000 MJ of thermal energy, but 60 gallons of ethanol would release only 4,800 MJ – about 32% of the chemical potential energy of the biomass.

This analysis ignores the external energy inputs to the biofuel refining process, such as electricity and natural gas, so the full energy-out/energy-in efficiency is much less than 32%.

US biodiesel consumption is small.

Biodiesel fuel is largely produced from canola or soybean sources. Annual production is under 1 billion gallons, compared to ethanol production over 10 billion gallons. Biodiesel is most commonly blended with 80% petroleum diesel and the resulting "B20" mixture is termed "biodiesel" at the pump. In cold weather B2 and B5 blends are sold because they do not gel. The energy content of pure biodiesel is about 9% lower than petrodiesel.

Energy Storage

Engineers have sought electric energy storage systems for decades. Electric power distribution systems do not have energy storage other than in the angular momentum of the turbo-electric generators. Any new power demand must be satisfied in seconds by increasing steam fed to turbines from boilers in nuclear, coal-fired, or wood-fired plants, water fed to hydroelectric plants water wheels, or natural gas supplied to gas turbines. Wind and solar power can not be dispatched like this.

Energy storage would benefit two situations: varying demand, and intermittent supply. Nuclear and coal plants have high capital costs and are designed to generally operate at full power. Electricity demand from consumers varies by a factor of two, so storing excess energy during periods of low demand could later provide supplemental energy when consumer demand rose beyond generating capacity.

The other benefit would be to store energy supplied from uncontrollable intermittent wind and solar power generators, making them more dependable sources and reducing the need for backup natural gas power generation.

Energy storage devices have three important parameters: power, energy, and efficiency. Power is the maximum sustained power that can be delivered, in MW. Energy is the total energy that can be stored, often in MWh. Efficiency is the ratio of the energy delivered to the energy stored.

Rechargeable storage batteries use chemical energy.

Batteries have two electrodes of different metals connected by a conductive electrolyte liquid or solid. When negatively charged electrons from the charger flow from anode to cathode there is a flow of positively charged anions from anode through elecdtrolyte to cathode. This changes the chemical states of the electrodes and the electrolyte, storing chemical potential energy. In discharging the flows are reversed and the chemical states revert.

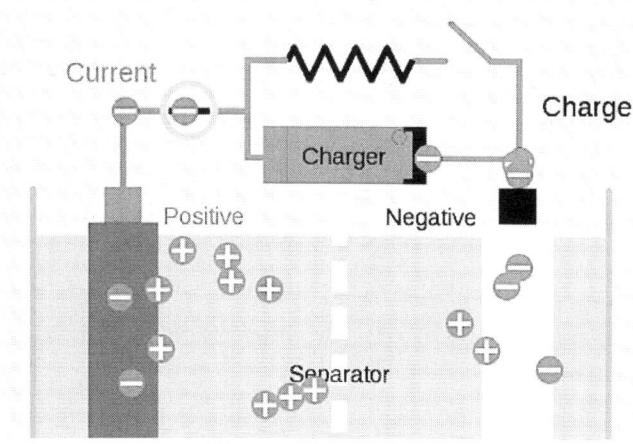

Storage battery

Batteries can be made of many combinations of metals and electrolytes, such as lead/sulfuric-acid/lead-oxide in common automobile batteries. Lithium ion batteries are common in consumer electronics and even the Tesla electric automobile. Sodium-sulfur batteries were developed for electric power applications. Fabricated from inexpensive materials, these batteries must operate at high temperatures, up to 350°C.

Charging today's electric vehicle batteries can take overnight. Flow batteries have liquid electrolytes that might be removed and replaced. In a future automobile the discharged electrolytes might be replaced with charged electrolytes at a filling station.

In Israel the company A Better Place is testing a different filling station method – a battery exchange station where a discharged one is exchanged for a charged one in 5 minutes.

Donald Sadoway's group at MIT has designed a liquid battery with molten antimony metal at the bottom, covered by a layer of molten salt electrolyte, topped with molten magnesium metal, operating at 700°C, and capable of scaling to meet utility needs. A battery that could deliver 1 GW of power for 48 hours would cost $1.8

billion just for the magnesium and antimony. That would be near the price of a LFTR that could deliver 1 GW continually.

Batteries meet some special electric utility needs.

Using batteries requires AC to DC to AC conversion and chemical potential energy storage, with round-trip efficiency up to 75%. The largest utility battery storage system is in Rokkasho, Japan, 245 MWh at 34 MW, used to store wind energy. It uses NGK's sodium-sulfur technology, reported costing $3/W.

In Fairbanks, Alaska, a 7 MWh, 27 MW nickel-cadmium battery system is used to stabilize power. A 5 MWh, 40MW, 1200 ton battery bank is used to supply that isolated city during a power failure while emergency diesel generators start.

A typical large 1 GW power plant supplies 24,000 kWh per day – ten times the capacity of the largest battery system ever built. Microsoft founder and energy investor Bill Gates pointed out that all the world's batteries could not supply all the world's electricity for more than 10 minutes.

Flywheels can store electric energy.

In New York state Beacon Power has installed a system of 200 high speed flywheels each capable of providing 25 kWh of energy at 100 kW of power for 15 minutes. The purpose of such systems is to stabilize the grid, so that small fluctuations in power demand do not require rapid changes to fossil fuel power plants' fuel burning, which lowers their efficiencies and so increases CO_2 emissions and air pollution.

Pumped storage hydroelectricity requires two reservoirs.

Electric energy can be stored with a relatively high round-trip efficiency of 75% by using excess generated power to pump water to a high elevation. The water later is drawn down to spin generators when demand rises. Such pumped hydro technology accounts for 99% of all stored electric energy capacity in the US.

For example, the illustrated Raccoon Mountain pumped storage plant in Tennessee pumps water from a reservoir up 1,000 feet into another reservoir built at the top of a mountain. The stored water can run the 1.6 GW power plant for 22 hours, expending the reservoir's 35 GWh of stored energy. Adjusting for inflation, the 1979 $300 million plant would cost $1 billion, or 3 cents/Wh of energy storage, or $0.63/W of generation capacity.

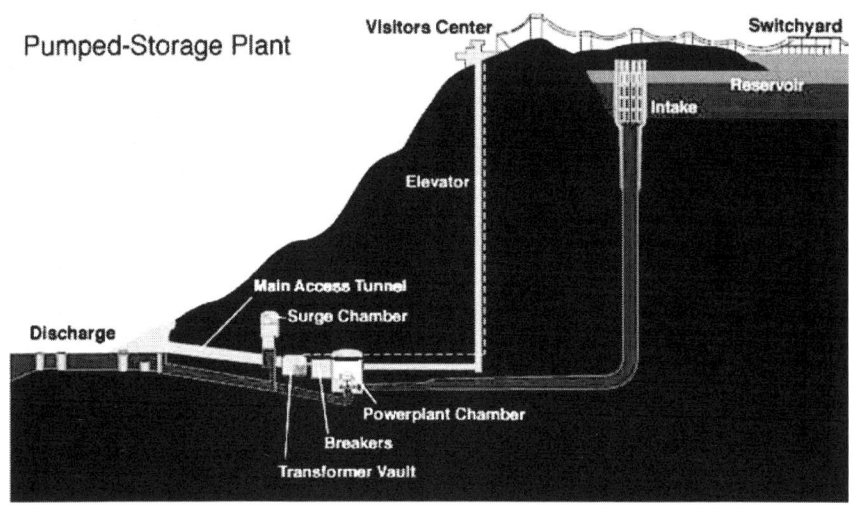

Raccoon Mountain pumped hydro storage

Pumped hydro storage requires two water reservoirs, and there are few possible sites remaining in the US, where permitting is also more difficult than in 1979. In contrast to Raccoon Mountain, cost estimates now are about $2/W of power generation capability, or 25 cents/kWh of energy storage.

Compared to pumped hydro, batteries are more expensive but more energy dense. A single AA battery can store nearly 10,000 joules of energy – about 2.5 watt-hours. Pumped hydro storage of that same energy would require pumping one liter of water to a height of 1,000 meters.

Compressed air energy storage also uses natural gas.

Using an electric motor and turbine or pump for compressing air into a tank raises the stored air pressure and temperature, rather like compressing a spring. When the air is released to flow back through the turbine and motor-now-generator, the energy stored in the compressed air is transformed back to electric energy.

Except for the motors and pumps, this process could be nearly 100% efficient if the tank of compressed air were perfectly insulated. You may have noticed that a tire air compressor heats up its tank; the tank then cools and loses heat; air released later is cold; compressed air energy storage is not 100% efficient.

Utility scale compressed air energy storage (CAES) has similar losses. The McIntosh, Alabama CAES plant stores compressed air in an underground salt cavern. Releasing it can deliver 2.6 GWh of energy at 100 MW for 28 hours. Because heat is lost from the compressed air, it must be reheated by natural gas when generating power. What is the round-trip energy efficiency?

The CAES output energy comes from two inputs: input electricity that compresses the air, and natural gas that reheats the air. EPRI reports that 1 kWh(e) of output requires 0.82 kWh(e) of input electricity and 1.34 kWh(t) of natural gas. A 60% efficient modern combined cycle natural gas turbine generator (CCGT) could have generated 1.34 x 0.60 = 0.80 kWh(e) with that same natural gas.

The comparable energy input is 0.82 + 0.80 = 1.62 kWh(e). So the round trip energy efficiency is 1.00/1.62 = 62%.

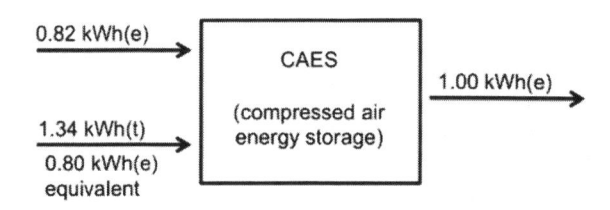

Compressed air energy storage efficiency

The McIntosh CAES plant cost $53 million in 1991, comparable to $89 million in 2012 dollars. That's $0.034/Wh energy storage capacity, or $0.89/W power generation capacity. The energy storage capacity cost is about triple EPRI's estimate in the following table, but this was a first-of-a-kind CAES plant in the US.

Energy storage costs vary with technology.

The Electric Power Research Institute tracks energy storage projects. In their report EPRI characterizes the capital costs in two ways: cost per unit of power capability, and cost per unit of energy storage. These costs don't add; they are just two different analyses.

Electric energy storage system estimated capital costs			
Storage technology	Efficiency	Power delivery capacity cost, $/W	Energy storage cost, $/Wh
Pumped hydro	80%	1.50 - 2.70	0.25 – 0.27
Advanced lead acid	90%	4.60 - 4.90	0.92 - 0.98
Lithium ion	90%	1.80 – 4.10	0.95 – 1.90
Compressed air	70%	0.96 - 1.25	0.06 – 0.12
Flywheel	85%	1.95 – 2.20	7.80 – 8.80
Sodium sulfur	75%	3.10 – 3.30	0.52 – 0.55
Zinc bromine flow	60%	1.45 – 1.75	0.29 – 0.35

The far right column is the capital cost for the capability to store (not generate) the electricity. Pumped hydro and below-ground compressed air stored energy costs are the lowest, under 20 cents/kWh. Lead acid battery stored power costs 40 to 60 cents/kWh.

Energy storage adds to the cost of electricity.

How does the cost of batteries or other energy storage devices add to the cost of electric power? Using our standard financial model, that capital costs be recovered over 40 years at 8%, assume that the storage devices cycle once per day. Doing the math on your spreadsheet or financial calculator, an investment of $1/Wh for storage capacity, repaid over 365 x 40 days, costs 23 cents/kWh, reflected in the following table, based on the midrange capital cost.

Added electricity costs for energy storage		
Storage technology	Energy storage cost, $/Wh	Capital cost recovery, cents/kWh
Pumped hydro	0.25 – 0.27	6
Advanced lead acid	0.92 - 0.98	21
Lithium ion	0.95 – 1.90	33
Compressed air	0.06 – 0.12	2
Flywheel	7.80 – 8.80	191
Sodium sulfur	0.52 – 0.55	12
Zinc bromine flow	0.29 – 0.35	7

Some caveats are in order. The table is a very rough guide. It does not represent the effects of efficiencies. It does not include the cost of electric power obtained to store. For compressed air it does not include the cost of natural gas. For flywheels, the cost is high because it is based on only one cycle per day.

Batteries are a very expensive solution to intermittent power.

To make intermittent power from wind or solar more reliable, generated energy could be stored in batteries. What would be the cost of storing one day's wind generation in batteries, to be used on a windless day? From the table above we estimate that $4.75/Wh is a reasonable estimate of the capital cost for a lead acid battery, so purchasing the storage for 24Wh would cost $114. This dwarfs the capital cost of wind turbine farms, at $5.80/W.

So installing a wind generation system with lead-acid battery capacity to store and deliver the energy a day later would cost $120/W. Because the batteries only smooth power availability, the average capacity factor of such a wind/storage system would still be 30%, and so the generated energy cost would be 360 cents/kWh. This is about 70 times today's cost for electric energy from coal, natural gas, hydro, or nuclear power.

Siemens proposes using hydrogen to store electric energy.

Siemens has developed new water-electrolyzing technology that can start and stop as intermittent wind or solar power varies. Based on proton exchange membrane technology used for fuel cells, the warehouse-sized units dissociate hydrogen from water at an energy conversion efficiency of 60%. The hydrogen would be stored, then burned by turbine generators to produce electricity at a maximum 60% efficiency. The round-trip, electricity-to-electricity energy efficiency would be no more than 35%, with the rest lost to heat. As inefficient as this is, Siemens sees this as the only way for Germany to store electric energy on a scale that would replace coal and nuclear power plants with wind and solar generation. One concern is the likely high costs of the technology, which has been too expensive for vehicles. Storing the vast amounts of hydrogen will also be a challenge. Some hydrogen can be compressed and added to the natural gas stored in pipelines that feed existing natural gas turbine electric power plants, replacing some burning of natural gas. However the small molecules of hydrogen can leak through metals, so pipelines and natural gas distribution systems would have to be lined with a material such as Teflon.

Molten salt can store pre-electric thermal energy.

Although molten salt can not store electricity like a battery, it can store thermal energy. At a concentrated thermal solar power generation station the sun's heat is normally used to make steam to run turbines that generate electricity. Alternatively, the thermal energy can be stored for many hours by heating an insulated tank of molten salt. Hours later that molten salt can make steam that makes electricity. The thermal-electric energy conversion is the same, just delayed. There will not be efficiency losses of battery charge-discharge cycles, only the loss of heat from the tank.

HYDROELECTRIC POWER

Hydro electric power opportunities are limited.

Hydro power is an attractive source for generating electricity. It is price competitive with coal, natural gas, and nuclear power (roughly 5 cents/kWh). It emits no CO_2. It is renewable, extracting energy from sun and rain. Capital costs can be recovered over a 50-100 year lifetime.

Worldwide hydro power generation averages 390 GW, supplying 16% of all electricity consumption.

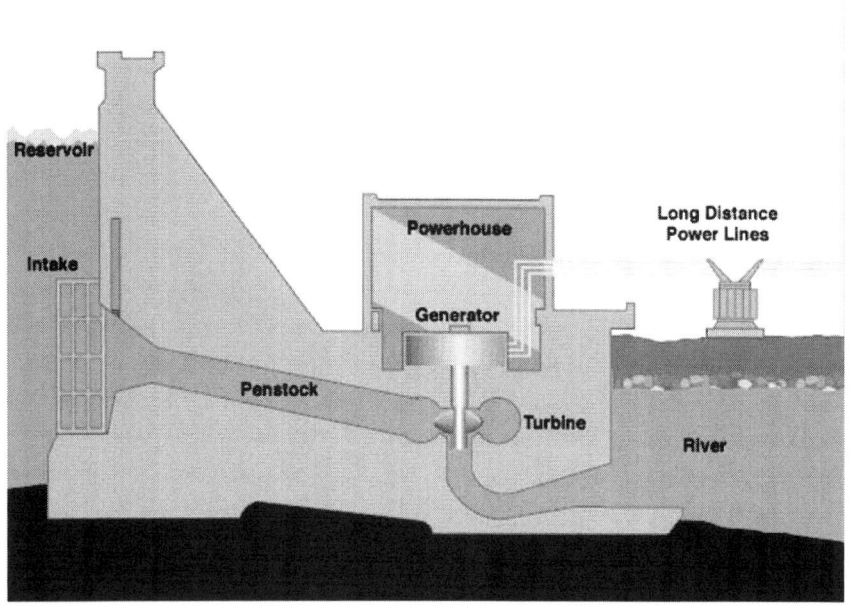

Hydroelectric power plant

Hydro electric power is controllable; it can be turned on or off gradually or in minutes. It can provide non-CO_2-emitting backup power for intermittent wind and solar power sources. Where coal and nuclear power plants usually run a full capacity, additional hydro power can be generated as needed to supply peak demand.

Hydro power plants generally use water from reservoirs faster than it is replenished, so hydro plants are often just used for peak demand periods, when market prices for electricity are highest. Capacity factors are typically under 50%. The Three Gorges Dam, the world's largest with a maximum capacity of 20 GW, is planned to provide an average power of 14 GW.

Sites for hydro electric power plants are limited to places where substantial rainfall flows into an area suitable for a reservoir. Such energy drives industrial development. Power from the Columbia River Grand Coulee Dam was at first dedicated to the aluminum smelting industry in Washington state. Aluminum for aircraft construction was critical to winning World War II. Alcoa and Boeing have plants there.

Worldwide, about 100 GW of capacity of hydro power plants were under construction in 2012. Building more hydro electric dams is difficult because most of the best sites have been already used. The environmental impacts of flooding so much land and displacing many people also limit expansion of hydro power. In the US some dams are being dismantled to restore natural water flow.

The proposed Grand Inga Dam in the Democratic Republic of Congo could generate 39 GW, nearly doubling Africa's electricity, at a capital costs of $8 billion, or about $2/watt. Political instability prevents building it. Excluding industrialized South Africa and the northern Mediterranean countries, this part of Africa has electricity energy poverty -- less than 30 W/capita.

Energy Conservation

Conserving energy and improving energy efficiency frees up energy to be used for new applications. Reducing electricity demand can forestall the construction of new electric power plants. Amory Lovins invented the idea of "negawatts" – supplying power by redirecting power saved through conservation and efficient electrical power use. Efficiency gains are beneficial not just for electric power, but also for transportation, industrial, and commercial applications.

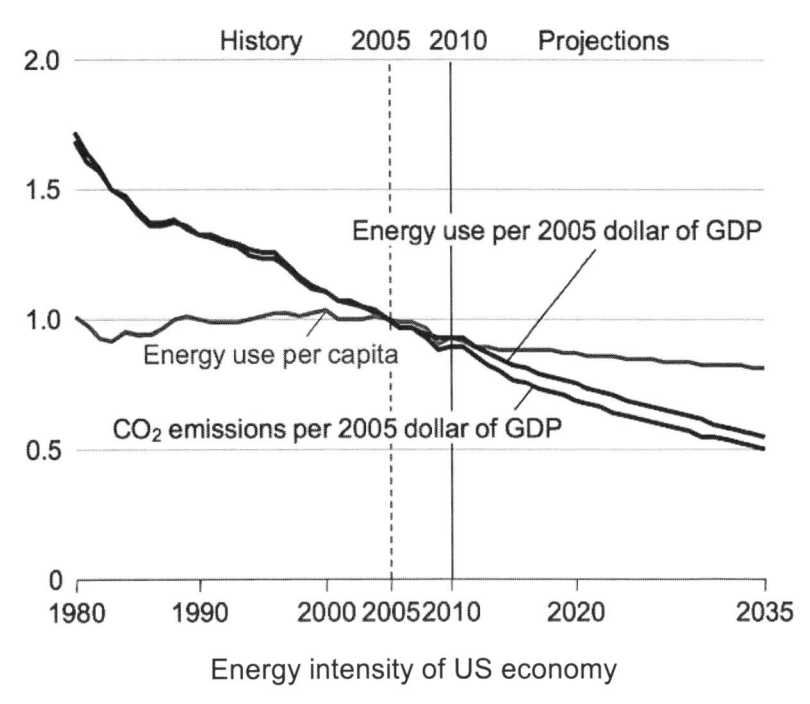

Energy intensity of US economy

This chart from the US DOE illustrates US improvements in energy efficiency. Energy use per capita is diminishing slowly, even as new energy uses arise. Square footage of homes has nearly doubled since 1950. Large vehicles such as SUVs have become popular. Television screens are bigger. Computers are faster. Although energy use per dollar of gross domestic product drops about 1% per year, US productivity, consumption, and GDP increase, countering efficiency gains.

US 2006 energy intensity was 2.6 kWh(t)/$GDP, about the same as the world average, but less efficient than Europe at 1.9 kWh/$GDP.

Many countries have legislative efforts to conserve electricity, such as ending sales of incandescent light bulbs in favor of three times more efficient fluorescent bulbs. Phasing out all incandescent light bulbs worldwide would reduce world electric power requirements by almost 50 GW, or 2.5%. Redirecting these "negawatts" would be the equivalent to constructing 50 large, 1 GW power plants. Exchanging all old refrigerators in Europe for 40% more efficient modern ones would cut electric power demand there by 2 GW.

Improving building design can save energy. Homes with reflective roofs need 40% less cooling energy than those with black roofs. For new construction, Passivhaus-designed 200 m² homes with super insulation and low air infiltration can be heated for 3 MWh(t) per year – an average of only 350 W(t).

Transportation is the biggest application for petroleum. Complex US corporate average fuel economy (CAFE) standards are designed to increase fleet average fuel economy by 2% per year.

However vehicle demand is increasing. GM sells more vehicles in China than the US, numbering 3 million per year. The $3000 Tata Nano is spurring vehicle sales in India.

Conservation and efficiency are not enough.

Conservation measures such as more efficient lighting and appliances can save electric power, but these "negawatts" are insufficient to solve the global problem.

Some environmentalists argue that we can solve the CO2 problem by consuming less, but the numbers don't bear this out. The US consumes electricity at an annual rate of about 434 GW, about 12,000 kWh/yr per person, or 1400 watts per person. (US electric power use dipped during the 2009 recession, then continued its historical growth.) Assume, for the sake of argument, that conservation and efficiency experts manage to cut US electric power consumption in half, to 6,000 kWh/yr per person – about 700 watts on average.

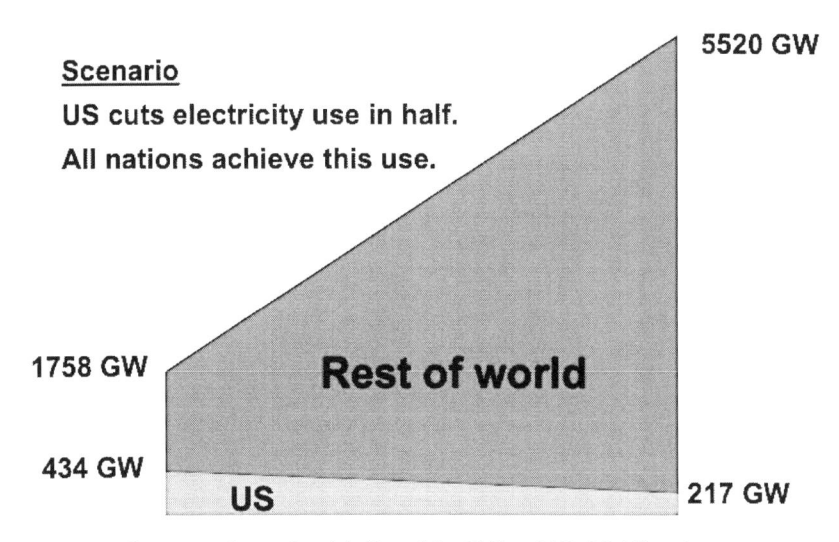

Scenario
US cuts electricity use in half.
All nations achieve this use.

5520 GW
1758 GW
Rest of world
434 GW
US
217 GW

Consuming electricity at half the US 2012 rate

The rest of the world aspires to achieve a lifestyle as prosperous as that of the US – requiring that same 6,000 kWh/yr per person. As the chart illustrates, even under this 50% conservation scenario, world electric energy use will triple as population grows to 9.2 billion people and developing nations improve their lifestyles.

Food choices impact energy consumption.

World livestock production produces more CO_2 than transportation. Raising beef cattle in the US requires 48 kWh(t) of energy per kilogram of meat. Two thirds of this is for raising and transporting food for stockyard animals, rather than having animals graze.

The birth to market time for chickens, hogs, and cattle varies. Growing one kilogram of meat requires sustaining the animal life for 50, 400, or 1000 days, requiring proportional amounts of food. So growing chicken meat costs 50/1000 of the food energy for growing beef meat. Humans on vegetarian diets consume even less energy.

50 days — 1 kg
400 days — 1 kg
1000 days — 1 kg

Relative energies to grow 1 kg of meat

Climate scientist and anti-CO2 campaigner James Hansen said,

"If you eat further down on the food chain rather than animals, which have produced many greenhouse gases, and used much energy in the process of growing that meat, you can actually make a bigger contribution in that way than just about anything. So that, in terms of individual action, is perhaps the best thing you can do."

We live in a world where many developing nations aspire to eating high-end meats such as in McDonald's hamburgers, increasing energy demand.

In conclusion, negawatts from conservation and efficiency will be overwhelmed by increasing demand for energy.

OTHER ELECTRICITY SOURCES

Oil powers cogeneration of electricity and desalinated water.

Oil is expensive for electric power generation. With oil at $100/barrel, the fuel cost for generating electricity is about 18 cents/kWh. This use of oil is being replaced with natural gas. Oil is the source for electricity in isolated regions of Alaska.

Cogeneration of heat and power is practical in compact communities where oil is used for heating, such as Dartmouth College. Edison's first power plant produced both heat and electricity in 1882 in New York City.

However the need for desalinated water enables economic co-generation of desalinated water and electricity in arid regions such as Australia, the Mid-East, or the Caribbean. Most such cogeneration occurs in the Mid East or North Africa, where petroleum resources are readily available. Cogeneration has even shaped the governments of Saudi Arabia and Kuwait, which each have a government organization called Ministry of Electricity and Water.

Desalination is a growing global market at 68 million cubic meters (Mm³) per day, worldwide, projected to be 120 Mm³ by 2020. The largest plant, in the United Arab Emirates, produces nearly 1 M m³ per day. Grand Cayman has 7 desalination plants producing 34,000 m³ per day, powered by oil.

Nuclear power can generate more, clean, safe electricity.

Nuclear energy supplies 14% of the world's electricity from 454 reactors. The nuclear power industry has 15,000 reactor-years of operating experience; naval reactors have a similar history. Nuclear power is well understood. In 2012 63 new reactors are under construction with 163 ordered or planned.

Electricity from nuclear power plants does not release the climate-changing CO_2 of coal or natural gas power plants, nor does it emit deadly particulates that cause the deaths of millions of people.

Nuclear power is by far the safest way to make electric power, counting Chernobyl, the only accident where radiation killed or injured people.

Existing uranium fuels are sustainable for decades at current consumption rates, and new thorium technologies provide inexhaustible fuels.

Nuclear wastes are hazardous, but they can be safely handled and securely sequestered.

The cost of nuclear power is less than that of renewable wind and solar power, which is intermittent and unpredictable.

New liquid fuel technology means nuclear power can even be cheaper than coal power, ending CO_2 emissions through economic self-interest. That is the subject of the rest of this book.

5 Liquid Fluoride Thorium Reactor

President John F Kennedy to the Atomic Energy Commission

"The development of civilian nuclear power involves both national and international interests of the United States. At this time it is particularly important that our domestic needs and prospects for atomic power be thoroughly understood by both the Government and the growing atomic industry of this country which is participating significantly in the development of nuclear technology. Specifically we must extend our national energy resources base in order to promote our Nation's economic growth." *March 17, 1962*

AEC Chairman Glenn T Seaborg to President Kennedy

"In contrast, our supplies of uranium and thorium contain almost **_unlimited amounts of latent energy_** that can be tapped provided "breeder" reactors are developed to convert the fertile materials, uranium-238 and thorium-232, to fissionable plutonium-239 and uranium-233, respectively."

"Among the most promising solutions ... is to use the ***fuel in fluid form***, thus permitting continuous extraction and reprocessing to remove the fission products. ... The currently most promising approach is the use of fused uranium salts which can be circulated, both for reprocessing purposes and for heat transport."

"Meanwhile, thorium-uranium-233 breeders will, if vigorously developed, **no doubt also become economic**. ... Initial economic pressures may well, however, tend to favor the uranium-plutonium cycle since plutonium will be an immediate product of the converters that will constitute the bulk of the initial power reactor installations."

Nov 20, 1962 (Emphasis added)

LFTR technology still answers Kennedy's request.

We still have the opportunity, missed a half century ago, to develop inexpensive, unlimited energy from thorium. The *"unlimited amounts of latent energy"* in thorium ores can power civilization for millennia. The *"fuel in fluid form"* is the key technology that means the liquid fluoride thorium reactor will *"no doubt also become economic"*.

A supernova created uranium and thorium energy.

A star near our Sun burned up its hydrogen about 5 billion years ago, cooling and then collapsing by gravitational force.

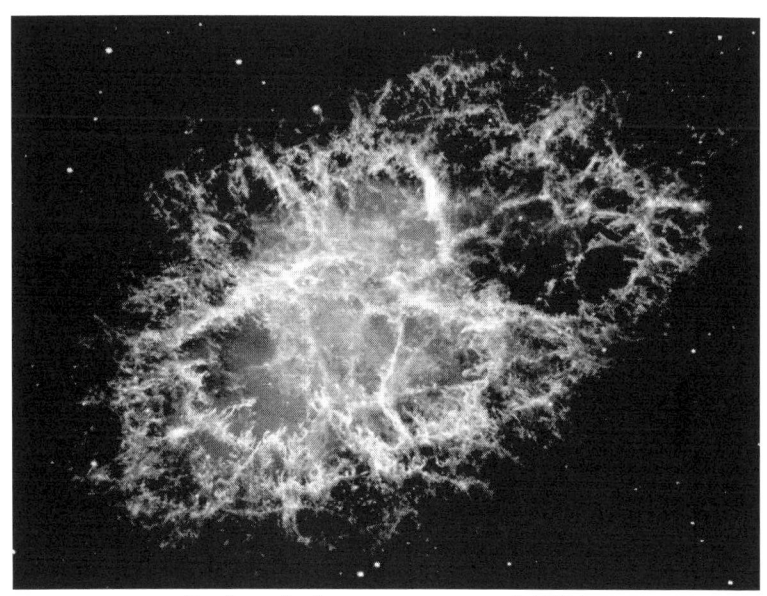

Crab nebula, supernova of 1054

This compressed the energy and simple atoms into new elements that fill the remainder of the periodic table, including the elements uranium and thorium.

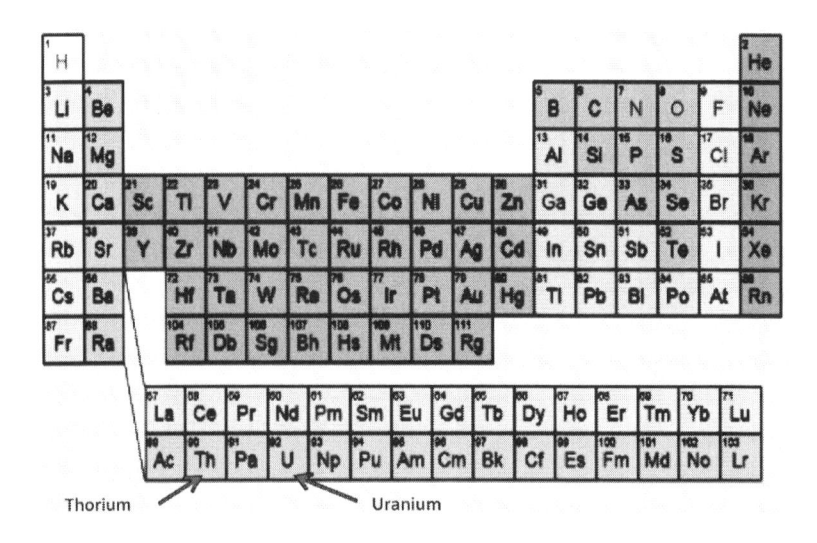

Thorium   Uranium

The matter blown into space and captured into orbits by our Sun's gravity coalesced into planets such as Earth.

The energy that was stored in the heavy metals can be released by nuclear fission.

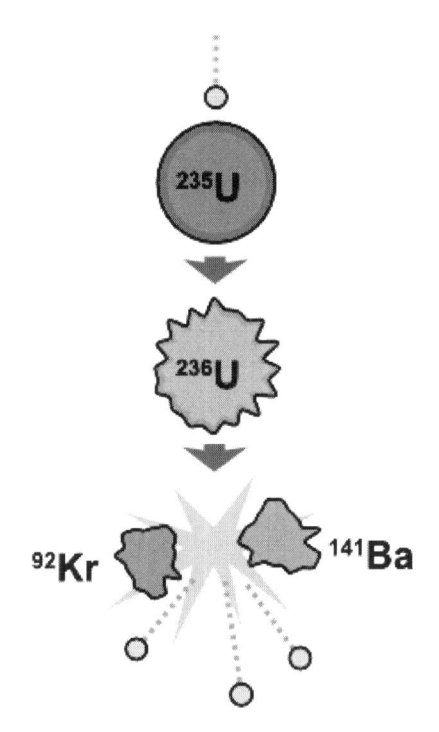

In this example a neutron impinges on a uranium atom with 235 protons and neutrons. The U-235 becomes U-236 for an instant and then splits into krypton and barium and three neutrons. The total mass of the resulting

> barium-141
> krypton-92
> 3 neutrons

is 166 MeV less than the mass of the original U-235 and neutron, immediately releasing 166 MeV of energy, later up to a total of about 200 MeV as the unstable fission products Kr-92 and Ba-141 decay.

200 MeV means 200 million electron volts. One eV is the kinetic energy obtained by one electron traversing an electric potential of one volt. MeV is also shorthand for a unit of mass because $E = m \times (c^2)$ and c is a constant.

One eV is a ballpark estimate of the chemical potential energy of a typical molecular bond. For example, the energy per molecule released by burning methane (CH4) is 9.6 eV, or about 2 eV per atom. The typical 200 MeV released by fissioning U-235 is 100 million times more per atom!

PRESSURIZED WATER REACTORS

Today US nuclear power reactors use solid fuel.

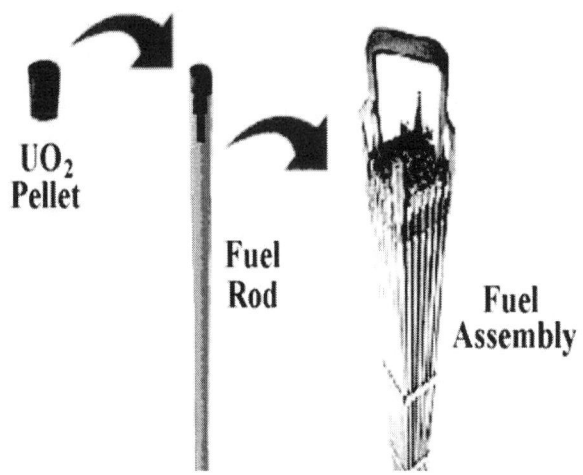

UO$_2$ Pellet

Fuel Rod

Fuel Assembly

Pellets of uranium dioxide in zirconium fuel rods are bundled into fuel assemblies. One cm diameter fuel pellets are sealed in zirconium tubes in 4 m long assemblies placed in the reactor core. Zirconium absorbs few neutrons. These assemblies are placed within the reactor vessel under water at 160 atmospheres pressure to keep the water liquid at 330°C. This hot water transfers heat from the fissioning fuel to a steam turbine that spins a generator to make electricity. 25,000 or more such fuel rods power the reactor for about 5 years.

Alvin Weinberg invented the pressurized water reactor (PWR) in 1946, the same year Hyman Rickover went to Oak Ridge to study whether nuclear fission might power a submarine. Weinberg convinced Rickover that the simple, compact PWR was the best technology choice for naval propulsion, even though Weinberg was then pursuing liquid fuel reactors for society's energy future. Rickover instilled PWR technology throughout the Navy. The naval reactors used metallic uranium fuel elements.

FIG. 3.

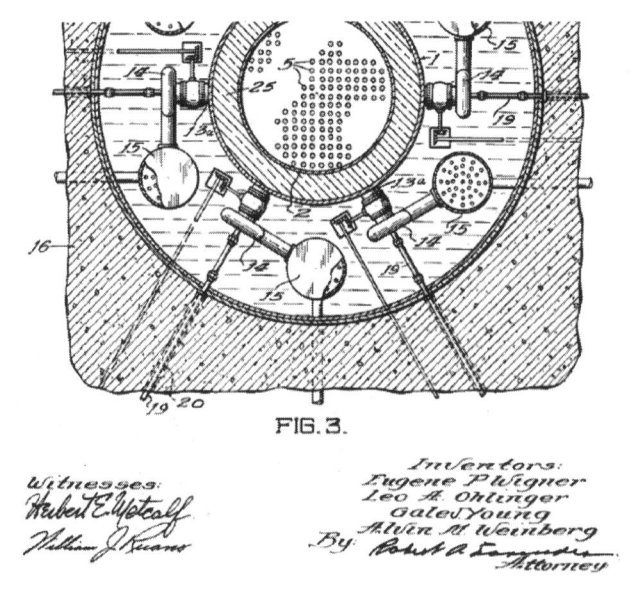

Alvin Weinberg signed this PWR patent.

After President Eisenhower's 1953 *Atoms for Peace* speech, Rickover's team then developed a uranium-oxide PWR at Shippingport PA, creating the first US nuclear power plant in 1957. The entire project took 37 months. The momentum carried on to all US power reactors, bypassing all other technologies, including liquid fuel reactors. Westinghouse sells the PWR. GE developed a lower pressure (60 atmospheres) variant, the boiling water reactor, BWR. The term LWR encompasses both BWR and PWR.

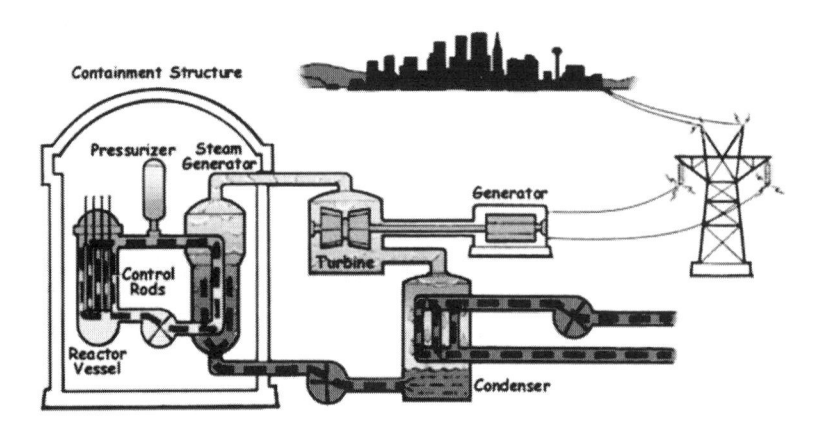

Pressurized water reactor, with heat-exchanging steam generator

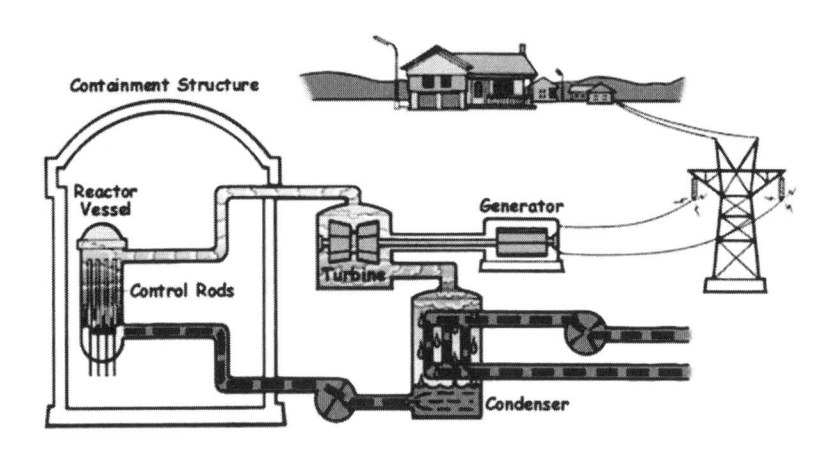

Boiling water reactor, with steam direct from vessel to turbine

Uranium fission produces energetic fission products and fast, high-energy neutrons. Multiple neutron collisions with light nuclei, such as the hydrogen in H_2O, moderate the neutron speed and energy so the neutron will more probably cause another U-

235 fission. Energetic fission products heat the fuel rods, which heat the water, which expands, lowering its density, thus lowering its moderation effectiveness, reducing the rate of fission. When the water cools, the reverse happens and the fission rate increases. This inherent negative feedback keeps the reactor at the point of criticality and prevents runaway. All US power reactors have this inherent stability. The water serves for cooling and thermal energy transfer, moderation, and reactivity stability.

The solid fuel form limits energy production.

The fuel pellets contain UO_2 with fissile uranium U-235 expensively enriched to 3.5% or more, the remainder being U-238. After about 5 years the fuel is spent and must be replaced; the spent fuel still has about 2% fissile material remaining. The reactor is stopped every 18 months or so and a third of the fuel assemblies are replaced. Although fresh fuel is not very hazardous, the spent fuel assemblies are intensely radioactive from fission products. During refueling the assemblies are carefully moved by remotely operated cranes and kept under water to keep from melting and to shield the operators. After a few years radioactivity decays enough that the spent fuel can safely be moved to dry cask storage.

The solid fuel form limits the amount of fissile material that can be placed in the fuel rods to be consumed. Noble gases such as krypton and xenon build up. Other fission products such as samarium accumulate and absorb neutrons, keeping them from sustaining the chain reaction. The solid fuel is stressed by internal temperature differences, by radiation damage that breaks covalent UO_2 bonds, and by fission products that disturb the solid lattice structure. As the solid fuel swells and distorts, the irradiated zirconium cladding tubes must continue to contain it and all fission products while in the reactor and for centuries thereafter in a waste storage repository. This limits the fissile U-235 than can be loaded into fuel rods to about 4%.

A cross section of fuel rod

Spent fuel damage

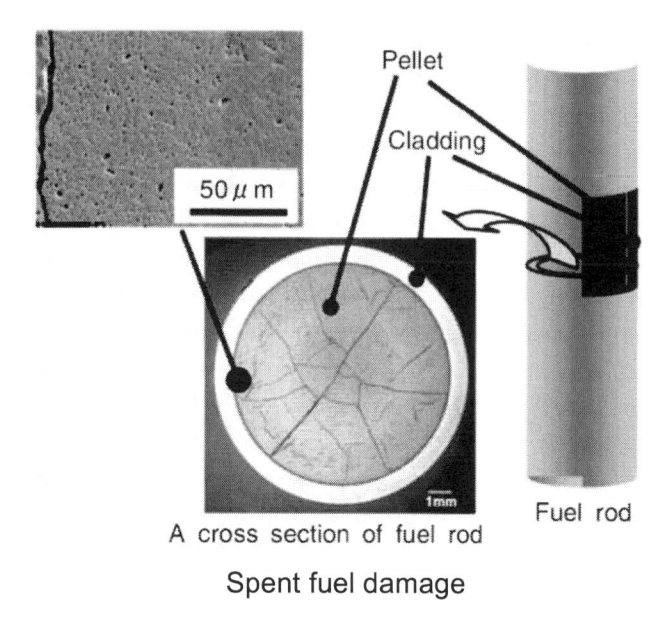

Spent fuel rods contain long-lived radioactive transuranics.

Transuranics are the elements of the periodic table beyond uranium (U). Spent nuclear fuel contains long-lived radioactive transuranic elements such as plutonium Pu-239, created after U-238 nuclei absorb neutrons. Some Pu-239 is fissioned, contributing as much as a third of reactor power. All such transuranics could eventually be destroyed in the neutron flux, either by fission or transmutation to a fissile element that later fissions, except that the solid fuel is removed long before this.

Discharged spent fuel also contains the radioactive fission products, which decay rapidly, reverting to radiation levels of natural uranium ore in a few hundred years. Nuclear waste storage concerns stem primarily from the long-lived transuranics, which could be consumed by leaving them in the neutron flux.

Liquid-Fuel Nuclear Reactors

Transuranics could continue to burn in a fluid fuel reactor.

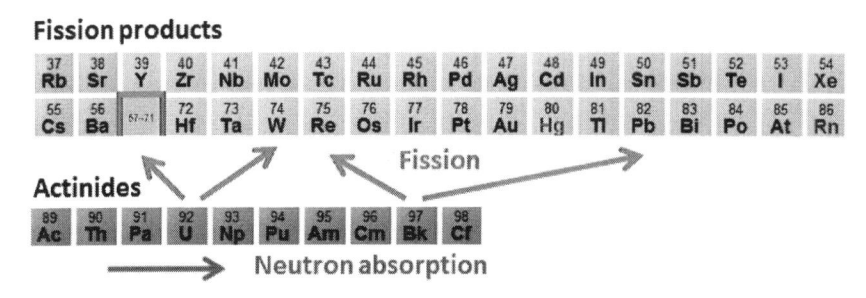

Fission products

37	38	39	40	41	42	43	44	45	46	47	48	49	50	51	52	53	54
Rb	Sr	Y	Zr	Nb	Mo	Tc	Ru	Rh	Pd	Ag	Cd	In	Sn	Sb	Te	I	Xe

55	56	57–71	72	73	74	75	76	77	78	79	80	81	82	83	84	85	86
Cs	Ba		Hf	Ta	W	Re	Os	Ir	Pt	Au	Hg	Tl	Pb	Bi	Po	At	Rn

Fission

Actinides

89	90	91	92	93	94	95	96	97	98
Ac	Th	Pa	U	Np	Pu	Am	Cm	Bk	Cf

⟶ Neutron absorption

Neutron-induced absorption and fission element changes

When irradiated and fissioned in a neutron flux, fluid fuel is not subjected to the structural stresses of solid fuel rods in LWRs. In a molten salt reactor the transuranics (plutonium, americium, curium, berkelium, ...) can simply remain dissolved in the fluid fuel salt. There they may absorb neutrons and either (1) fission, releasing energy, or (2) become another, heavier transuranic isotope, also subject to neutron absorption. Fission products like xenon gas can bubble out, so they do not absorb neutrons from the chain reaction. Fission product noble metal solids like silver precipitate out. Others remain in the fuel solution until transmuted or chemically removed.

Liquid-fueled nuclear reactors bypass many disadvantages of solid-fuel reactors. If the fissile material is in a liquid, there are no heat and neutron induced stresses and strains such as build up and distort UO_2 fuel pellets in their zirconium tubes. There is no zirconium to react with water and release hydrogen at high temperatures during a cooling failure. The heat transfer capabilities of flowing fluids exceed those of solids. Fissile material can flow in fluids in and out of the reactor vessel, as needed, rather than having years of it pre-stored within the reactor. The fluid containing the fissile material is also the heat-transfer agent, eliminating two heat exchange step in LWRs: (1) at the interface between the fuel pellets and the containing zirconium tube, and (2) from the zirconium tube to the pressurized water in an LWR. Fluid fuels allow continuous chemical processing, only accomplished with LWR solid fuels by chopping up spent fuel rods and then chemically dissolving the solids.

Fermi started up the first of the fluid-fuel nuclear reactors.

Enrico Fermi created the first nuclear reactor in a pile of graphite and uranium blocks at the University of Chicago. Fermi also started up the world's first liquid-fuel reactor, which used uranium sulfate fuel dissolved in water. As in the solid-fuel reactor, the water moderates neutron speeds, but expands as the reaction heat increases, lowering moderation and reaction rate. Another inherent negative feedback arises because some dissolved uranium sulfate fuel expands out of the reactor.

A hydrogen nucleus sometimes absorbs a neutron, preventing it from continuing the chain reaction, so an aqueous reactor doesn't quite reach criticality unless fueled with uranium that has been enriched beyond the natural 0.7% isotopic abundance of U-235. A fix is to use deuterium, hydrogen that already has an extra neutron, in heavy water, D_2O, which absorbs few neutrons, so heavy water aqueous reactors can use inexpensive, unenriched uranium.

The aqueous reactor at Oak Ridge National Laboratory fed 140 kW of power into the electric grid for 1,000 hours. In operation it successfully removed xenon fission products. The inherent reactivity control was effective, and shutdown was accomplished simply by turning off the turbine generator. Aqueous reactors turned out not to be practical for efficient electric power generation, which required temperatures exceeding 300°C, because of two-phase instability of the uranium sulfate solution.

In 1940 scientists at England's Cavendish Lab had pioneered a low power reactor with uranium oxide powder slurry in heavy water. In the 1970s Netherlands scientists experimented with such a 1 MW aqueous reactor with uranium and thorium particles in liquid suspension. Babcock & Wilcox is again developing a low power aqueous reactor to manufacture the fission product molybdenum-99, which decays to technitium-99m used in medical imaging and treatment.

Los Alamos operated a molten plutonium metal reactor.

Later US national laboratories experimented with liquid metal fuels. Bismuth melts below 300°C and has a low cross section for absorbing neutrons. Brookhaven Lab designed liquid metal fuel reactors with circulating molten bismuth and uranium in the 1950s. This fluid fuel also had the advantages of easy fuel handling and inherent criticality control from thermal expansion. But because of difficult corrosion control, low heat capacity of bismuth, limited solubility of uranium in bismuth, and requirements for enriched uranium, no liquid metal fuel reactors of this type were completed.

Planning for the time when world supplies of U-235 might become depleted, Los Alamos Lab developed a molten plutonium reactor. It had a 600°C core of molten plutonium and iron contained in tantalum thimbles cooled by liquid sodium. This 1 MW reactor ran from 1961 until 1963.

Oak Ridge scientists conceived molten salt reactors.

Oak Ridge scientists proposed the idea of a fluid fuel reactor with UF_4 dissolved in molten fluoride salts. A mixture of LiF and BeF_2 salts is fluid at temperatures as low as 360°C. The Li-7 and Be-9 in the salt and a graphite moderator reduce neutron kinetic energies so they fission uranium. Reactivity is stable because expanding hot salt dilutes the moderator and also expresses some uranium fuel from the critical core. The hot salt circulates and transfers thermal energy out of the reactor. The strong ionic bonds of the fluoride salts are stable under irradiation at high temperature. Although fluorine gas is highly corrosive, fluoride salts are not.

In the Cold War the US Air Force wanted bombers that could continuously circle the Soviet Union without landing to refuel, leading to the aircraft reactor experiment (ARE). Oak Ridge built the first molten fluoride salt reactor, which ran for 100 hours in 1954 at temperatures up to 860°C – red hot! The ARE demonstrated inherent reactivity stability and automatically adjusted power, without control rods, as the 2.5 MW heat exchanger airflow varied. The Hastelloy-N metal vessel and piping withstood corrosion.

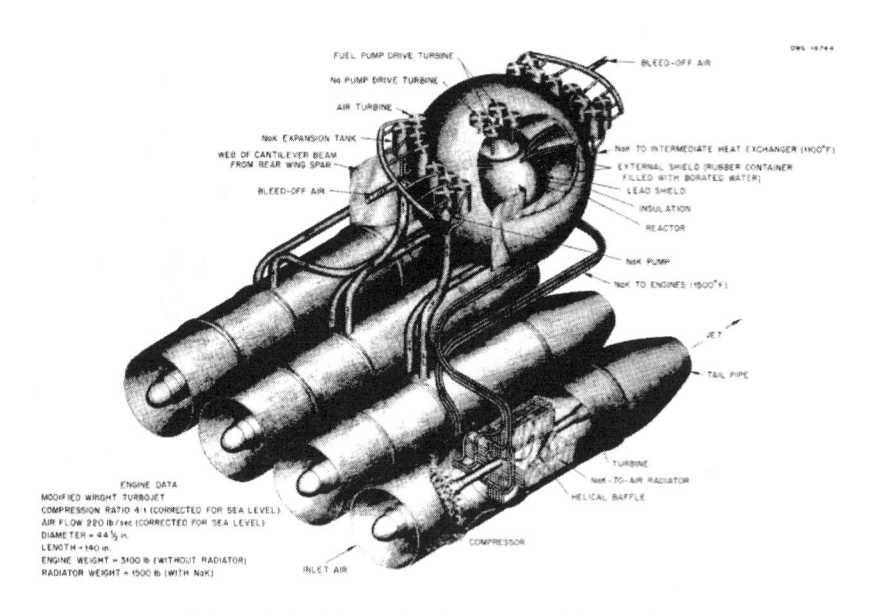

The *Fireball* reactor to power jet engines.

This success led to design of the compact, 1.4 m diameter *Fireball* reactor containing a fluid core of UF$_4$ dissolved in molten salt in a beryllium metal sphere. It heated liquid NaK (sodium and potassium) metal to transfer 200 MW of thermal power to aircraft turbine jet engines. The reactor sphere was only 1.4 meters in diameter and generated 200 MW(t) to heat the air in the jet engines. This aircraft nuclear engine project was cancelled before Fireball testing because practical in-flight refueling allowed sustaining a fleet of airborne bombers, backed up by ICBMs in submarines and on land.

Thorium is a mildly radioactive element, and a possible fuel.

Thorium is interesting because it is more plentiful than uranium and can be transmuted into uranium in a nuclear reactor. Thorium is a heavy, silvery metal about as abundant as lead -- four times more abundant than uranium, and 500 times more abundant than the fissile U-235 isotope. Thorium is not very radioactive because it decays so slowly; its half-life is 14 billion years – about the age of the universe. It alpha-decays in a chain of ten elements ending in lead, releasing heat. The heat released by thorium within the earth is the primary source of geothermal energy. Thorium's heat also sustains the liquidity of the earth's molten iron core, where convection currents create the geomagnetism responsible for the earth's magnetic field. That magnetic field diverts the stream of charged particles from the sun. Were it not for the earth's magnetic field, that solar wind would strip away earth's atmosphere and water, much like Mars

Thorium was first burned in solid-fuel reactors.

Thorium can become uranium in a nuclear reactor, where neutrons not only cause fissions but some are absorbed to create new elements. In today's solid-fuel LWRs some U-238 becomes U-239. It decays to neptunium and then plutonium via beta decay -- the ejection of an electron, making a new element. The resulting Pu-239 is fissile, and some is consumed to help power the reactor. Near the end of a LWR fuel cycle, about 1/3 of the reactor power comes from Pu-239.

The columns in the next table represent the heavy metal actinides, thorium, protactinium, uranium, neptunium, plutonium, and americium, labeled with their element abbreviation and atomic number – the number of protons in the nucleus. The rows correspond to isotopes of each element, labeled by their total number of protons plus neutrons – their atomic weights.

nucleons	Th 90	Pa 91	U 92	Np 93	Pu 94	Am 95
241						
240						
239			↑	→	→ 💥	fertile
238			●			
237						💥 fission
236						
235			💥			→ beta decay
234						
233	↑	→	→ 💥			↑ neutron absorption
232	●					

Changes to elements exposed to neutrons

Similarly to what happens with U-238, if thorium is placed in a nuclear reactor some Th-232 becomes Th-233. It beta decays to protactinium Pa-233 and then to U-233, which is also a fissile fuel. Relatively little plutonium is produced from the Th-232, because six more neutron absorptions are required than from U-238. Thorium-232 and uranium-238 are called fertile because they can be transformed to fissile elements by neutron absorption and beta decay.

Fuel of thorium combined with uranium was successfully tested at the Shippingport power reactor from 1977 to 1982; end of life analysis showed that the reactor produced about 1% more fissile material than it consumed. Thorium was used in the German THTR-300 (thorium high-temperature reactor 300 MW) pebble bed reactor between 1983 and 1989. Alvin Radkowsky founded Thorium Power (now Lightbridge), which designed fuel rods to use thorium in existing reactors, but the concept was not commercialized. Nobel prize winner Carlo Rubbia at CERN designed an accelerator-driven thorium reactor. Since 1996 India has operated its experimental Kamini 30 kW(t) reactor with U-233 fuel created by an adjacent 40 MW(t) fast breeder test reactor in which thorium is irradiated to produce U-233; India's national strategy to produce 30% of its electricity from thorium by 2050. China and Canada are testing thorium in heavy water moderated CANDU reactors.

But all these reactors use solid fuel forms.

The molten salt reactor realizes thorium's true potential.

Yet in 1943 Eugene Wigner and Alvin Weinberg had designed the aqueous reactor as a first step to a liquid fuel thorium-uranium breeder reactor! Oak Ridge director Alvin Weinberg, who had convinced Rickover to use PWRs for submarines, led development of the liquid fluoride thorium reactor, convinced that "humankind's whole future depended on" this inexhaustible energy.

The thorium/uranium fuel cycle depicted below converts fertile thorium-232 into fissile uranium-233, which fissions and releases energy.

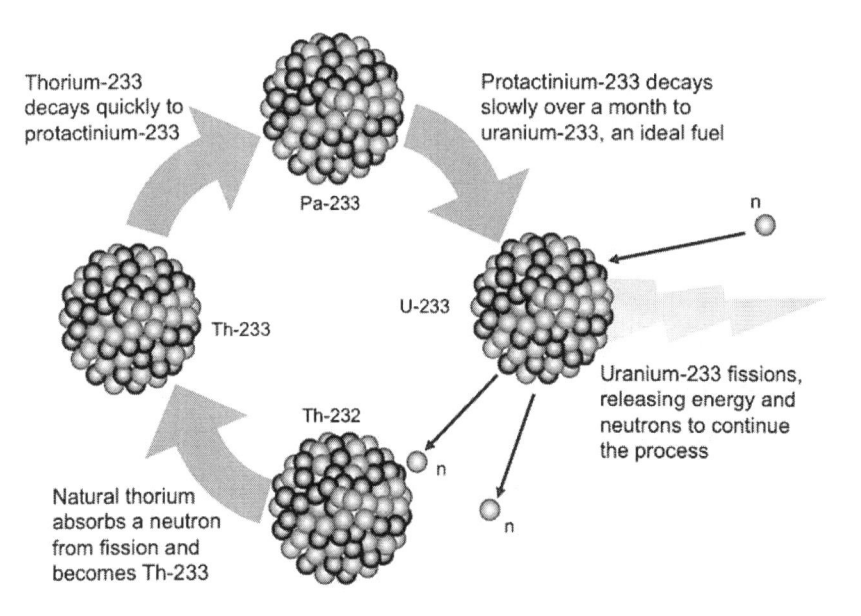

Thorium-233 decays quickly to protactinium-233

Protactinium-233 decays slowly over a month to uranium-233, an ideal fuel

Pa-233

Th-233

U-233

Th-232

Uranium-233 fissions, releasing energy and neutrons to continue the process

Natural thorium absorbs a neutron from fission and becomes Th-233

Neutron-induced breeding of thorium to fissile uranium

ORNL's Molten Salt Reactor Experiment was a success.

The Oak Ridge Lab molten salt reactor experiment (MSRE) drew on the 1950s experience with the aircraft reactor experiment. The MSRE reactor operated successfully over 4 years through 1969. It was initially fueled with uranium enriched to 33% U-235. After six months of operation the uranium was removed from the molten salt by exposure to fluorine gas, converting the dissolved UF4 to gaseous UF6. Fluoride salts of U-233 were dissolved in the recycled molten salt and the MSRE then demonstrated that U-233 was also a viable energy source.

To simplify engineering and testing, the Th-232/U-233 breeding step was separate; the U-233 came from other reactors breeding Th-232. A secondary molten salt loop was heated by a heat exchanger designed to keep radioactive materials confined to the primary loop. No turbine generator was attached; the thermal energy was dissipated to the air through another heat exchanger heated by the secondary loop of (clean) molten salt.

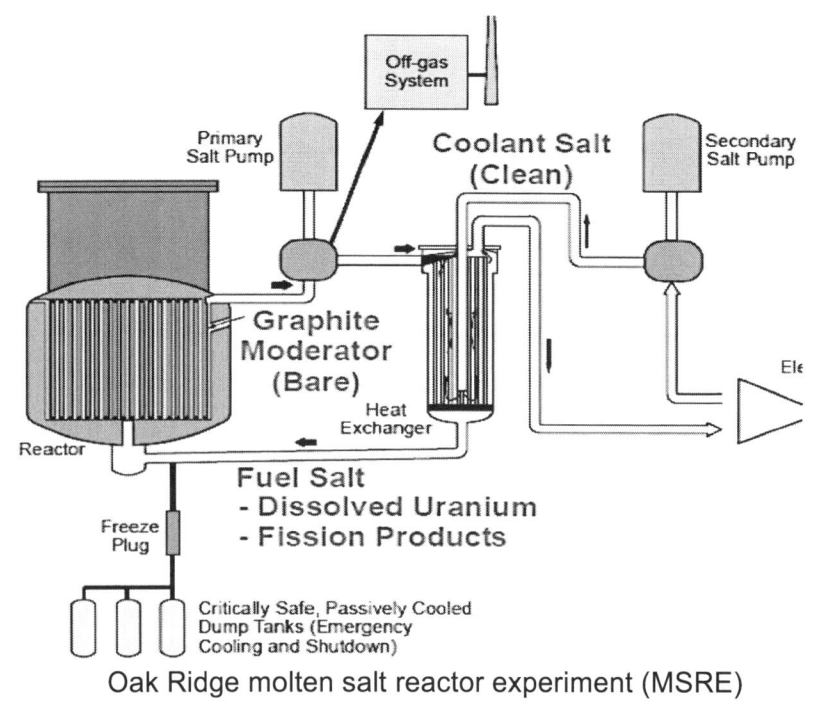

Oak Ridge molten salt reactor experiment (MSRE)

The MSRE was successful. Fission product xenon gas was continually removed to prevent unwanted neutron absorptions. Online refueling was demonstrated. Graphite structures and corrosion resistant Hastelloy metal for vessels, pipes, and pumps proved suitable. Oak Ridge developed chemistry for separation of thorium, uranium, and fission products in the fluid fluorine salts. For example, UF_4 (in solution) + F_2 (gas) → UF_6 (gas), so bubbling fluorine gas through the molten salt could remove the bred fissile uranium, leaving the thorium fluoride behind.

The MSRE was a single fluid molten salt reactor. Thorium can be bred to uranium in a single fluid reactor or in a two-fluid molten salt reactor with the fissile uranium and fertile thorium separate, illustrated below.

LFTR makes its own fissile uranium from thorium fuel.

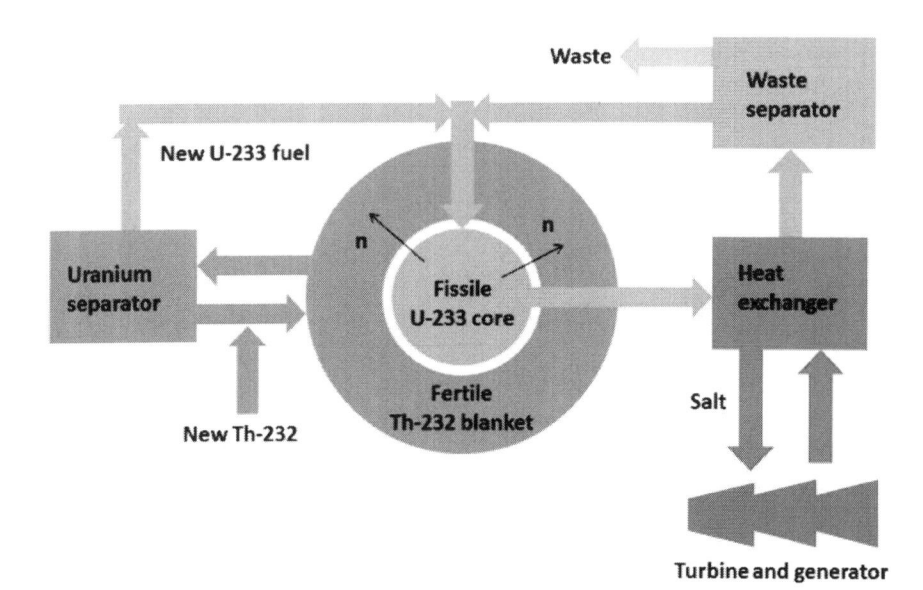

Two-fluid LFTR concept

In the liquid fluoride thorium reactor (LFTR) concept above, the chain reaction in the fissile core heats the molten salt in which it is dissolved. That radioactive, heated molten salt flows through a heat exchanger that transfers the thermal energy to clean, non-radioactive salt. That salt flows to an energy conversion system to generate electricity. Waste fission products in the core salt are removed. Some neutrons from the U-233 fission enter the blanket where they convert Th-232 to U-233, which is separated chemically and flows into the core salt, to replace the U-233 lost to fission. New Th-232 is added to replace that converted to U-233.

The molten salt is a mixture of fluorides of beryllium and lithium fluorides (LiF and BeF2) termed flibe (F_3LiBe). It is eutectic, meaning that mixture's melting point is lower than that of LiF or BeF2 alone, depending on the mixture. LiF+BeF2 melts at 360°C; 2LiF+BeF2 melts at 460°C and is less viscous. Flibe is transparent.

In operation the fission reaction heats the molten salt to about 700°C before leaving the reactor core, passing through a heat exchanger, and returning to the core at about 560°C. The heat exchanger transfers that thermal energy to a stream of molten salt that is nonradioactive, so that the power conversion system remains nonradioactive to facilitate maintenance. That molten salt at 620°C heats a gas (helium, CO2, or air) that runs a turbine generator.

Molten salt does not boil at temperatures less than 1400°C, so the LFTR operates at atmospheric pressure. Unlike a conventional LWR, there are no pressurized radioactive isotopes that can be propelled by steam to escape to the environment in an accident.

Neutrons from fission are fast, with kinetic energies near 1 MeV. Fissioning U-233 requires slower, less energetic neutrons with kinetic energies under about 1 eV – roughly the same as that of the heat motion of the molecules of molten salt, so they are called thermal neutrons. Slowing the neutrons is accomplished through collisions with the light atoms of Li-7, Be-9, and F-19 in the molten salt, and C-12 in a graphite moderator.

LFTR molten salts can be continuously reprocessed.

The two-fluid LFTR core salts and blanket salts can be continuously reprocessed with small integrated chemical systems that process the salt inventory once every 10 days or so. Thus the reactor only has a few days of excess fissile material, not several years as required by LWRs. The radioactive fission products can similarly be removed from the reactor in days, rather than storing them for years in zirconium-cladded fuel rods of LWRs. (Single fluid reactors can avoid chemical processing for years.)

The uranium separator moves new U-233 to the core salt.

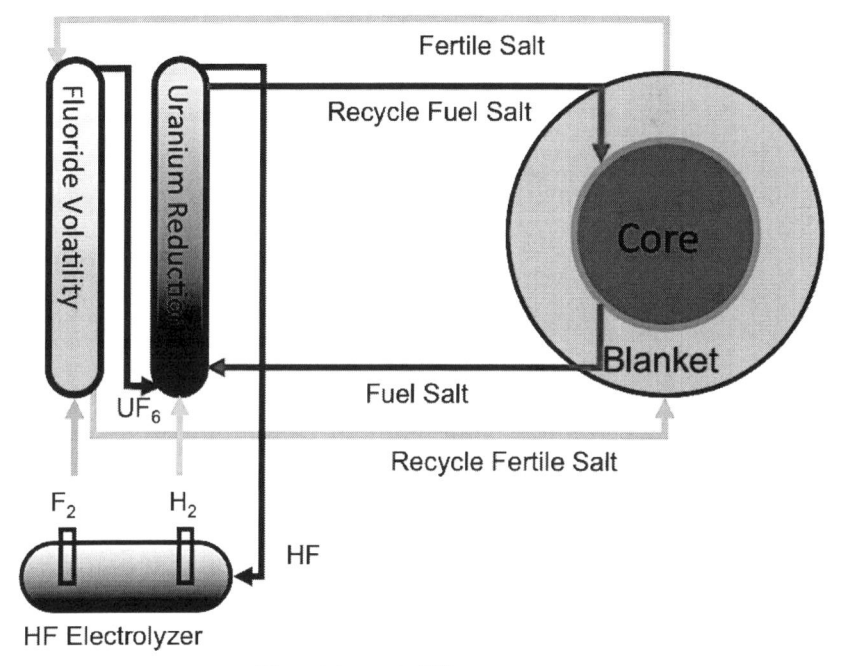

Fluoride volatility process

The blanket salt containing newly created U-233 is exposed to fluorine gas in the fluoride volatility vessel, converting the dissolved uranium fluoride to uranium hexafluoride gas, by $UF_4 + F_2 \rightarrow UF_6$. That gas is then exposed to hydrogen in the uranium reduction vessel producing U-233 in the fuel-salt-soluble UF_4 form, by $UF_6 + H_2 \rightarrow UF_4 + 2\ HF$. Hydrogen fluoride is separated by electrolysis and used over again.

The waste separator uses chemistry and physical properties.

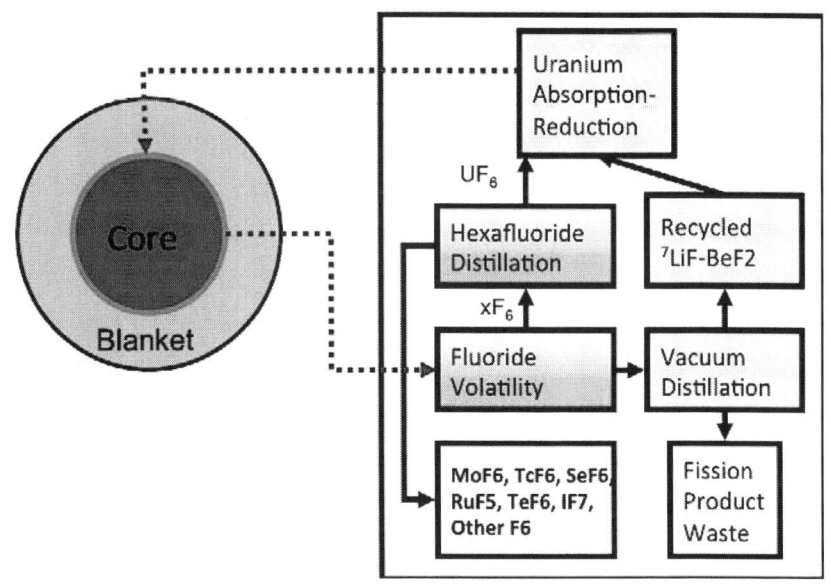

Waste separation concept

ORNL did not build a continuous waste separator such as illustrated above. In addition to the fluoride volatility and other chemical processes, distillation can be used to physically separate molecules with different boiling points. Removal of fission products from the core molten salt is complicated by the variety of chemical elements that must be separated from the flibe. There is considerable engineering needed to perfect the process.

LFTR has inherent safety.

Today's LWRs achieve safety through defense in depth – multiple, independent, redundant systems engineered to control faults. LFTR's inherent safety keeps such costs low.

Pressure: LWRs have strong reactor vessels pressurized to 160 atmospheres; they have large containment domes to contain any radioactive materials propelled by steam in an accident. A molten salt reactor operates at atmospheric pressure, so radioactive materials can not be dispersed this way.

Stability: The reactor power is inherently stable. If the reactivity increases and generates more heat, some molten salt expands out of the critical core into pipes where chain reactions can not be sustained. The reduced amount of fissile U-233 in the core reduces the reactivity and heat generation, keeping the reactor stable. As temperatures rise the neutron fission rate also decreases with increased neutron energy, and more neutrons are absorbed by U-238 or Th-232, contributing to MSR thermal stability.

Disconnect: If an electrical transmission line disconnects, so the electric generator and heat exchanger can not remove generated heat, the molten salt expands to reduce power, stablely.

Backup safety. ORNL invented a freeze plug -- salt kept solid by a cooling fan. Should control systems lose power or the molten salt temperature somehow rise the plug melts and the salt flows out of the reactor into a drain tank where nuclear fission is impossible.

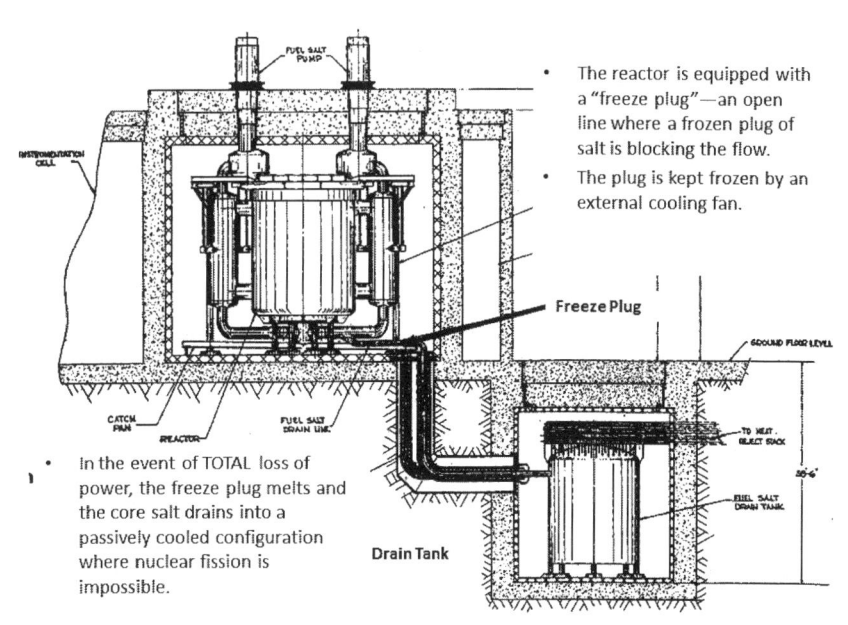

- The reactor is equipped with a "freeze plug"—an open line where a frozen plug of salt is blocking the flow.
- The plug is kept frozen by an external cooling fan.
- In the event of TOTAL loss of power, the freeze plug melts and the core salt drains into a passively cooled configuration where nuclear fission is impossible.

ORNL's single fluid MSR with freeze plug and drain tank

Melt down. A molten salt reactor can't melt down because the core is already molten -- its normal operating state. The salts are solid at room temperature, so if a reactor vessel, pump, or pipe ruptured the salts would spill out and solidify.

LFTR gains its efficiency from high temperature.

Efficiency goes up as heat flows through a greater temperature difference. The limit is

$$\text{Efficiency} \leq \frac{T_H - T_C}{T_H}$$

where temperatures are in degrees Kelvin, °K above absolute zero. The higher temperature of molten salt compared to pressurized water contributes to LFTR high electric/thermal efficiency. LFTR safely operates at high temperatures. Salt remains liquid below 1400°C; internal graphite core structures maintain integrity even above this. Molten salt heat capacity exceeds that of the water in PWRs or liquid sodium in LMFBRs, allowing the most compact, lower cost heat transfer loops. The molten salt heat exchange loop components of high-nickel metals such as Hastelloy-N are qualified for use up to 750°C.

Brayton cycle power conversion is efficient and compact.

LFTR molten salt temperature is 700°C, compared to LWR steam at 315°C, enabling new, more efficient electric/thermal power conversion technology. The triple-reheat closed-cycle Brayton gas turbine achieves a 45% efficiency of conversion from thermal to electric power, compared to 33% typical of existing nuclear and coal power plants using traditional Rankine steam cycles. The working fluid can be helium or nitrogen gas.

The Brayton rejected heat to power ratio is 1.2 (55/45) rather than Rankine's 2.0 (67/33). So the cooling requirements are nearly halved, reducing cooling tower costs, water evaporation, or heating of cooling water borrowed from rivers, lakes, or the sea. This compact Brayton turbine machinery is a quarter the mass of a steam turbine, suggesting a similar cost reduction.

The supercritical CO2 turbine is another emerging technology that may also lead to an even more compact and less expensive LFTR power conversion cycle.

LFTR high temperature allows dry air cooling.

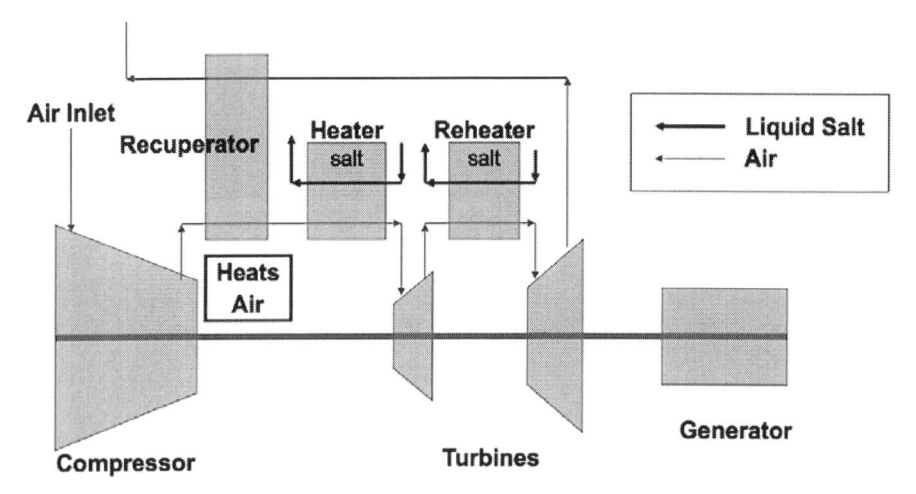

Open air Brayton cycle turbine

The 700°C molten salt from LFTR can heat the compressed air in a turbine similar to an aircraft jet engine, not heated by burning fuel but heat transferred from molten salt. Such air-cooled LFTRs will be practical in arid regions or where water is scarce. In the schematic above the recuperator, heater, and reheater transfer heat from the reactor molten salt to hot air that spins the turbine and generator. The electric/thermal efficiency of 40% is a bit lower than 45% for the triple reheat closed cycle version.

The Nixon administration stopped LFTR development.

Weinberg had invented the PWR used by the Navy, but raised concerns about its safety compared to the molten salt reactor, creating a dispute with AEC deputy director Milton Shaw. Shaw was Rickover's single-minded, forceful protégé, who relied on well-documented procedures and naval management discipline to carry them out.

The Oak Ridge work was stopped; the Nixon administration decided instead to fund only the solid-fuel LMFBR (Liquid sodium Metal cooled Fast Breeder Reactor. This bred plutonium-239 faster than the LFTR bred uranium-233. Weinberg argued for the LFTR and was also critical of LWR relative safety. Weinberg was fired; funding was ended, briefly restored in 1974, and then finally ended in 1976.

Retired ORNL project manager Paul Haubenreich recounts: "Milton Shaw ... was working for Rickover, the Navy was still pursuing the sodium-cooled reactor which went in the Seawolf submarine and the pressurized-water reactor that went into the Nautilus. And so, by the late 60s Milt Shaw still had it in his mind that the sodium-cooled reactor, which was the type of reactor EBR-I (Experimental Breeder Reactor I) out at Idaho, was still viable. But it needed more money to develop it, and so he said 'well we can get some money from shutting the molten salt program down', and as far as I know, that was his idea. And the fast breeder persisted for quite a while, as you know."

Later Weinberg said "It was a successful technology that was dropped because it was too different from the main lines of reactor development."

Weinberg credits the concept of dissolving uranium fluorides in molten salt to his teammates: Ray Briant, Ed Bettis, and Vince Calkins. LFTR development requires deep chemistry expertise, and the liquid fuel technology is unfamiliar to most nuclear engineers today. This is why LFTR is sometimes called "the chemists' reactor".

One motivation for LFTR and LMFBR was the concern about limited uranium reserves. Subsequently new uranium ore discoveries diminished interest in breeder technologies such as LFTR or LMFTR that overcame the 0.7% scarcity of U-235 and high enrichment costs. Three fast reactors were built at national labs; Detroit's Fermi plant was the first such commercial power plant, followed by the Clinch River plant, which never operated. No US LMFBRs are operating.

Nuclear power expansion waned after the Three Mile Island and Chernobyl accidents and the NY state 1983 disapproval of evacuation plans for the completed $6 billion Shoreham NY nuclear power plant. Bond interest rates reaching 17% discouraged nuclear power plant capital investments. Antinuclear activists were able to delay or stop construction, increasing financing costs. Since 1980 total worldwide CO2 emissions from coal rose from 6.6 to 12 billion tons per year.

LFTR ADVANTAGES AND FLEXIBILITY

LFTRs can be started with U-233, U-235, or Pu-239.

A 100 MW LFTR requires 100 kg of fissile material to start the chain reaction and the neutron flux that converts fertile Th-232 to fissile U-233. Uranium-233 can start a LFTR, but U-233 is not found in nature because its half-life of 159,000 years is short compared to the time of its creation by a supernova 5 billion years ago. The US government owns over 500 kg of U-233, which could be used for starting up a few experimental LFTRs. Unfortunately the Department of Energy is about to destroy this asset by diluting it with U-238 and burying it, at a cost of $511 million.

It is possible to design LFTRs that can be started with uranium enriched to 20% U-235. Because such fuel contains 80% U-238 it will initially make long-lived radioactive transuranics such as plutonium.

With a different LFTR design, plutonium-239 can be another possible start-up fissile material, and it can be obtained from the stored spent fuel rods produced by LWRs. All the troublesome transuranics (neptunium, plutonium, americium, californium) can be used. The world now has 340,000 tonnes of spent LWR fuel containing approximately 3,400 t of fissile plutonium, enough to start one 100 MW LFTR each day for 93 years.

Fast MSRs can convert LWR waste to U-233 for LFTRs.

A double-benefit approach to start up LFTRs may be to make U-233 from the plutonium and other transuranic elements in spent PWR fuel. This technique can both destroy the long-lived, radiotoxic materials in nuclear waste and also start up a fleet of LFTRs.

In 1944 Manhattan Project scientists discovered that the Pu-239 they bred for weapons also contained Pu-240, which spontaneously fissions and might cause predetonation of the weapon. To avoid this Wigner designed a reactor that fissioned plutonium rather than uranium. Neutrons from plutonium fission could convert a blanket of Th-232 into U-233 for weapons use. The reactor was never built because Robert Oppenheimer succeeded in building the spherical implosion "gadget" that compressed the plutonium rapidly enough to set off a chain reaction explosion without a predetonation fizzle.

We can use that conversion idea today. A plutonium reactor to transmute thorium into uranium can be a liquid chloride fast reactor (LCFR) – a cousin to the liquid fluoride thorium reactor. The LCFR is better for keeping more plutonium in solution. It is "fast" so more neutrons fission Pu rather than being absorbed by it. Ordinary NaCl and KCl salts can be used. Excess neutrons can transmute a blanket of Th-232 to U-233 used to start up LFTRs. A 1 GW LCFR could generate about 1 tonne of U-233 a year.

The US defense department has excess weapons-grade plutonium to be disposed of. The US and Russia have agreed by treaty to dispose of 34 tonnes of plutonium each by 2014. The current US plan is to mix the Pu-239 with UO2 to make MOX (mixed oxide) fuel rods for LWRs. This project is behind schedule and utilities are reluctant to burn MOX fuel. Instead, LCFRs could consume this excess weapons-grade plutonium. This 78 tonnes of plutonium could be used directly in LFCRs to start up about 780 100-MW LFTRs.

Fusion reactors might someday produce startup U-233.

Beyond plutonium availability, a source of startup U-233 could be a future fission-fusion hybrid reactor, another interest of Ralph Moir. Such a reactor could produce 8 t of U-233 per year A fusion reactor with a molten salt breeding blanket could supply start-up fissile U-233 for 2-4 LFTRs of similar power output each year. Alternatively it could supply the annual makeup fissile material for 19 similar power DMSRs, discussed later.

Such uranium produced by fission reactions can contain 5% U-232, whose decay daughters' decay radiation make the uranium highly proliferation resistant, as discussed later.

A handful of thorium can provide a lifetime of energy.

Just 100 grams of thorium can provide all the energy you need for the whole of your life. The resulting golf ball of waste is largely benign in a few hundred years. A LFTR can burn 100% of the thorium, while a LWR burns less than 1% of mined uranium. And thorium is three times more plentiful.

Using a LFTR the illustrated one tonne ball of thorium would generate 1 GW of power for a year, enough energy for a small city. The fuel cost would be less than $300,000.

LFTR energy from thorium is inexhaustible.

At about 12 parts per million in the earth's crust, thorium is distributed throughout all the world, with known large amounts in the United States, Australia, Turkey, Brazil, and India. The common thorium ores are water insoluble and remain where deposited by geology. Thorium is often found with the chemically similar rare earth elements. The recoverable amounts of thorium vary with the cost of extraction, but at $80/kg the World Nuclear Association estimates worldwide reserves exceed 2 million tonnes. Equally rare lead costs $2/kg, so the thorium price could be much lower, too. Currently thorium is a nuisance byproduct of rare earth element mining, so thorium would be inexpensive.

3,752 tons of thorium in US desert

268,000 tons of thorium in Lemhi Pass

Thorium Energy claims 1.8 million tons of ore containing 268,000 tons of thorium as ThO2 on 1,400 acres of Lemhi Pass, Idaho. The US Geological Survey estimates reserves of 300,000 tons.

On average the earth's crust contains 26 grams of thorium per cubic meter. A LFTR can convert 26 g of thorium to over 250,000 kWh of electricity worth $7,500 at 3 cents/kWh. By contrast, a cubic meter of coal, 1400 kg worth $230, makes roughly 13,000 kWh of electricity worth only $700 at today's 5 cents/kWh typical prices.

Suppose all the energy used by the whole world came just from thorium. The world consumes about 500 quad per year – about 500 EJ (exajoules) = 500,000,000 GJ. The energy coming from thorium in a LFTR is 80 GJ/g, so world demand would be just 500,000,000/80 g per year, or 6,250 tonnes/year. World Nuclear Association's conservative estimate of 2 Mt of thorium reserves implies a 300 year supply.

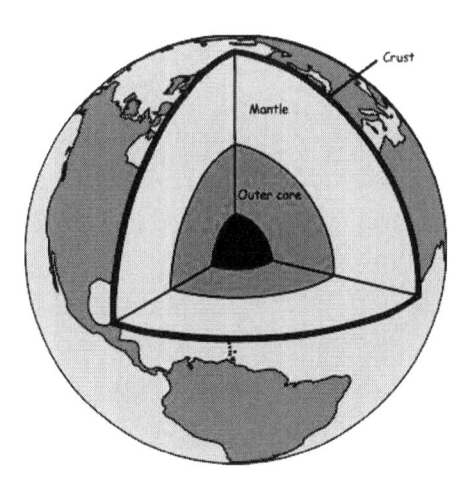

After this time civilization could mine thorium distributed throughout the earth's crust, which contains 12 parts per million. Obtaining the 6,250 t of thorium would require mining 500 Mt of material per year. In comparison, world coal mining is 8,000 Mt per year, with reserves of about 150 years. The earth's continental crust contains over 4,000 Gt of thorium, nearly enough for a million years of energy from thorium.

LFTR produces < 1% of long-lived radiotoxic waste of LWRs.

LFTRs reduce nuclear waste storage issues from millions of years to a few hundred years. The radiotoxicity of nuclear waste arises from two sources: the highly radioactive fission products from fission, and the long-lived actinides from neutron absorption. Thorium and uranium fueled reactors produce essentially the same fission products, whose radiotoxicity in 500 years drops below even that of the natural uranium ore that would power a PWR or BWR.

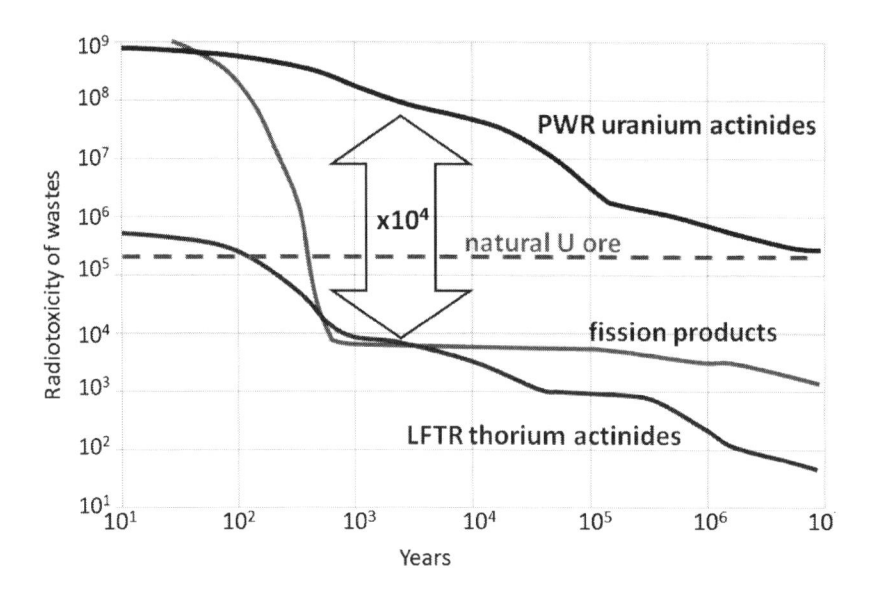

Radiation dose from ingestion of waste from 1 GW(t) reactor

LFTR creates far fewer transuranic actinides because Th-232 requires 7 neutron absorptions to make Pu-239, whereas U-238 requires just one. After 300 years radiation from LFTR waste would be 10,000 times less than radiation from LWR waste. Chemical separation processes are not perfect, so 0.1% of the LFTR transuranics might pass through the waste separator instead of being retained to be burned in the LFTR. LFTR waste radiotoxicity would be 1/1000 that from PWRs. Geological repositories smaller than Yucca mountain would suffice.

A single fluid LFTR has simpler plumbing.

The single fluid LFTR contains both the fertile Th-232 and fissile U-233 dissolved in the same molten salt. There is no separate breeding blanket.

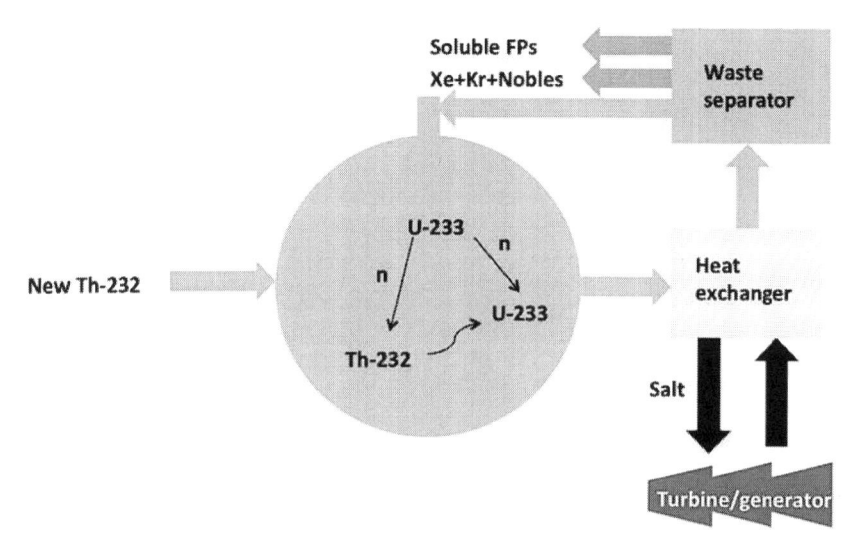

Single fluid LFTR

The diagram illustrates some neutrons fissioning U-233 atoms and some being absorbed by Th-232, eventually decaying to U-233. There is no need for an external uranium separation chemical process facility. Noble metals and noble gases can be physically removed. The difficulty is in the design of the waste separator, which must separate fission products that are chemically similar to thorium. If a waste separator could be perfected, perhaps based on both physical and chemical properties, this single fluid reactor could be very attractive.

ORNL's ARE and MSRE were single fluid reactors, but did not breed thorium to uranium. Molten salt reactors can be designed for a wide variety of nuclear fuels. The two-fluid LFTR uses thorium via the Th-232/U-233 fuel cycle. Fast MSRs can consume plutonium and other transuranics in spent LWR fuel. The denatured MSR uses a mixture of thorium and enriched uranium.

DENATURED MOLTEN SALT REACTOR

The denatured molten salt reactor (DMSR) is a single fluid reactor. Both fissile uranium and fertile thorium are dissolved in the molten salt. The word "denatured" means the fissile U-235 is diluted with at least 80% U-238, so the uranium is unsuitable for weapons.

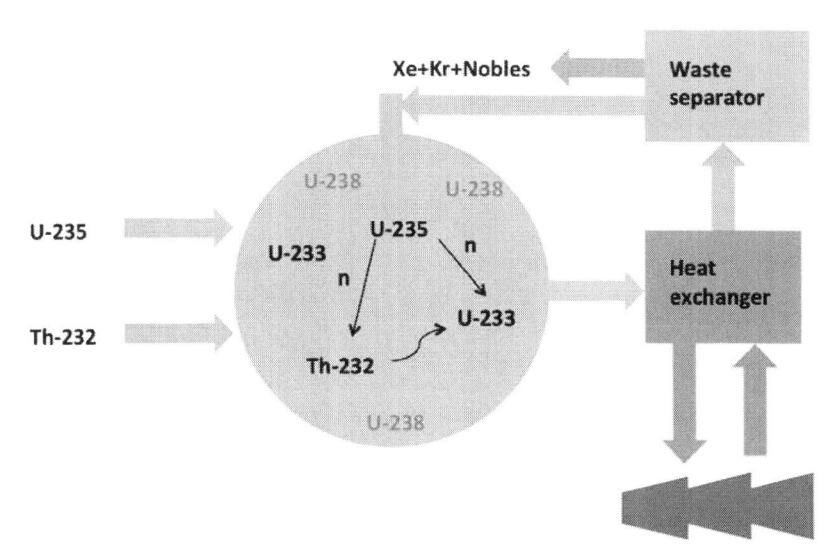

DMSR: U-235 fission, Th-232 breeding, U-233 fission

DMSR is started with fissile U-235. Neutrons from fission can either continue the fission chain reaction by interacting with a uranium atom nucleus, or can be absorbed by a Th-232 nucleus and then decay to (Pa-233 to) U-233. All this happens within the molten salt. Of the fission products, the noble gases and noble and semi-noble metals are removed by physical processes. The remaining fission product elements become fluorides that remain dissolved in the molten salt for up to 30 years.

DMSR fuel additions are 75% thorium and 25% uranium.

The DMSR molten salt core contains Th-232, fissile U-233, fissile U-235, and U-238. Enough neutrons are absorbed by the plentiful U-238 (becoming fissile Pu-239) that the process can not be self-sustaining. Consequently more fissile U-235 must be continually added to the molten salt core along with fertile Th-232. The Moir-Teller DMSR design has a conversion ratio of 0.75, meaning that 75% of the fissile material is U-233 converted from supplied Th-232, and the other 25% is from U-235 that must also be added continuously. David LeBlanc and early ORNL designs propose 80% Th-232 and 20% U-235 fuel additions.

DMSR fuel salt can be reprocessed after 30 years.

The DMSR waste separator removes some of the nonreactive fission products, namely the noble gases (xenon and krypton) that bubble out of the salt using the helium sparge technology demonstrated at ORNL.

Excerpted below from ORNL's 1 GW DMSR design are inventories of some isotopes at start-up and 30 years later, when the fuel salt is discarded or reprocessed.

DMSR fissile and fertile material		
Isotope	Start kg	End kg
Th-232	110,000	92,900
U-233	0	1,910
U-235	3,450	1,250
U-238	14,000	28,600
Pu-239	0	231
Pu-other	0	505

Most of the fission products become fluoride salts and remain dissolved in the molten salt. Without chemical processing, the

DMSR accumulates radioactive fission product fluorides in its fuel salt for the 30 year lifetime of the fuel and salt charge.

Then the salt can be reprocessed at a chemical plant, where the valuable uranium can be extracted for re-use. Left behind in the salt will be the dissolved fission products, plutonium, and other actinides. Possibly the valuable flibe salt might be extracted for re-use. The remaining actinides and fission products will be prepared for sequestration. Recovered flibe and uranium can be used to restart the DMSR, with addition of LEU uranium and thorium.

Alternatively the reprocessing can be avoided and the salt including its thorium, uranium, and transuranics can be sequestered as waste. The DMSR can be recharged with new flibe, thorium, and LEU and operated for another fuel cycle.

DMSR electricity will be cheaper than coal.

Compared to the two-fluid LFTR, the DMSR has only a single molten salt fluid containing both fertile and fissile materials, and therefore DMSR should have an even lower capital cost.

Unlike LFTR which runs entirely on abundant thorium, the DMSR does require some expensive fissile U-235 fuel, but it uses only 1/4 the U-235 of a standard LWR. World uranium supplies are substantial if the price is high enough. Current uranium prices are about \$100/kg, but at \$1000/kg it even becomes economic to separate the UO_2 dissolved in seawater at 3 mg/tonne. At that price uranium for a DMSR would only be 0.5 cents/kWh.

DMSR can recycle LWR spent fuel.

Per Peterson points out that DMSR also provides a simple, low-cost technology to recycle LWR spent fuel. The entire fuel rod, including its zirconium cladding, could be converted to fluoride salts with hydrogen fluoride (HF). The zirconium would become part of the ZrF_4-NaF molten solvent salt, such as was used in the first ORNL molten salt reactor, instead of flibe. The fluoride volatility process can remove the uranium leaving the fissile plutonium and other actinides as fuel for the DMSR.

If designed to use a two-fluid configuration, such a DMSR could also breed Th-232 into U-233 that could be used to start up two-fluid LFTRs that could then operate completely on thorium.

The simplest spent DMSR fuel processing technique would be to discharge depleted salt to waste and recharge with fresh salt derived from processed LWR spent fuel.

The Denatured Molten Salt Reactor will be first to market.

DMSR will likely be the first thorium molten salt reactor to enter commercial service.

1. DMSR has minimal fuel processing, requiring just xenon off-gas removal and noble metal plate-out.
2. No structural boundary layer is required between fissile fuel salt and fertile blanket salt of LFTR.
3. Less R&D must be accomplished before commercialization.
4. End-of-life salt reprocessing R&D can proceed in parallel with 30 years of commercial operation.
5. DMSR LEU (low enriched uranium) fuel is compatible with current licensing requirements.
6. Using 1/4 the uranium of LWRs, DMSR fuel will be available for centuries.
7. DMSR is highly proliferation resistant, more so than any other nuclear power technology.
8. DMSR is less costly than LFTR, because it has fewer components.
9. DMSR can make energy cheaper than coal sooner, achieving the benefits sooner.

Using DMSRs instead of two-fluid LFTRs foregoes these benefits:

1. No fissile material need be transported to or from the LFTR after startup.
2. LFTR's 100% thorium fuel obviates the need or excuse for uranium enrichment plants.
3. Worldwide availability of thorium creates energy security for all nations.
4. Inexpensive thorium fuel will last for may thousands of years.

PEBBLE BED MOLTEN-SALT-COOLED REACTOR

PB-AHTR is a molten-salt-cooled solid-fuel reactor.

The pebble-bed advanced high-temperature reactor (PB-AHTR) uses solid fuel, but in a pebble form much different from the fuel rods of today's LWRs. A bed of such pebbles forms a critical mass that generates heat carried away by a molten salt coolant. The pebbles contain thousands of sand-sized particles of uranium fuel.

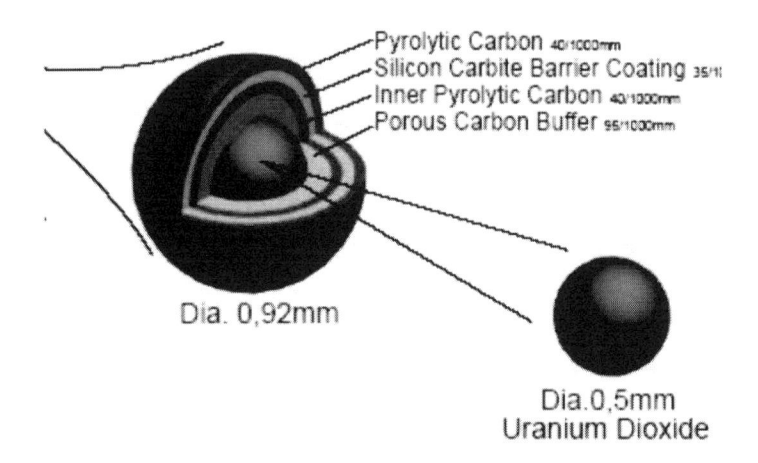

TRISO fuel particle

These sand-sized particles of uranium fuel, coated with three impermeable barrier layers, are a key technology that contains both the fuel and its fission products. The porous carbon buffer layer provides moderation and a place to contain fission product gases. The three successive impermeable layers provide a triply redundant containment for all the radioactive materials. The three redundant layers (pyrolytic carbon, silicon carbide, pyrolytic carbon) maintain their structural integrity at temperatures over 1600°C. These so-called TRISO particles have three isolating layers.

LIQUID FLUORIDE THORIUM REACTOR 213

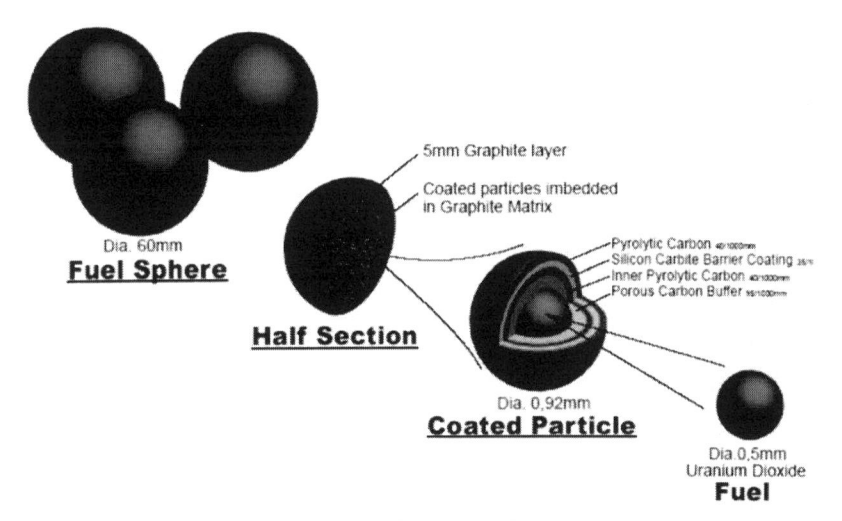

TRISO fuel pebble

Over 10,000 of TRISO particles are embedded in a billiard-ball-sized graphite pebble.

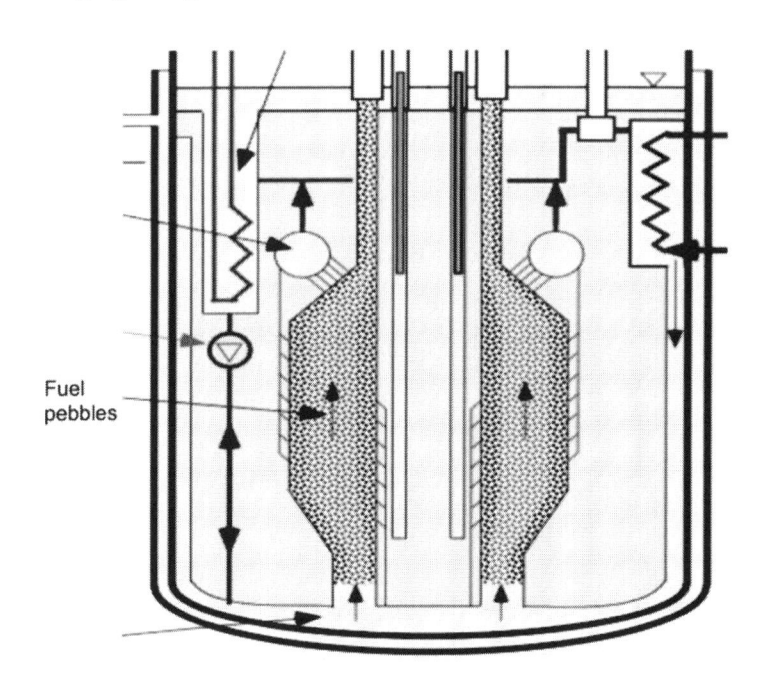

Fuel pebbles in PB-AHTR core

Thousands of these TRISO fuel pebbles form the pebble bed that achieves a critical mass of fissile uranium. The fuel pebbles are packed closely together in an elongated toroid container. A cross section is shown in the illustration. The pebbles are cooled by molten salt that flows through an isolating heat exchanger represented by the zigzag line at the upper right. That molten salt then flows to a power conversion system to make electricity.

The pebbles flow upward, but very slowly, a few per hour. They are examined by automatic machinery that measures the remaining useful fissile fuel content. Spent pebbles are set aside for ultimate waste disposal and replaced with fresh fuel pebbles. The pebbles are not quite close-packed, because they reposition themselves somewhat randomly as they move up. The pebbles do tend to maintain their relative positions as they rise, allowing cylindrical layering of reflectors or fertile thorium fuel pebbles. The lubricity of the molten salt helps pebbles maintain their relative positions.

Initial plans for the PB-AHTR are to use uranium oxide fuel. The organized flow of pebbles may allow for controlled insertion of an auxiliary blanket of thorium TRISO pebbles. As in LFTR, the Th-232 absorbs a neutron, becoming Th-233, which beta decays to Pa-233, which decays to fissile U-233 with a half-life of 27 days. Pa-233 is a strong neutron absorber, diminishing the production of U-233. In PB-AHTR the flow of pebbles may allow freshly irradiated thorium pebbles to be held outside the reactor core where the Pa-233 can not absorb neutrons. After Pa-233 decay to U-233 the pebbles can be returned for further breeding or as fuel.

The pebbles are strong and hard, already in a form suitable for waste disposal. Reprocessing would be much more difficult than for LWR fuel rods, because of the requirement for crushing the pebbles and also the TRISO particles.

PB-AHTR has many of the advantages of LFTR.

The PB-AHTR is also cooled by high-temperature molten salt with high heat capacity. This means it can be compact, leading to lower capital costs, contributing to low electric power costs. The high temperature, up to 900°C, enables high efficiency helium Brayton or supercritical CO2 power conversion systems. The high 1400°C boiling point of molten salt provides a large safety margin in the event of overheating. Dry air cooling is possible. The reactor also runs at atmospheric pressure, reducing potential forces that could propel radioactive material in an accident.

One advantage of PB-AHTR is that the TRISO fuel form is well understood and has already been used in other types of nuclear reactors, namely high-temperature helium cooled pebble bed reactors. Germany operated a uranium fueled 15 MW(e) pebble bed reactor for 15 years. Germany built and ran a 300 MW thorium fuelled reactor, the THTR-300 for six years. China operated a 10 MW HTR-10 demonstration pebble bed reactor at Tsinghua University and is now constructing the first of several such commercial 250 MW reactors. The US has capabilities for TRISO fuel manufacturing.

PB-AHTR fuel is not dissolved in the molten salt, but kept separated in TRISO particles. There are no fission products in the salt that could interact with the materials of the vessel, piping, and pumps, simplifying the materials requirements. The PB-AHTR might be developed more rapidly than LFTR.

The design of the PB-AHTR is being advanced actively by UC Berkeley, MIT, and U Wisconsin, with modest government funding support, $7 million over 3 years. Much of the research and operational findings will be directly applicable to LFTR and DMSR work. The molten salt is heated to over 700°C temperatures in both reactors. This would allow sharing of technology developed for high-temperature power conversion systems, such as closed Brayton cycle helium or nitrogen gas turbines, open air Brayton cycle turbines, or supercritical CO2 turbines.

ENERGY CHEAPER THAN COAL

LFTR will provide energy cheaper than coal.

Taxing carbon seeks to encourage energy sources that do not emit CO2, yet this has not been effective. Developing countries will not agree to carbon taxes and forgo the cheap coal energy advantage they perceive led to prosperity in OECD nations. Alternatively, a source of energy cheaper than coal would dissuade all nations from burning coal, without imposing tariffs or taxes that reduce economic productivity. Affordable electric power can also help developing nations reach modest levels of prosperity and lifestyles that include sustainable birthrates.

Decisions about coal-fired electricity generation versus nuclear power are made at the time of construction of a new plant. The Chapter 4 "Energy sources" cost models for new-build electric power plants are summarized in the table below.

Electricity costs from alternative sources, cents/kWh					
	Coal	Gas	Wind	Solar	Biomass
Capital cost recovery	2.8	1.0	24.4	22.5	4.0
Fuel	1.8	2.8	0	0	4.7
Operations	1.0	1.0	1.0	1.0	1.0
Total	5.6	4.8	25.4	23.5	9.7

Examining the table derived in Chapter 4, it is clear that wind, solar, and biomass fuel will not undersell electric power generated from fossil fuels coal and natural gas. The table also suggests that

CCGT natural gas generators will partially replace coal, because of gas's economic advantage, reduction in air pollution, and 2/3 reductions in CO_2 emissions. The economic incentives for generating electric power favor both CO_2-emitting fossil fuels.

Thorium energy must be cheaper than coal to reduce the largest source of CO_2 emissions from power plants. Thorium energy cheaper than from natural gas can also replace that lesser source of CO_2 emissions and fugitive methane leaks. LFTR must produce electricity for less than 4.8 cents/kWh to competitively displace burning both coal and natural gas as a source of energy. To replace fossil fuels for power generation, the LFTR cost objectives are a capital cost of $2/watt of generating capacity and $0.03/kWh for electric energy. How can liquid fluoride thorium reactors produce energy cheaper than from coal?

Molten salt reactor cost estimates have been about $2/watt.

Costs of 7 molten salt reactor proposals			
Estimate	Year	$/watt	2012 $/watt
Sargent & Lundy	1962	0.65	4.95
Sargent & Lundy ORNL TM1060	1965	0.15	1.09
Kasten, MOSEL reactor	1965	0.21	1.53
ORNL-3996	1966	0.24	1.70
McNeese et al, ORNL-5018	1974	0.72	3.36
Engel et al, ORNL TM7207	1978	0.66	2.33
Moir	2000	1.58	2.11

The table above presents seven independent cost estimates to build experimental molten salt reactors. The $/watt is the cost of research, development, construction, and testing of the proposed experimental reactor divided by the power produced. The last column is inflation-adjusted to 2012 dollars. This suggests that $2/watt is a reasonable goal for commercially produced power reactors that do no bear the R&D costs. New, up-to-date designs can furnish more accurate cost estimates.

Following are additional reasons that LFTR can produce energy cheaper than coal.

The compact LFTR operates at atmospheric pressure.

All radioactive materials in LFTR are at atmospheric pressure. There is no need for high pressure piping, valves, and pressure vessels such as an LWR requires. This reduces costs for Hastelloy piping, fittings, valves, pumps and other materials. It also simplifies safety engineering, for there are no pressurized radioactive materials that could be propelled into the environment in a severe accident.

This leads to a compact reactor, reducing mass and costs. The genesis of LFTR was a nuclear reactor small and light enough to sit on an airplane wing. Airplane jet engines are already examples of compact Brayton cycle power conversion systems.

Inherent thermal stability lowers control costs.

As flibe heats and expands, the density of fissile material is reduced and the chain reaction slows. Rising temperatures also increase neutron absorption and lower fission probabilities, slowing the reaction. Neutron-absorbing control rods are not necessary, lowering costs in comparison to LFTR.

The simple backup safety freeze plug melts at high temperatures or control failures, dumping the fuel salt into special tanks where the chain reaction stops.

Decay heat removal systems are passive.

When a reactor stops the fission products continue to decay and the heat must be removed. Because LFTR operates at higher temperatures than LWRs, and because the liquid flibe conducts and convects heat efficiently, heat transfer is more rapid. Molten does not boil under 1400°C, so there is a large safety margin.

LFTR's high temperature increases its efficiency.

The high 700°C outlet temperature enables 45% efficient Brayton cycle power conversion, compared to LWRs at about 33%. This means that LFTR can deliver 45/33 the electricity for a thermally comparable LWR. Also, the rejected heat dissipated by the reactor cooling system is also reduced by 39% for the same electrical power, reducing costs for cooling towers or alternatively permitting dry air cooling.

The high heat capacity of molten salt reduces size.

High molten salt heat capacity exceeds that of the water in PWRs or liquid sodium in LMFBRs, allowing compact geometries and heat transfer loops that make the reactor more compact, requiring less material such as Hastelloy-N or SAE 316 stainless steel, lowering materials cost.

New power conversion systems are smaller.

Two new power conversion systems are candidates for LFTR uses. The triple-reheat closed Brayton cycle turbine mass is smaller than a comparable steam turbine by a factor of about four. Open cycle Brayton turbine engineering has been developed to a high art for the aircraft industry. A $24 million GE90 turbine delivers 83 MW -- only $0.29/W. After perfecting, the Brayton closed sycle helium turbine costs should similarly drop relative to massive steam turbines used in LWRs. The newer supercritical CO_2 turbine is even smaller, and requires more engineering to perfect.

Waste disposal costs are smaller.

LFTR produces less than 1% of the long-lived radioactive transuranic isotopes produced by LWRs. Their heat production is the cost driver for long-term geological storage sites such as Yucca Mountain.

Small modular LFTRs can be mass produced.

Commercialization of technology leads to lower costs as the number of units increase. Experience benefits arise from work specialization, new processes, product standardization, new technologies, and product redesign. Business economists observe that doubling the number of units produced reduces cost by a percentage termed the learning ratio, seen in the early aircraft industry to be 20%. Today Moore's law in the computer industry illustrates a learning ratio of 50%. In *The Economic Future of Nuclear Power* University of Chicago economists more conservatively estimate the learning ratio is 10% for nuclear power reactors.

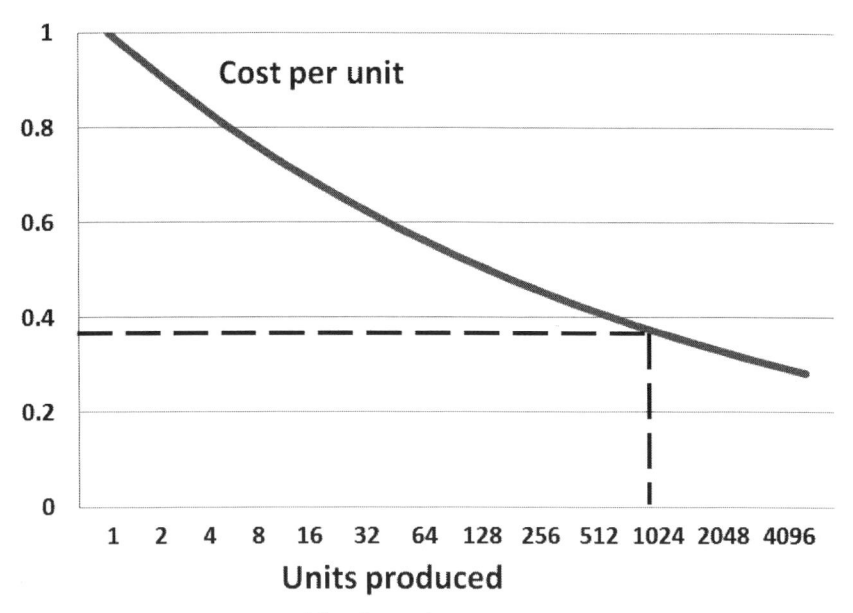

The learning curve

In this illustration, the cost of the 1024th LFTR would be about 35% the cost of the first commercial LFTR. Some engineers advocate economy-of-scale to justify large reactors, but this analysis shows that 100 MW units would have a 30% cost advantage over 1000 MW units because of the ten times more production experiences.

Boeing 737 production line

Boeing made 477 airplanes in 2011 costing up to $330 million each. Boeing, capable of manufacturing $200 million units daily, is a model for LFTR production. Airplane manufacturing has many of the same critical issues as manufacturing nuclear reactors: life safety, reliability, strength of materials, corrosion, regulatory compliance, documentation, design control, supply chain management, and cost, for example.

Reactors of 100 MW size costing $200 million can similarly be factory produced. Manufacturing more, smaller reactors traverses the learning curve more rapidly. Producing one per day for 3 years creates 1,095 production experiences, reducing costs by 65%.

Documentation control integrated with manufacturing saves costs and increases accuracy. New manufacturing techniques are enabled with CAM (computer aided manufacturing), automatically converting designs to manufacturing instructions for machine tools and industrial robots. CAM can vary manufacturing parts and processes to produce a variety of units on one production line.

In the Boeing photograph above, observe that the wing tips are not identical on all units.

Ongoing research will lead to lower LFTR costs.

Cost reductions are presaged by current engineering research. Compact, thin-plate heat exchangers may reduce fluid inventories, size, and cost. Possible new materials include silicon-impregnated carbon fiber with chemical vapor infiltrated carbon surfaces, and higher temperature nickel alloys. Operating at 950°C can increase thermal/electrical conversion efficiency beyond 50%. Such high temperatures can improve efficiency for water dissociation to create hydrogen, to lower manufacturing costs of synthetic fuels such as methanol or dimethyl ether that can substitute for gasoline or diesel oil.

Initial fissile material quantities and costs are low.

A 100 MW LFTR requires only about 100 kg of fissile material, such as U-233 or U-235, to start up. Thereafter it is fueled by thorium, or thorium and enriched uranium in DMSR. A LWR or LMFBR requires 5 times this, adding to capital costs.

Thorium fuel is plentiful and inexpensive.

One ton of thorium can power a 1,000 megawatt LFTR for a year – enough power for a city. Just 500 tons would supply all US electric energy for a year. Fuel costs at $300,000 per ton for thorium would be $0.00004/kWh, compared to coal at $0.03/kWh.

Uranium enrichment costs are low.

The expanding worldwide fleet of LWRs increases demand for uranium and also for the enrichment services to convert it from 0.7% to 4% U-235. Some LFTRs may require enriched uranium only for startup. Designs such as DMSRs will require a continued supply of enriched uranium, but less than 25% of the amount used by LWRs.

Fuel fabrication costs are low.

Unlike LWRs, there are no costs for producing high quality zirconium tube fuel rods to contain UO2 pellets and their fission products for centuries. Unlike pebble bed reactors using TRISO particle fuel, there is no cost for triple-coating millions of UO2 particles designed to retain fission products within the three redundant layers. The LFTR fuel supply form might be solid UF4 crystals or gaseous UF6, which are already intermediate, steps in the production of solid UO2 used in LWRs.

New control system technologies can reduce labor costs.

The number of people required to operate today's LWRs is higher than for other forms of power production. Nuclear power plants operate 24x7, and each job employs 6 people: 4 for the 4 work shifts per week, 1 for vacation and sick leave, and 1 for training time, so labor costs mount up. In my visits I observed there are more than 1000 employees per GW of power output, adding about 1 cent/kWh to electricity costs.

Information systems and control systems technologies have improved immensely since LWRs were designed in the 1970s. Safety critical software techniques enable low-labor-cost operation of aircraft, helicopters, and rapid transit. Reducing direct operator control of reactors can also avoid mistakes, such as the series of operator errors that led to the Chernobyl disaster. Security guard costs should be proportional to the possible damage threat, much lower with a non-pressurized LFTR. Even US ICBMs in missile silos were guarded with remote electronic surveillance.

Transmission line costs are less with distributed LFTRs.

Much of the costs associated with multi-GW power plants are for transmission lines to transport power hundreds of miles on low-loss high-voltage direct-current (HVDC) lines. Fewer transmission lines are required when 100 MW power sources such as LFTRs are near cities and manufacturing centers. Costs for

HVDC lines are roughly $1 million per mile, so the costs for energy transmission over 1,000 miles is roughly 1 cent/kWh.

The program *objective* must be energy cheaper than coal.

For all the above reasons, low costs of $2/W and 3 cents/kWh is an achievable objective. A $2/watt capital cost contributes $0.02/kWh to the power cost, assuming a 40 year life, 8% interest rate, and 90% capacity factor. With plentiful, inexpensive thorium fuel, LFTR can generate electricity at <$0.03/kWh, underselling power generated by burning coal.

Producing one LFTR of 100 MW size per day could phase out all coal-burning power plants worldwide in 38 years, ending 10 billion tons of CO_2 emissions from world coal plants now supplying 1,400 GW of electric power.

Low LFTR costs are crucial to this coal replacement strategy, achievable if cost objectives are maintained at every design choice. Less expensive electric power will check global warming by dissuading all nations from burning coal. It will also help developing economies to improve their prosperity, encouraging lifestyles with sustainable birthrates. Keeping LFTR energy costs cheaper than coal is critical to achieving the social and environmental benefits.

Cost challenges can be met at the R&D stage.

There are cost challenges for LFTR development. Meeting the production cost objectives of $2/W and 3 cents/kWh requires a well-executed research and development program. Corporations with deep pockets may develop advanced nuclear power, as evidenced by Bill Gates' investment in Terrapower's LMFBR reactor, building on prior US $16 billion R&D expenditures. There is an opportunity for substantial government or philanthropic investment in LFTR R&D to keep ultimate production costs low by removing amortization of imprecise R&D costs. Public investment in energy R&D is a much more effective public policy than ongoing alternative energy production subsidies being paid today.

LFTR DEVELOPMENT ENGINEERING

Commercialization of LFTR or DMSR power reactors is a multi-fold engineering task. There are no show-stoppers; ORNL has already demonstrated two molten salt reactors; chemical separation processes are known. Component technologies are in different states of maturity. Engineering is required to bring each technology from laboratory to pilot plant to commercialization. Here is a sampling of major components of a liquid fluoride thorium reactor.

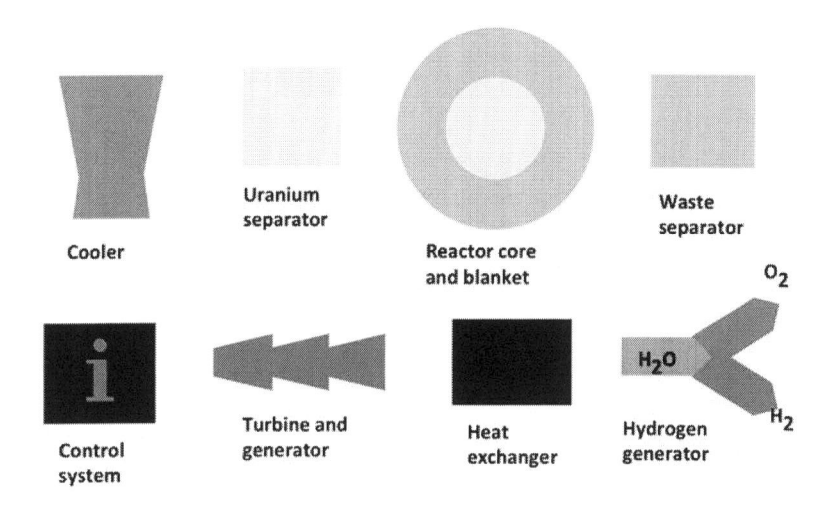

Overview of LFTR components

The reactor core is the locus of the heat-generating U-233 fission; the blanket contains the Th-232 converted to U-233 through neutron absorption.

The uranium separator and waste separator must process the molten salt slowly and reliably within the high radiation environment of the primary containment cell.

A heat exchanger in the containment cell must transfer thermal energy from the fuel salt to nonradioactive molten salt for the modules outside the containment cell.

Electrical generator technology is off the shelf. Open cycle Brayton *turbines* are well developed in the aircraft and natural gas power industries, but closed cycle helium turbines have not been demonstrated in a power plant. Supercritical CO2 *turbines* have only been developed at laboratory scale.

Air cooling for the turbine power conversion system can eliminate water consumption, but it is an uncommon technology for power generation.

New computer-based control systems will make reactors safer and less expensive, reducing labor cots and the potential for human error. Safety-critical software systems engineering is well developed for industries such as aeronautics, rail transport, medicine, and spaceflight.

Hydrogen generation by water dissociation at utility scale is an engineering challenge required for vehicle synfuel production.

LFTR DEVELOPMENT TASKS

The work outlined below is extensive. An actual, functioning, commercial nuclear power plant is highly sophisticated and complex. LFTR's design, development, and deployment require an extensive investment of talented human capital. It is a challenge that is very appropriate to modern engineering knowledge, design tools, and new materials. The construction cost of nuclear power plants is not so much from the commodities used in construction but in human capital -- engineering. Much of the expensive engineering labor expenditure ned not be repeated once LFTRs are in commercial production.

What better environmental investments could there be? Invest in human capital rather than the extractive industries. Meet the climate challenge. Reduce pollution. Improve global prosperity. Conserve resources.

Build a LFTR technology reference database.

Much of what is known about the molten salt reactor is decades old. New knowledge is scattered in many publications, email chains, and discussion forums. There are a dozen centers of excellence in molten salt technology with physical, computational, and analytical data, ranging from the US National Institute of Standards and Technology to the Institute for Transuranium Elements in Karlsruhe, Germany.

Develop the program plan, budget, and schedule.

Oak Ridge National Laboratories developed a rather complete technology tutorial and program plan in 1974. The document ORNL-5018, *Program Plan for Development of Molten Salt Breeder Reactors* forms an excellent starting point. It could be updated to account for new knowledge of materials, experience with fast breeder reactors, experience with high-temperature gas-cooled reactors, and current costs. ORNL published a wrap-up review (TM-6415) of LFTR and DMSR in 1979 outlining the work to be accomplished.

Design an appropriate neutron economy.

This diagram is a hypothetical neutron economy for a two-fluid LFTR. The starting point is the average 252 neutrons from 100 fission events. The LFTR uses neutrons to continue the U-233 fission reaction and to convert Th-232 to U-233, but loses neutrons to unproductive absorption by graphite, salt, Pa-233, and fission products.

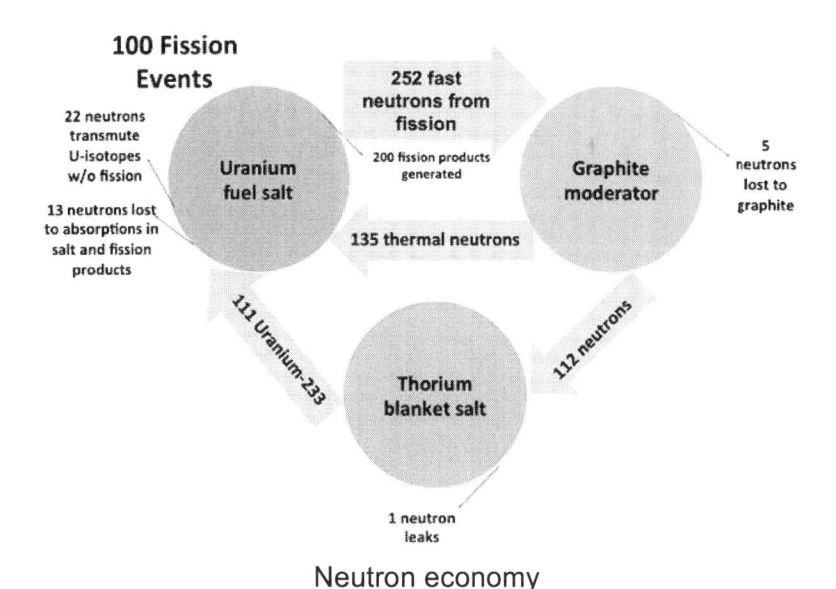

Neutron economy

This example shows 111 U-233 atoms created for every 100 fissioned. An appropriate neutron economy would produce just as much U-233 as LFTR consumes; such a reactor is termed an isobreeder.

The absorbed neutron creates uranium in three steps: Th-232 to Th-233 to Pa-233 to U-233. ORNL's designs included separation and isolation of the Pa-233 because it parasitically absorbed neutrons unproductively, but this is not necessary for an isobreeder. Indeed such Pa-233 becomes U-232, which adds proliferation resistance. U-233 absorptions occasionally lead to production of some transuranic elements, but these are eventually fissioned with an adequate neutron economy.

The neutron economy choices interact with MSR design type: two-fluid LFTR, one-fluid LFTR, or DMSR. For DMSR the neutron economy does not permit an isobreeder, leading to the compromise of feeding external fissile U-235 or Pu-239.

Control reactivity and power output.

A feature of molten salt reactors is the negative temperature coefficient, meaning that as fuel salt temperature rises the fission rate decreases. This is best accomplished through immutable physical materials properties rather than fallible control systems. In the LWR it happens because as heated water expands neutron moderation from hydrogen collisions drops. In LFTR it happens because the molten salt carrying the U-233 expands, squeezing that fissile material out into pipes away from the critical mass, lowers neutron fission cross-sections, and increase neutron absorption by U-238 and Th-232.

Reactivity control for LFTR must be validated for many a range of temperatures and mixtures of fission products. Redundant control rods to absorb neutrons may be added, although they are not needed to control excess reactivity, as in LWRs, which contain years of supply of fissile U-235 fuel.

If the electrical grid disconnects from an operating power plant the power conversion turbines can not convert heat into electricity, so a LFTRs molten salt temperature will rise and reactivity will decrease. This provides an opportunity for LFTR to be inherently load-following, decreasing its power output to match the power demand, or load. Today's LWR plants can slowly change power output with manual controls. Reportedly, load-following happened unexpectedly with the MSRE reactor when half its load (a thermal radiator) stopped functioning. Load-following is operationally flexible, but not economically attractive since revenue decreases with power output and LFTR operational costs continue virtually unchanged.

Another proposed load-following scheme is to operate LFTR at full power and store excess heat in large, external, insulated molten salt tanks for later conversion to electric power when demand rises. This has been done with concentrated solar power. Yet another is to use excess heat, when available, for heating oil shale to further the years-long in situ process of converting the fossil kerogen to oil that can be pumped out. And another is to use excess heat and power for desalination to produce and store fresh water.

Control molten salt chemistry.

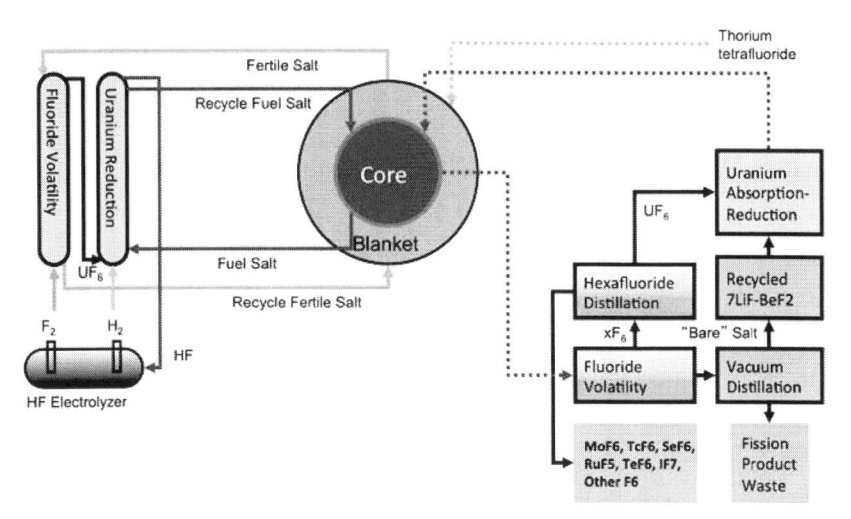

Molten salt chemical processing

Managing the chemistry of the molten salts is important to LFTR operations. There is good reason LFTR is termed "the chemist's reactor". The two-fluid LFTR sketch above shows on the left the process for separating uranium from the blanket salt and on the right the more complex separation of fission products from the fuel salt.

Uranium has several valence states in which it can make compounds, such as in UF4 and UF6. The fluorination process UF4 + F2 → UF6 changes the dissolved uranium salt into gaseous uranium hexafluoride. Thorium has only the +4 valence state so remains behind. The hydrogen reduction process then changes UF6 + H2 → UF4 + 2HF. Managing fluoride ions in the salt is important to control corrosion. The electrolyzer makes HF into H2 and F2.

The fission product separator on the right side is more challenging, because there are more elements to deal with, arising from all fission products. Most fission products will combine with fluorine and dissolve in the molten salt. Tellurium fluorides were the cause of minor crevice corrosion discovered in the autopsy of the MSRE. Separation techniques include chemistry and distillation, which relies on different boiling point of different compounds. Even centrifuge separation has been suggested to make use of the mass density differences.

The single fluid LFTR has more challenging chemistry, because thorium is chemically similar to many of the fission products. Alternatively, fission product fluorides can be allowed to remain dissolved in the molten salt, requiring reprocessing or replacement after 10 to 30 years. The DMSR operates this way.

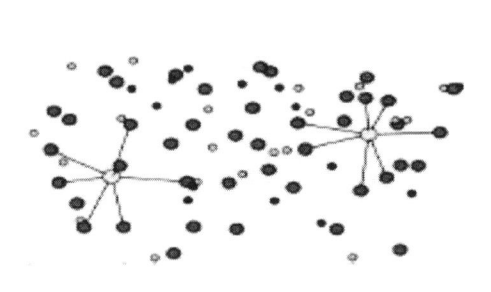

Virtual chemistry is a new tool enabled by very fast computers, used by Paul Madden at Queens College. Theoretical chemistry and fast computers can predict physical properties of liquids, such as heat capacity and viscosity of Flibe.

Remove noble fission products.

The noble gases from fission are xenon and krypton. The fission product metals that do not form fluoride salts are the noble and semi-noble metals: molybdenum, ruthenium, silver, tin, tellurium, and sometimes niobium.

Xenon-135 results from 6% of the fissions of uranium or plutonium. Xenon-135 is a prolific neutron absorber and will stop the chain reaction in a reactor like LFTR with little excess fissile material. However Xe-135 has a half life of only about 9 hours, decaying to cesium, ending the neutron absorption and allowing a stopped reactor to restart. The on-off behavior puzzled the nuclear pioneers. Failure to manage increasing reactivity from xenon decay contributed to the Chernobyl reactor power overshoot.

Xenon is a noble gas, not forming any fluorides and not dissolving in the molten salt, but remaining as gas bubbles. ORNL discovered that these could be removed by injecting a stream of small helium bubbles into the salt and removing them after they had taken up the xenon. The process is called sparging. Krypton is similarly swept out

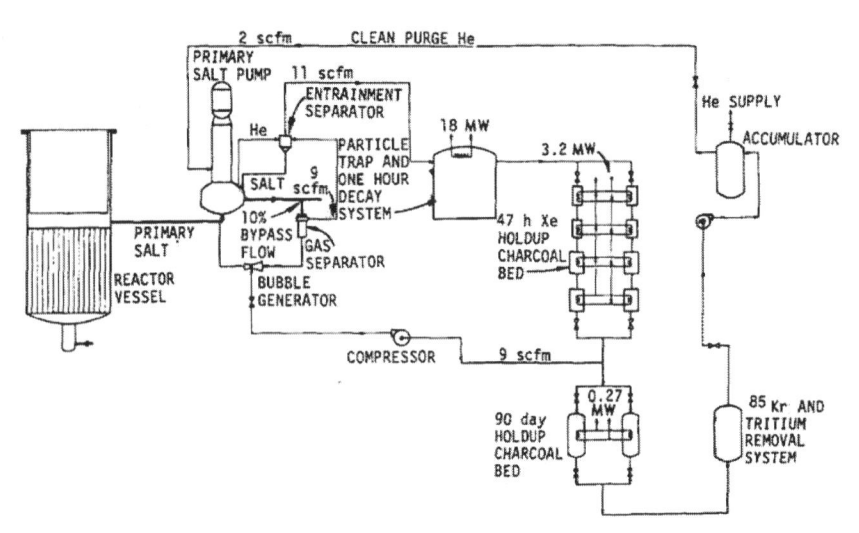

Off-gas processing system

ORNL accomplished this, shown in the diagram of the off-gas processing system is taken from their report. Sparging was also able to remove krypton noble gases, as well as some of the noble metal fission product particles. This sparge reduces neutron losses due to xenon absorption to less than 0.5%.

Helium sparging and off-gas processing will be part of any MSR design, including the two-fluid LFTR, the single-fluid LFTR, and the DMSR. Some fission product particles of the noble and semi-noble metals (Nb, Mo, Ru, Sb, Te) can partially be removed by the helium sparge and also can plate out on special metal adsorbers.

ORNL found these metals on Hastelloy metal pipes and pumps, and on graphite structures, and at liquid-gas boundaries The LFTR designer must provide a way to capture or plate-out the noble metals. Noble metal plating has been used for corrosion control in LWRs. Ralph Moir has also suggested investigating centrifuge separation.

Neutron-irradiated graphite swells, then shrinks.

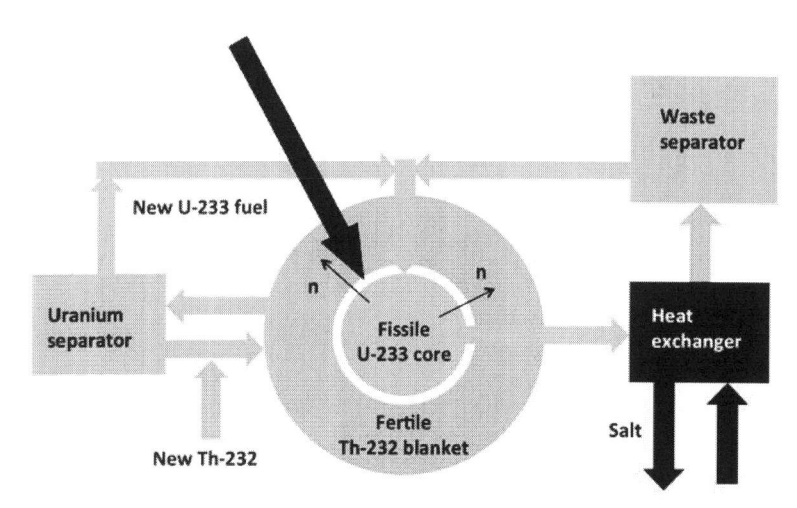

Graphite structural material

Graphite is form of carbon used in molten salt reactors. It moderates neutrons because of its low atomic weight. It also reflects neutrons. A high purity form is used as a structural material, for example in the barrier between the fissile fuel salt and the fertile blanket salt. Neutron irradiation of graphite causes it to swell while further irradiation causes it to shrink. This makes the mechanical design difficult. LFTRs may have to be designed with the ability to change the graphite after perhaps ten years. Other proposed designs do not use graphite as a structural material, but do moderate neutrons with graphite in a form that lets be replaced if necessary. Fast reactors have no graphite.

Metals must withstand heat, irradiation, and corrosion.

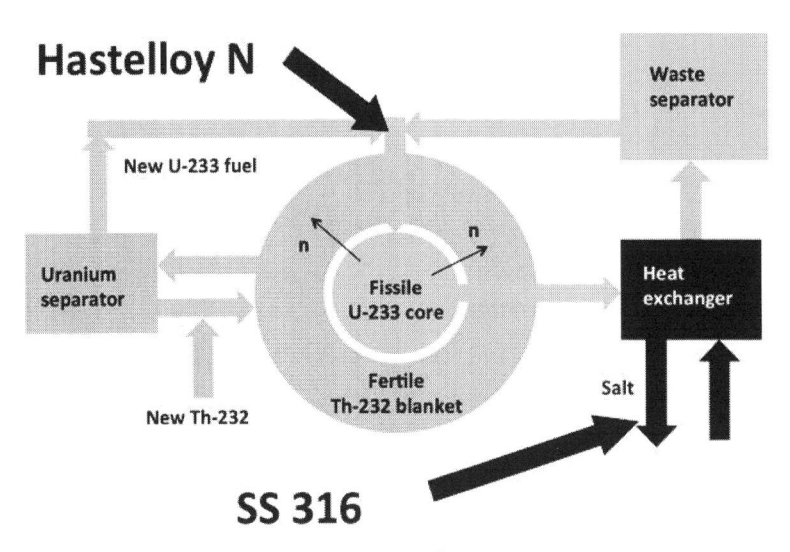

Applications of metal alloys

Most components of a LFTR may be metals, used for vessels, pipes, tubes, and pumps. Standard metals such as 316 grade stainless steel can be used for many purposes. Hastelloy N is a nickel alloy used in nuclear power applications because of its resistance to corrosion, erosion, and high temperature. It is an alloy of nickel, molybdenum, chromium, iron, silicon, magnesium, manganese, cobalt, and other metals. The properties of such metals placed in service for a lifetime of 60 years or more must be confirmed, tested, and validated.

Carbon composites might replace metal materials.

With metals such as Hastelloy N the molten salt reactor can operate to temperatures up to about 760°C. Increasing the temperature to near 1000°C would improve the thermal-to-electrical conversion efficiency and importantly permit the direct thermochemical dissociation of water to make hydrogen at a thermal-to-chemical efficiency near 50%. Other applications of high process heat include in situ conversion of oil-shale-embedded kerogen to oil.

New carbon composite materials have already replaced metal in modern aircraft, lowering weight and increasing fuel efficiency. Carbon fiber-reinforced carbon (C/C) composites can withstand temperatures to 2000°C. Carbon fiber-reinforced silicon carbide (C/SiC) is another possible high-temperature, high strength material. Incorporating such materials in LFTR requires not just R&D but validation that the materials can survive the reactor lifetime in a high-temperature, high radiation environment while in contact with molten salts with dissolved fluorides of thorium, uranium, and fission products. Carbon is a neutron moderator.

Heat exchangers isolate fluids, at a temperature loss.

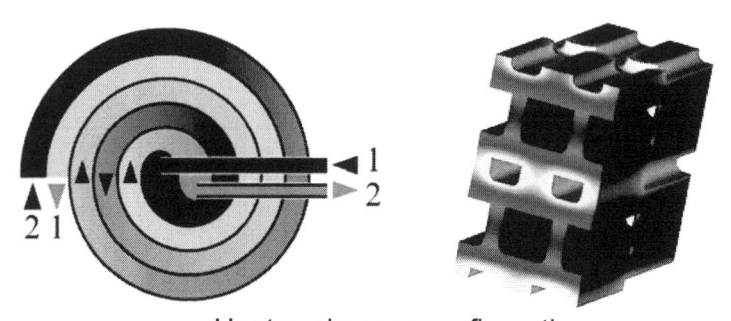

Heat exchanger configurations

LFTR and DMSR designs have two heat exchangers. The first isolates the radioactive fuel salt from the secondary salt used to transfer heat outside the radioactive hot cell. At the power conversion system that clean salt transfers heat to a gas to drive a turbine-generator. That gas might be steam, air, or helium.

Heat exchangers typically have lots of tubes or channels to increase the surface area isolating the two fluids, to increase the heat flow rate. The steam generator of a PWR is a water-to-water heat exchanger that isolates the radioactive matter from the water-steam circuit that powers the steam turbine-generator, simplifying its maintenance.

The engineering challenge is to keep the heat exchanger small, minimize the thickness of the walls separating the two fluids to minimize heat flow resistance, and maximize the surface area of the walls separating the fluids so more heat can flow. All this must take place in a system with high temperatures, changing temperatures, ionizing radiation, and exposure to molten salts with dissolved uranium and fission product fluorides.

Lithium-6 must first be removed from Flibe molten salt.

The molten salt in LFTR may be a mixture of lithium fluoride (LiF) and beryllium fluoride (BeF2), forming a eutectic that melts at 460°C, a lower temperature than either of its components. Lithium is composed of two stable isotopes, Lithium-6 (7%) and Lithium-7 (93%). Unfortunately the Li-6 isotope absorbs too many neutrons, ruining the neutron economy, making chain reaction fission impossible. The reaction is n + Li-6 → H-3 + He-4. So the LFTR needs lithium with the Li-6 isotope removed. The Li-6 absorbs neutrons so strongly that lithium needs to be enriched to perhaps 99.999% Li-7.

Commercially Li-6 has been separated using mercury, because Li-6 has a greater affinity for mercury than does Li-7. Unfortunately the separation process leaked mercury into the environment so it has been discontinued in the US. There is now a shortage of Li-7, which is used as lithium hydroxide to control pH and corrosion in LWRs. Other possible techniques are vacuum distillation and laser isotope separation. Reportedly the GE Silex laser isotopic separation process will be tested with lithium.

Tritium must be continually removed.

Tritium is H-3, hydrogen with two extra neutrons. Tritium is unstable, with a 12-year half life, beta decaying to He-3, and releasing and electron with about 6 keV of energy. This low energy electron can be stopped by 6 mm of air or the dead epidermis of human skin. Tritium is potentially dangerous to health if ingested because tritium is hydrogen that can form water and then become part of human cells, where tritium decay energy could cause damage. However the biological half-life of tritium in humans is less than 10 days because water is continually ingested and excreted. So only a small fraction $10/(365 \times 12)$ of any ingested tritium will decay internally.

A LFTR makes tritium from lithium-6 by n + Li-6 → H-3 + He-4. Although most Li-6 was previously removed from the fuel salt, more is continuously generated by n + Li-7 → Li-6 + n + n. Tritium also comes from n + Li-7 → He-4 + H-3 + n. A 100 MW LFTR would generate 25 mg of tritium per day, responsible for 240 curies of radiation. The US legal emissions limit is 10 curies per day (although 5200 curies per day is allowed in Canada), so the tritium should be removed and sequestered where it can decay harmlessly with a 12-year half life.

Some tritium can be removed from the fuel salt by the sparging process used to remove xenon gas fission products. Tritium (hydrogen) molecules can dissociate into atoms on a metal surface, especially at elevated temperatures. The tritium atom (^{3}H) can share its electron with the free electrons of the metal, allowing the triton (^{3}H$^+$) to pass among metal atoms and through materials such as the primary heat exchanger that transfers thermal energy to the secondary salt loop. Consequently the secondary salt loop will also contain tritium, another place that tritium removal systems could be developed. Tritium will also migrate through the secondary heat exchanger that heats the gas that powers the turbine generator. If it is powered with a closed loop Brayton cycle turbine, then the build-up of tritium there does not matter much because not much will escape though the lower temperature gas cooling loop.

To reduce tritium generation, alternatives to flibe molten salt can be considered. Neutron irradiation of the flibe lithium makes tritium, which is hard to control. Beryllium is toxic. Lithium and beryllium are expensive. Other salts will not moderate neutrons as well as Flibe. Salts to consider include NaF and ZrF4.

Select a high-temperature power conversion turbine.

LFTR designers will want to benefit from the high thermal-electric power conversion possible with the 700°C temperature molten salt. For example, the triple-reheat Brayton cycle pictured above uses a closed circuit of helium gas that is serially expanded in three turbines (T) at high, medium, and low pressure.

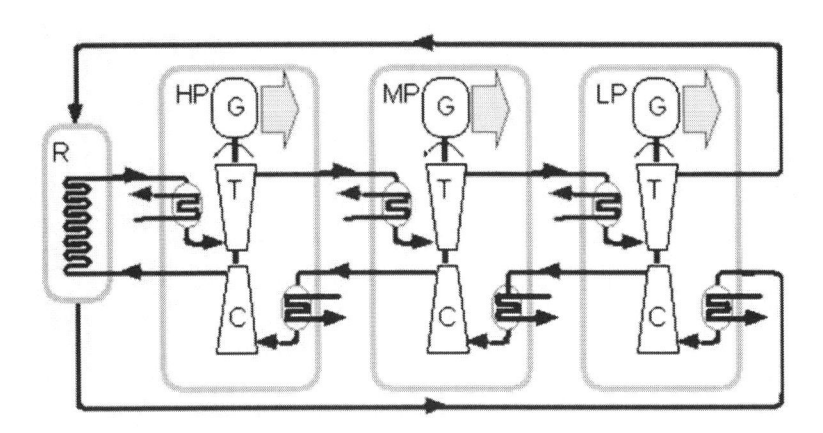

Triple reheat closed Brayton cycle gas turbine

The gas is heated three times by the molten salt as it transfers heat energy to turbines HP, MP, and LP. The turbines spin the generators (G) and compressors (C). There are 7 heat exchangers in this scheme. It's impressive thermodynamic engineering. Such power conversion schemes have not been demonstrated at scale of a nuclear power plant. South Africa's pebble bed reactor project was developing such a turbine with Rolls Royce and Mitsubishi, but unfortunately suspended work for lack of funds.

Compared to steam turbines in today's LWR power plants, the high temperature and high efficiency mean the rejected heat is nearly halved. For water-cooled plants this reduces the heating of rivers or lakes. For plants with classic waist-shaped cooling towers this reduces the water lost to evaporation.

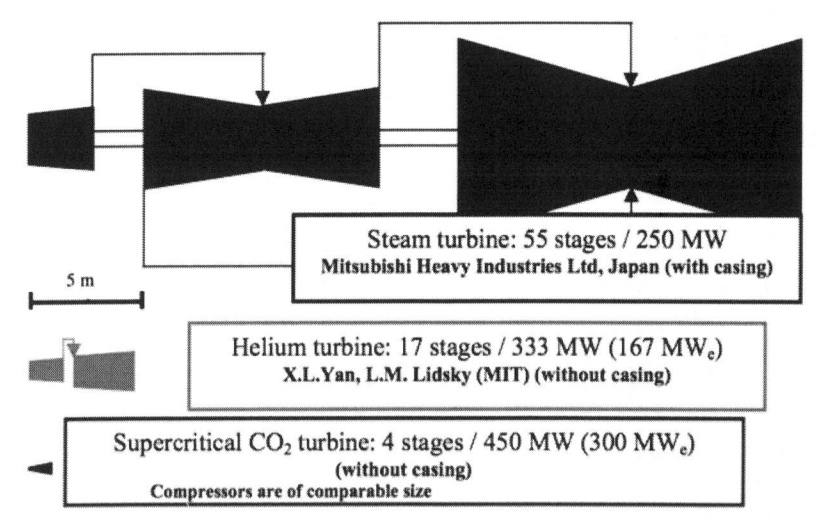

Steam turbine: 55 stages / 250 MW
Mitsubishi Heavy Industries Ltd, Japan (with casing)

5 m

Helium turbine: 17 stages / 333 MW (167 MW$_e$)
X.L.Yan, L.M. Lidsky (MIT) (without casing)

Supercritical CO_2 turbine: 4 stages / 450 MW (300 MW$_e$)
(without casing)
Compressors are of comparable size

Relative sizes of steam, helium, and SCO2 turbines

Another advanced power conversion cycle uses supercritical CO2 (SCO2) instead of gas in the closed loop, acting much like a jet engine running on a hot liquid. A 300 MW(e) SCO2 turbine would be only 1 meter in diameter. The illustration above from a paper by Dostal, Driscoll, and Hejzlar at MIT shows how much more compact the SCO2 turbine can be.

Steam turbines are a low-development-risk power conversion technology already developed and sold to the power industry by GE and Siemens. For example, Siemens today sells 46% efficient 620°C steam turbines and will be testing a 51% efficient 700°C steam turbine after 2015.

Implement passive waste heat dissipation.

With the lessons of Three Mile Island and Fukushima, all future nuclear reactors will have passive decay heat removal. When any operating nuclear reactor is shut down by stopping the fission chain reaction, existing unstable energetic fission products continue to decay and release considerable heat. One minute after shutdown the reactor still generates 4% of its full power heat. After one day 0.5% of full power heat must continue to be dispersed.

In normal operation both direct fission and fission product decay thermal energy is absorbed by the power conversion turbine-generator that makes electric power and also the related heat-rejection cooling system. After shutdown this energy transfer system does not function, so fission product decay heat must be removed another way. In today's LWRs this is accomplished by pumping water through an auxiliary cooling system, but this requires electrical power, which was not available at Fukushima.

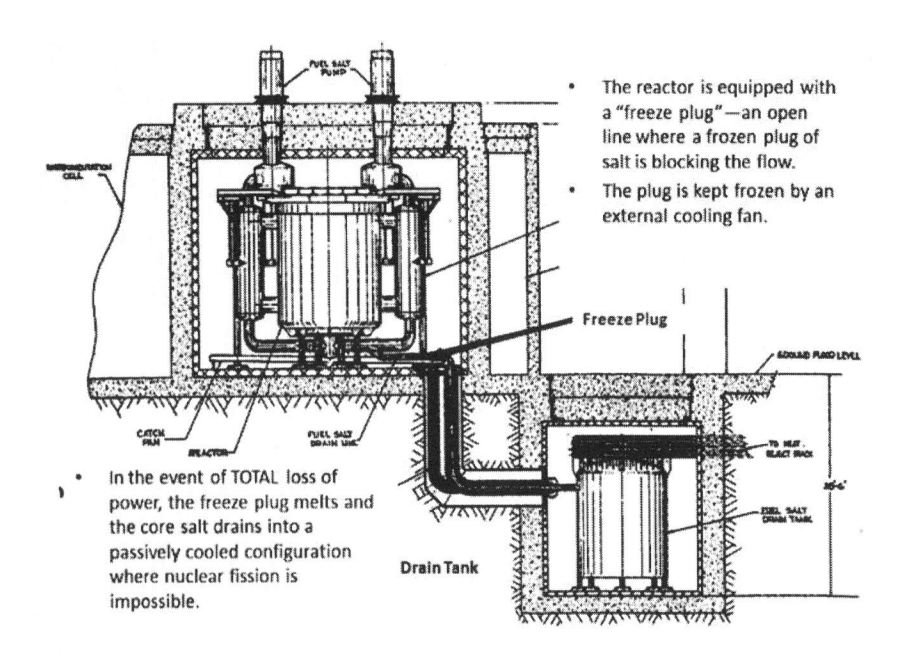

- The reactor is equipped with a "freeze plug"—an open line where a frozen plug of salt is blocking the flow.
- The plug is kept frozen by an external cooling fan.
- In the event of TOTAL loss of power, the freeze plug melts and the core salt drains into a passively cooled configuration where nuclear fission is impossible.

ORNL MSR with passive cooling

The LFTR and DMSR will have passive decay heat removal systems to allow passive transfer of heat by convection and conduction.

The ORNL drawing above illustrates a molten salt drain tank (with no fission-inducing moderator) on the lower right into which the molten salt flows in case of overheat or shutdown. It would be air cooled. Molten salt temperature can rise to 1400°C without boiling, so cooling high-temperature molten salt is simpler than cooling LWR fuel rods, which must stay below 800°C to avoid damage; at 1200°C the zirconium oxidizes releasing hydrogen gas.

Design a safe, maintainable plant.

Today's designs for small modular nuclear power plants employ undergrounding of the reactor vessel and associated radioactive equipment. This provides protection against terrorist attacks. Undergrounding can provide radiation shielding in accidents.

Maintenance of the salt processing units and off gas system will require remote handling within hot cells that shield personnel from radiation. It should be possible to replace any component. Maintenance observations may benefit from the transparency of the molten salts. Nuclear engineers know how to implement seismic isolation. The LFTR design must accommodate in-plant storage of fission products removed from the fuel salt. The DMSR design leaves most of them in the salt.

Develop nuclear materials safeguard systems.

Deploying LFTR and DMSR nuclear power plants will not proliferate nuclear weapons, because it is too difficult and expensive for a weapons-aspiring nation to attempt to pervert the technology, especially in comparison to methods already demonstrated by poor nations. Centrifuge uranium enrichment or plutonium production from carbon-moderated or heavy-water-moderated special purpose reactors have been used by Pakistan, North Korea, and India.

Safeguards applicable to all nuclear power activities help prevent misuse of nuclear materials by nations, revolutionary organizations, terrorists, criminals, thieves, and nuclear reactor operations personnel. Safeguards are basically accounting and control systems for all inventories and transfers of nuclear materials. Nuclear materials will include all fissile isotopes and fertile isotopes that may be transmuted into fissile material. Protective safeguards also apply to fission products that might be misused as contaminants. Safeguard regulations are set by individual countries such as the US with guidance from the International Atomic Energy Agency (IAEA).

Thus LFTR and DMSR designs must provide for remote measurement and observation of plant operations, including video and digital monitoring, with unscheduled audits. The security of the safeguard system must be strong. Safeguards have not yet been developed for fluid fuel forms, so this will be an area for development and regulatory cooperation. LFTR fissile U-233 may be designated as special nuclear material, with more stringent safeguard mechanisms than for DMSR or LWR power plants.

A proposed additional LFTR safeguard is the potential for remote controlled release of U-238 into the fissile U-233 reactor core, ruining that material for weapons use, but also ruining it for future LFTR power production.

Separate and immobilize waste.

Wikipedia's chart of radioactive isotopes resulting from nuclear fission is wonderfully information rich.

Actinides				Half-life	Fission products		
^{244}Cm	^{241}Pu f	^{250}Cf	^{243}Cm^f	10–30 y	^{137}Cs	^{90}Sr	^{85}Kr
^{232}U f		^{238}Pu	f is for	69–90 y			^{151}Sm nc→
4n	^{249}Cf f	^{242}Am^f	fissile	141–351 y			
	^{241}Am		^{251}Cf f	431–898 y	No fission product		
^{240}Pu	^{229}Th	^{246}Cm	^{243}Am	5–7 ky	has half-life 10^2		
	^{245}Cm^f	^{250}Cm	^{239}Pu f	8–24 ky	to 2×10^5 years		
4n	^{233}U f	^{230}Th	^{231}Pa	32–160 ky			
	4n+1	^{234}U		211–290 ky	^{99}Tc	^{126}Sn	^{79}Se
^{248}Cm		^{242}Pu	4n+3	340–373 ky	Long-lived fission products		
	^{237}Np			1–2 My	^{93}Zr	^{135}Cs nc→	
^{236}U		4n+2	^{247}Cm^f	6–23 My		^{107}Pd	^{129}I
^{244}Pu	4n+1			80 My	>7% >5% >1% >.1%		
^{232}Th		^{238}U	^{235}U f	0.7–12 Gy	fission product yield		

The rows contain isotopes of similar half lives. The left columns are the transuranic actinides, with each column having the elements of the four independent alpha decay chains. The isotopes in the right columns have similar production yields from fission. Fission products with half lives less than 10 years are not shown. LFTR produces few of the radioactive actinides in the left of the table. On the right hand side, the long-lived radioactive fission products of concern are technetium-99, tin-126, selenium-79, zirconium-93, cesium-135, palladium-107, and iodine-129.

To explore the components of spent fuel as the radioactive isotopes decay with time, experiment with Kirk Sorensen's excellent Java program at the *Energy from Thorium* website, http://www.energyfromthorium.com/javaws/SpentFuelExplorer.jnlp

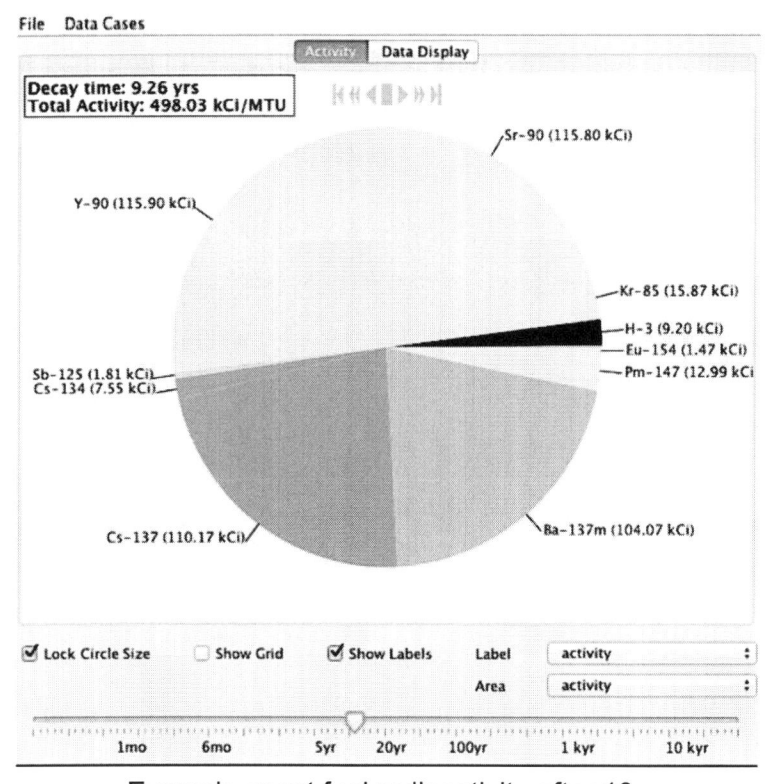

Example spent fuel radioactivity after 10 years

Some public anxiety about today's LWR nuclear power arises from the unspecific plans to manage the waste. The spent solid-fuel rods contain radioactive fission products and radioactive transuranic elements, plus twenty times as much benign U-238, much increasing the mass and volume of material to be sequestered. The most favored long-term disposal strategy seems to be to isolate complete fuel rod assemblies underground. France alone reprocesses these spent fuel rods, separating useful fissile U-235 and Pu-239, and dissolving the fission products in solid glass to be isolated forever.

Public acceptance of thorium energy cheaper than coal requires a plan for the waste. It starts at the plant. The non-noble fission products will be fluorides. Moir and Teller recommended dissolving them in fluorapatite, $Ca_5(PO_4)_3F$, the mineral found to have naturally immobilized for a billion years the fission products from the ancient natural nuclear reactors discovered in Africa.

The waste form should not be dispersible vapors, particulates nor liquids. Darryl Siemer is experimenting with glasses that can contain the fission product fluorides exiting from LFTR. Suitable glasses may include borosilicate (alkali+borate+silica) chosen for Yucca Mountain, aluminophosphate (alkali+Al2O3+phosphate) chosen by Russia, or iron phosphate (alkali+Fe2O3+phosphate). Siemer finds the iron phosphate glasses can contain the fission products if fluorides are converted to nitrates by boiling in dilute nitric acid and capturing the off-gases. LFTR waste immobilized in this glass would occupy about 9 cubic meters per GW-year of operation.

Integrate licensing process with design process.

Regulatory bodies such as the Nuclear Regulatory Commission have the duty to protect the public and to assure the public that protection is adequate. Commercial success of LFTR requires not just a license, but also a continuing, open process of interaction between regulators and designers.

In the US the NRC has expertise in LWR safety management. NRC has some dated experience in novel technologies from pre-application interactions with developers of the LMFBRs and the high-temperature gas-cooled reactor. NRC staff will need to develop deep expertise in safety performance of LFTR. NRC had embarked on establishing a technology-neutral licensing framework, which needs to be continued.

Manage for success.

A complex, billion-dollar project such as LFTR development requires experienced, motivating leaders and managers. LFTR advocate Joe Bonometti has such experience with NASA and provides us this advice. Manage the technology but satisfy the public. Educate and motivate every new team member about the LFTR goals. "The critics said it couldn't be done: heavier than air flight, landing on the moon, flying faster than sound, a stealth plane, a terabyte for $200, LFTR for $2/watt." Motivate with magic – the liquid fuel form. Picking a small, expert, empowered team is essential. Here is more Bonometti advice.

1. Collect extensive baseline data to be able to judge effects.
2. Build competency for fluid handling, filtering, storing, and technology support.
3. Understand, use, and validate safety-critical software both for operations and design.
4. Have a separate, redundant monitoring and diagnostic and health monitoring system.
5. Maintain an active, full-scale thermal/mechanical working model and modeling center to recreate anomalies.
6. Manage risk reduction by tackling hard problems early. Include diagnostic sensors, wiring, and imagery, with access to all parts.
7. Develop a long-term vendor/technology base.
8. Never defer correcting a defect.
9. Never tolerate an unexplained mystery in any test result.

Maintain cost priorities.

Many design decisions will be made during the course of a LFTR development project. For example, designers must consider safety, maintainability, longevity, reliability, energy security, proliferation resistance, waste management, and cost. LFTR energy cheaper than coal requires continuing focus on cost. The design objective of $2/W capital cost and 3 cents/kWh electric power cost can be achieved if cost impacts are considered at every design decision.

After design requirements are settled the project must not be burdened with opponents' add-on requirements such as more radiation protection, more aircraft impact resistance, more earthquake resistance, more evacuation routes, or more taxes. The project must anticipate and defend against injunctions, regulatory delays, and lawsuits that appear to protect people and the environment, but are in reality tools to increase costs. The tipping point for LFTR costs is "energy cheaper than coal". Weinberg's group nicknamed the MSR the "3P" reactor – a pot, a pipe, and a pump. Keeping the design simple keeps costs low.

Shorter projects cost less and have less cancellation risk.

I recommend a high priority, 5-year program to complete prototypes for the LFTR and the simpler DMSR. It might take an additional 5 years of industry participation to achieve capabilities for commercial production. Nuclear engineers and government regulators in today's nuclear industry would say this schedule is too aggressive. Yet nuclear projects have been done faster, at a time when engineers did not have today's advanced technologies for computing, materials, thermodynamics, and nuclear chemistry. Admiral Hyman Rickover developed the first ever nuclear power plant in 5 years from 1949 to 1954 and installed it on the Nautilus submarine. He actually built two power plants nearly in parallel; the prototype in Idaho was a few months ahead of the production power plant being installed in the Nautilus. The first US land-based electric power plant was built at Shippingport PA in 39 months.

Developers

Heightened public concerns about nuclear waste, global CO2 emissions, and nuclear power cost have led scientists and engineers to revisit the liquid fuel technologies bypassed in the 1970s. LFTR can generate carbon free, low waste, low cost power with the added benefit of consuming existing PWR spent fuel.

A number of LFTR initiatives are currently active around the world. France supports theoretical work by two dozen scientists at Grenoble and elsewhere. The Czech Republic supports laboratory research in fuel processing at Rez, near Prague. Design for the FUJI molten salt reactor continued in Japan. Russia is modeling and testing components of a molten salt reactor designed to consume plutonium and actinides from PWR spent fuel. LFTR studies are underway in Canada and at Delft University of Technology in the Netherlands. US R&D funding has been relatively insignificant.

United States

US scientists rejuvenated 21st century LFTR interest.

In 2004 Lawrence Livermore scientist Ralph Moir and Edward Teller, a Manhattan Project veteran and developer of the hydrogen bomb, called for the construction of a prototype thorium-burning molten salt reactor, but it was never funded. See Appendix B for this paper.

Oak Ridge had meticulously documented its research and in 2002 the reports were scanned for a NASA program investigating power plants for a manned mission to Jupiter. In 2006 graduate student Kirk Sorensen indexed and posted these documents at the www.energyfromthorium.com forum. A worldwide research community of scientists and engineers collaborate there, proposing ideas and designs online and receiving analytical comments within hours. Forum members post links to new research in the US, Canada, France, Russia, Netherlands, Czech Republic, UK, and Japan. Google has assisted the forum, producing five video presentations about LFTR now posted on YouTube.com as Google tech talks.

US R&D funding for liquid fuel reactors is nil, except that in 2012 MIT, UC Berkeley, and U Wisconsin were awarded $7 million over three years for related studies of solid-fuel, molten salt cooled reactors. In contrast to modest US research, France, the Czech Republic, Japan, Russia, and the Netherlands support MSR research.

The US is destroying its U-233, valuable to LFTR R&D.

U-233, at the core of the reactor, is important to LFTR development and testing. With a half-life of only 160,000 years, it is not found in nature. The US has 1,000 kg of nearly irreplaceable U-233 is now slated to be destroyed by diluting it with U-238 and burying it forever, at a cost of $511 million, which would far better be invested in LFTR development, making good use of the U-233.

Several people estimate that with adequate national laboratory support, a prototype could be operational in 5 years, for approximately $1 billion, as estimated by the Generation-IV International Forum and the Moir-Teller paper. It may take an additional 5 years of industry participation to achieve capabilities for mass production. If these timescales seem aggressive, note that the Shippingport 1957 PWR was built in 39 months, and Weinberg's 1943 Oak Ridge reactor in 9 months!

The Nuclear Regulatory Commission would need funding to train staff qualified to work with this technology. Today the NRC is a roadblock to permitting development of advanced nuclear technologies such as LFTR. As important as energy is to the US, the NRC 2012 budgeted congressional appropriation is only $129 million dollars. It receives another $910 million in fees charged to the existing nuclear power industry. Any company applying to the NRC for licenses must be prepared to pay over $250/hour for all staff work on the application, amounting to hundreds of millions of dollars, with an uncertain ruling by the politically appointed NRC commissioners.

Once LFTR is developed, the nuclear industry and utilities would also be shaken by this disruptive technology that changes the whole fuel cycle of mining, enrichment, fuel rod fabrication, and refueling.

US National Laboratories are capable of developing LFTR.

The US has many national laboratories that have thousands of scientists and engineers, along with physical capabilities to develop a modern LFTR. The US has the technical resources in these laboratories, but not the mission, direction, and funding.

Importantly, the national laboratories still retain the right to self-regulate the construction of research nuclear reactors. Electric utility companies do need NRC-licensed reactors, so the lengthy NRC licensing process could be pursued in parallel with LFTR development at the national laboratories.

Oak Ridge National Laboratories (ORNL) in Tennessee built 13 nuclear reactors including two molten salt reactors. ORNL has close ties to the University of Tennessee nuclear engineering department. Today it operates the HFIR (high flux isotope reactor) used for materials testing and NCSS (national center for computational sciences) providing the world's most powerful computers for unclassified computational research in subjects such as molten salt reactors. In 2011 ORNL scientists hosted meetings and published studies of fast molten salt reactors and salt cooled solid-fuel reactors.

Argonne National Laboratories grew out of Fermi's pioneering work at the University of Chicago. Argonne built over 28 reactors from the 1940s to 2004. Most recently Argonne scientists designed the integral fast reactor, a liquid metal cooled fast breeder reactor that burns plentiful U-238 via the uranium-plutonium fuel cycle. The story is told by Charles Till and Yoon Il Chang in *Plentiful Energy: The story of the integral fast reactor*.

Bettis Atomic Power Laboratory and Knolls Atomic Power Laboratory work exclusively on naval propulsion nuclear reactors. They have developed 20 different kinds of reactors for US submarines and 8 reactor types for aircraft carriers and other surface ships.

Idaho National Laboratory is slowly developing the NGNP (next generation nuclear plant) high-temperature gas-cooled reactor. They also operate the ATR (advanced test reactor) which can subject test materials to very high neutron fluxes to determine material lifetimes. Over 50 reactors were built at the site, including the prototype for the Nautilus submarine.

Lawrence Livermore National Laboratory conducts research and development of fusion reactors as well as supporting theoretical work in molten salt reactors. Edward Teller, Ralph Moir, and Robert Steinhaus worked there.

Los Alamos National Laboratory in the past worked on nuclear power projects such as the molten plutonium nuclear reactor. Today the lab is largely dedicated to military and weapons work.

Sandia National Laboratory is principally involved in classified military activities, but also supports open research and development such as the project Green Freedom to make gasoline from air and water using nuclear power.

Savannah River National Laboratory once operated five nuclear reactors. Today it is constructing a mixed oxide (MOX) plant to make LWR fuel rods from surplus plutonium and U-238. It is also hosting the construction of three SMRs (small modular reactors) being developed by US commercial ventures.

In summary, the US has the capability but not the will nor leadership nor funding to develop advanced nuclear power such as the liquid fuel nuclear reactor – energy cheaper than coal. Because the benefits are global, other nations may step in to lead development, as is happening in China and France.

US Energy Secretary Chu discounted LFTR potential.

United States Secretary of Energy Stephen Chu expressed historical criticism of the technology in a letter to NII Senator Jeanne Shaheen answering questions at his confirmation hearings.

> "One significant drawback of the MSR technology is the corrosive effect of the molten salts on the structural materials used in the reactor vessel and heat exchangers; this issue results in the need to develop advanced corrosion-resistant structural materials and enhanced reactor coolant chemistry control systems."

The corrosion of the MSRE vessel from fission products was analyzed by ORNL and solutions have been developed.

> "From a non-proliferation standpoint, thorium-fueled reactors present a unique set of challenges because they convert thorium-232 into uranium-233 which is nearly as efficient as plutonium-239 as a weapons material."

Proliferation resistance is contributed to by U-232 contamination of U-233. The DMSR is even more proliferation resistant.

Secretary Chu also recognized, however, that

> "Some potential features of a MSR include smaller reactor size relative to light water reactors due to the higher heat removal capabilities of the molten salts and the ability to simplify the fuel manufacturing process, since the fuel would be dissolved in the molten salt."

Flibe Energy is preparing to develop LFTR for the US military.

Flibe is the short name for the molten salt of LiF mixed with BeF2 – one of the key LFTR technologies. Alabama-based Flibe Energy was founded in 2011 by Kirk Sorensen, a former NASA employee who was researching LFTR for a moon-base power plant when he realized its potential on Earth. Sorensen also runs the *Energy from Thorium* blog and forum, where much of the initial LFTR interest was aroused.

The military has a need is for robust electric power plants in remote operational sites as well as for secure independent power generations for military bases in the continental US. LFTR can be configured in readily transportable modules to permit rapid installation. The US military has independent regulatory authority, separate from NRC, allowing R&D to proceed more predictably without indefinite regulatory delays and costs. The long path of NRC licensing of this new technology can proceed in parallel with military deployment, providing that experience to the NRC, so that the civilian sector will also benefit from LFTR.

Flibe Energy proposes that it build a pilot plant to be operated under US Army regulatory authority near Huntsville, AL The first demonstration reactor will be 40 MW(e) capability, and designed to operate for ten years. The next step would be 240-400 MW(e) utility-class reactors, although Sorensen says the technology could readily be scaled from 1 MW to 1 GW.

The company is raising funding of several hundred million dollars, with the objective of privately funding a LFTR development and demonstration by 2016. Author Robert Hargraves is an unpaid advisor to Flibe Energy.

Transatomic Power features MSR waste burning capability.

Mark Massie and Leslie Dewan are PhD candidates in MIT's department of nuclear engineering. With advisors from MIT and ORNL, they founded Transatomic Power, which features the waste-burning capability of their WAMSR (waste annihilating molten salt reactor) concept.

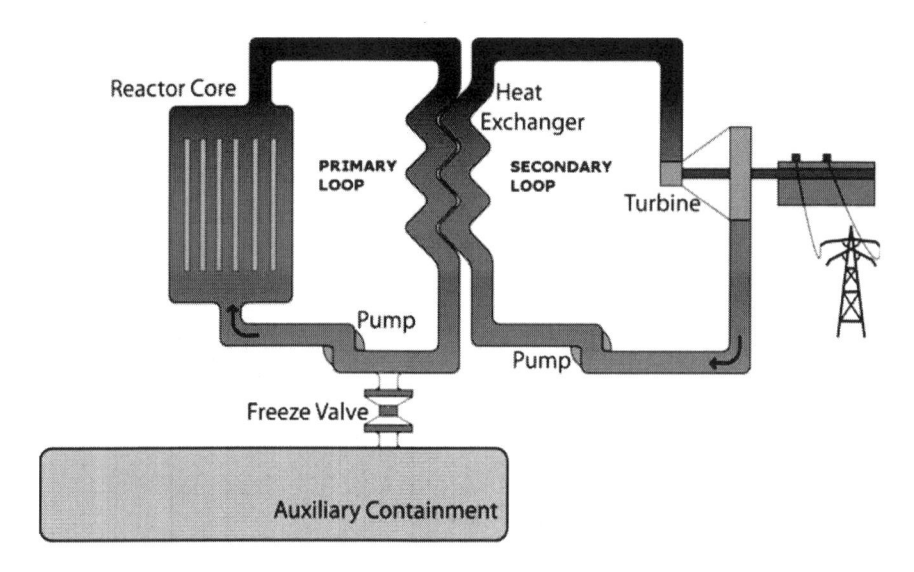

Transatomic Power molten salt reactor

A fleet of such 200 MW(e) reactors could consume 98% of the world's existing 270,000 tonnes of spent nuclear fuel rods and provide the world's growing electric power needs for 72 years. The WAMSR design passes 650°C molten salt through a heat exchanger to make steam for a conventional steam turbine generator. WAMSR does not use thorium, but U-238 is plentiful; there are 270,000 t of spent fuel worldwide containing 95% U-238. There is nearly ten times this amount of stranded U-238 left at uranium enrichment plants that separate it to concentrate the

0.7% naturally abundant U-235 to 4% for LWR fuel. Transatomic Power raised $763,000 in seed money in 2012.

Thorenco has a fast neutron LFTR design.

Thorenco's founder, Rusty Holdren, presented a design of a pilot LFTR at the 2012 conference of the Thorium Energy Alliance. This is a pool type reactor, in which the heat generated by U-233 fission in the core is transferred to the large thermal mass of the immersing, circulating molten salt coolant in the pool vessel.

A second heat exchanger at the top of the pool transfers the heat to steam or other gas used to drive a turbine generator.

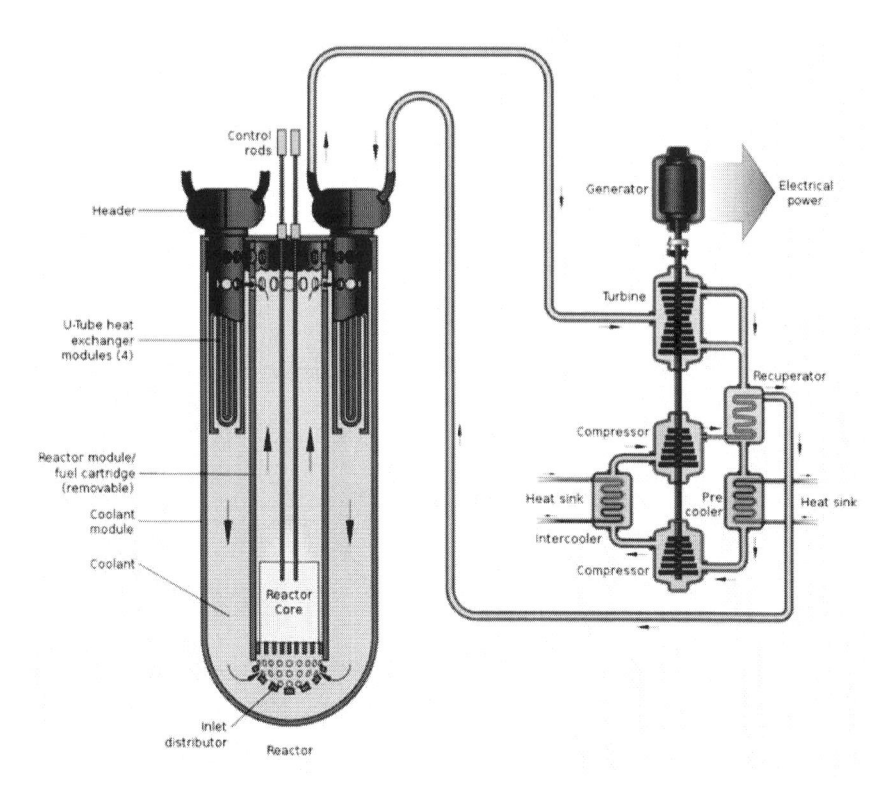

Pool-type molten salt reactor

Holdren invented a honeycomb core containing the fuel salt, which does not circulate. The fuel salt is cooled by molten salts of the pool circulating by convection through the hexagonal channels. Hastelloy metal hexagonal tubes separate fuel salt from cooling salt. Because of neutron absorption by nickel in Hastelloy, the structural material may need to be replaced frequently.

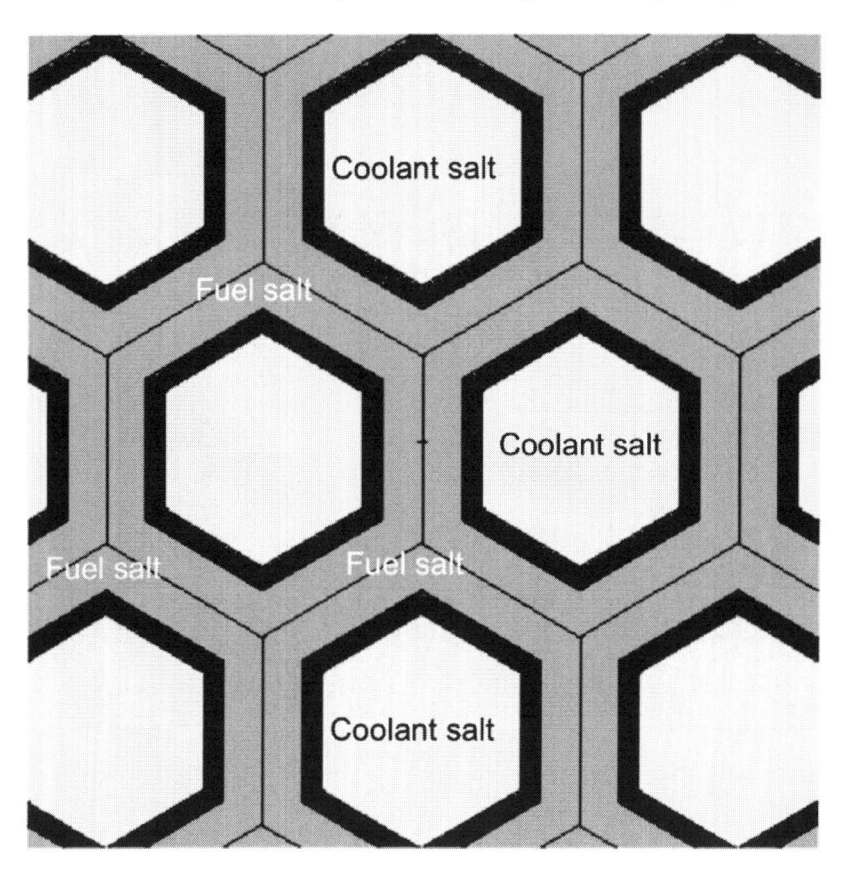

Coolant salt channels in molten salt fuel

The coolant salt is 57% NaF, 43% BeF2. The fuel salt composition in one study was 7% UF4, 7% ThF4, 53% NaF, 33% BeF2. Only the Be provides much neutron moderation, so fission cross sections are less than with thermal neutrons. This will require more uranium than a graphite moderated LFTR.

The 40 MW(t) reactor would operate for 10 years with a charge of 1,600 kg U-233 and 9,000 kg Th-232. In that decade the reactor produces 100 kg of U-233 but consumes 141 kg of it, reducing the initial U-233 stock from 1,600 kg to 1,559 kg.

Like other thorium MSRs, the reactor produces an insignificant amount of plutonium and other transuranics and exhibits strong proliferation resistance because of the U-232 production and its consequent 2.6 MeV gamma radiation.

CHINA

China is moving to reduce its dependence on coal for energy. Since 2006 China has shut down many small, inefficient, polluting coal plants that had generated 71 GW of power and released 165 million tonnes of CO_2 per year. China is aggressively expanding its electric power generation using several new, advanced nuclear power technologies. These include the light water reactor technologies used in all US reactors, the heavy-water-moderated CANDU technology developed and used in Canada, the gas-cooled high-temperature pebble bed reactor first operated in Germany, and the liquid sodium metal cooled fast reactor being obtained from Russia.

China has 14 nuclear power plants in operation and 25 under construction, with a 2020 capacity of 60 GW(e), growing to 200 GW by 2030. For scale comparison, the Three Gorges hydropower project generates 18 GW.

China bases its nuclear expansion on Generation III LWRs.

China also has a domestic nuclear reactor and fuel industry, the China National Nuclear Corporation, which built LWRs. China has contracted with Areva to build four of EPR (European Pressurized Water) reactors, two of which are under construction in Guangdong province, to deliver 1.66 GW, beginning operation in 2014.

China has employed Westinghouse to build four of its AP-1000 reactors, each capable of 1.1 GW net electric power generation. Two of the four Westinghouse AP1000 reactors are nearing completion. The World Nuclear Association reports the capital cost of $2/watt is expected to drop to $1.60/watt for further units. Eight more AP-1000 reactors are planned and thirty more proposed. China is also gaining intellectual property rights to this advanced technology, with the intention to become self-sufficient and an exporter of nuclear technology.

China is building commercial pebble bed reactors.

PBR technology was first developed in Germany, where the THTR-300 thorium-fueled pebble bed reactor operated from 1983 to 1989. South Africa's established PBMR Pty, Ltd to develop such a commercial reactor, but ran out of funds in 2010 and the project ended. One attraction of the PBR is the inherent safety; at high temperatures U-238 absorbs more neutrons, enough to stop the fission chain reaction. Passive air-cooling removes fission product decay heat.

The first pebble bed reactor (PBR) in China became operational at Tsinghua University in 2003, based on technology from Germany's AVR reactor experiments in the 1960s; China purchased AVR components and reassembled them. It is a 10 MW(t) high-temperature research reactor cooled by helium gas; the gas heats steam for a turbine generator. The Australian Broadcasting Company visited the HTR-10 pebble bed reactor in China to video its operation. Professor Zhang Zuoyi described the events as the reactor's helium cooling system was purposefully shut down to demonstrate on television the inherent safety of the pebble bed reactor fission. The temperature rose, causing U-238 in the fuel to absorb enough neutrons to stop the chain reaction, and the reactor vessel was then passively cooled by convection.

Pebble bed reactors planned at Rongcheng

China is now building a 190 MW demonstration reactor power plant at Rongcheng. If successful, a total of 19 pebble bed reactors generating 3,600 MW will be constructed at that site.

Russia is selling two fast reactors to China.

China has experimented with fast neutron breeder reactors at the China Institute of Atomic Energy. An experimental 20 GW(e), sodium-cooled, pool-type reactor first went into production operation in 2011. The $350 million project aimed to accumulate experience in fast reactor operation and to be a facility to irradiate fuels and materials at high neutron energies.

Russia has operated its BN-600 sodium cooled fast neutron reactor successfully since 1980, and is now constructing an improved BN-800 880 GW reactor, slated for operation in 2012. The advantage of the fast neutron reactor is that it can consume fertile U-238, which is over 100 times more plentiful than fissile U-235 used in standard LWRs. China and Russia agreed to build two such BN-800 units at Sanming, with construction starting in 2013 and completing in 2019.

China imports 95% of its uranium ore, but has 8.9 million tons of thorium associated with its rare earth reserves. China is testing thorium in its CANDU and pebble bed reactors.

China is undertaking a LFTR R&D project.

In January 2011 the Chinese Academy of Sciences (CAS) announced the launch of its project to develop a thorium molten salt reactor (LFTR). CAS vice-president Dr. Jiang Mianheng left that position in late 2011 to lead the new LFTR project. Jiang Mianheng, the son of the former president Jiang Zemin, is an electrical engineering graduate of Drexel University in the US.

After the July 2010 publication of an article about LFTR in *American Scientist*, Jiang led a delegation to visit Oak Ridge National Laboratories, where molten salt technology had been conceived and tested in two reactors. Thereafter the Chinese team received the endorsement and funding from the CAS to begin the

LFTR development project. ORNL shared information with 1,894 visitors from China in 2011, yet China intends to acquire and control LFTR intellectual property for itself.

Chinese Academy of Sciences

The work is underway at the Shanghai Institute of Applied Physics, where researcher Wen Wei Po announced the project in January 2011 in the Wen Hui Bao news article and posted it on the *Energy from Thorium* forum. The R&D scope also includes Brayton cycle power conversion, hydrogen production, and methanol synthesis from CO_2 and H_2.

In 2012 the TMSR (thorium molten salt reactor) project reportedly employs 432 people, expected to rise to 750 by 2015, with a budget of $350 million over 5 years. The project will proceed in four stages, starting the first two at once:

1   2 MW(t) 660°C PB-AHTR by 2015
2   2 MW(t) MSR by 2017
3   10 MW(e) MSR by 2020
4   100 MW(e) MSR by 2030

Unknown outside the country, China had investigated molten salt reactor principles shortly after the 1965-1969 molten salt reactor operation directed by Weinberg at Oak Ridge. China scientists built a zero-power experimental dry-salt reactor with lithium and beryllium fluoride salts containing enriched U-235 and thorium salts, reaching criticality in the early 1970s. Recently they contacted scientists in the Czech Republic and Japanese scientists familiar with the 1970s MSRE project at ORNL.

At the end of 2010 Dr. Xu Hongjie, lead researcher at the Shanghai Institute of Applied Physics, described the breakthrough molten salt reactor using molten fluoride salts similar to magma carrying nuclear fuel [in my edited Google Translate transcript]. He said the molten salt reactor was selected from six advanced nuclear power concepts proposed by the Generation IV International Forum because MSR uses liquid fuel, is small, has a simple structure, runs at atmospheric pressure, can consume several kinds of fuels, and will generate as little as one thousandth the waste of existing technologies.

The Chinese Academy of Sciences and the Shanghai Institute of Applied Physics are collaborating with nuclear engineering experts at UC Berkeley, MIT, and U Wisconsin, especially with regard to safety assessment and licensing. One of the two 2 MW research reactors will be a molten salt cooled pebble bed reactor similar to the PB-AHTR design conceived at UC Berkeley. China already has capabilities to manufacture TRISO fuel pebbles such as used in their HTR-10 experimental reactor and Rongcheng pilot plant.

The university collaborators will develop independent models to predict the neutronic and thermal hydraulic behavior of the CAS reactor design such as reactivity, fuel and coolant temperatures, temperature reactivity feedback, and shutdown control rods. US students may spend time as interns at the CAS, where they may construct molten salt flow loops for materials testing.

The Chinese Academy of Sciences and the US Department of Energy have a Nuclear Energy Cooperation Memorandum of Understanding, with executive committee co-chairs Mianheng Jiang and Pete Lyons, DOE undersecretary for nuclear energy.

Participants include scientists from INL, MIT, UC Berkeley, and ORNL.

The CAS and SIAP are hosting the 4th annual International Thorium Energy Organization conference, in Shanghai, October 29 to November 1, 2012. The iTheo announcement states:

> China is taking the lead in exploring fresh approaches to nuclear fission in its quest for sustainable, environment-responsible energy that can be delivered reliably and in quantity. The Chinese initiated action to find viable energy sources significant enough to wean the country off its dependence on carbon-based energy. The large amounts of thorium being produced as a by-product of China's rare earth mining operations is a further incentive.

Chinese Premier Wen Jiabao says in a government report published on March 5, 2012, that China will accelerate the use of new-energy sources such as nuclear energy and put an end to blind expansion in industries such as solar energy and wind power.

FRANCE

Grenoble scientists are designing fast neutron thorium MSRs.

Although France is not currently building a molten salt reactor, France's national laboratory scientists in Grenoble have been investigating molten salt reactors and thorium since the 1990s. Initial studies looked at using molten salt reactors to burn up plutonium and other actinides in spent LWR fuel assemblies. The thorium breeder also became a focus of their work.

Current research publications deal with a graphite-free, unmoderated, thorium-blanketed molten salt reactor, termed MSFR (molten salt fast reactor). Such fast reactors require more fissile matter for the fast neutron to interact with nuclei before being lost from the core.

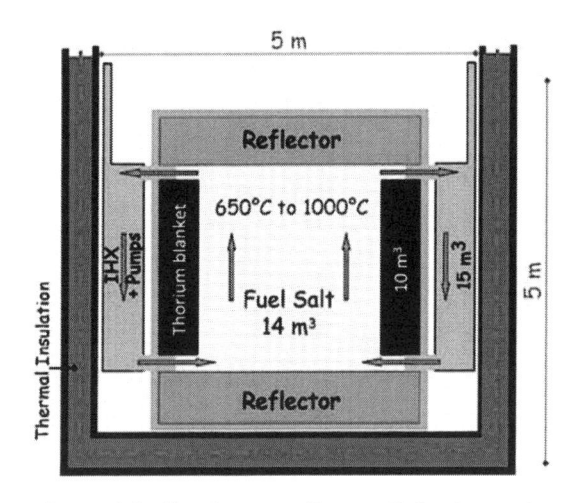

Grenoble thorium molten salt fast reactor

The MSFR salt is 78% LiF with dissolved $^{233}UF_3$ and $^{232}ThF_4$. The fissile U-233 can be replaced with U-235, Pu-239, or a mixture of transuranics found in spent LWR fuel. A 1,000 MW(e) reactor would also produce 95 kg of U-233 per year that could start up other reactors. The MSFR requires a large U-233 inventory of 3,400 kg to capture the fast neutrons.

The cylindrical core is 2.3 m by 2.3 m, containing half the 28 m^3 of molten salt, the rest being in pipes, pumps, and heat exchangers. Only 40 liters of salt must be reprocessed per day.

France is a member of the Europe's project EVOL (evaluation and viability of liquid fuel fast reactor systems), along with Netherlands, Germany, Italy, United Kingdom, Czech Republic, Hungary, and Russia.

OTHER EMERGING LFTR DEVELOPERS

Czech Republic and Australia may develop LFTR.

The Czech Republic has supported research and development in molten salt reactors for years. Jan Uhlir is a leader in research on thorium MSRs at the Nuclear Research Institute Rez, near Prague.

Molten salt test loop at Rez

The institute has theoretical and laboratory experience in chemical partitioning of actinide fluorides, transmutation of actinides for destroying long-lived transuranics, and converting Th-232 into U-233. The laboratory is conducting limited theoretical and experimental development MSR technology, mainly in fuel cycle processes, development of structural materials (nickel alloys) and some system studies. Their molten salt test loop is pictured here.

In November 2011 Australia's Thorium Energy Generation Pty, Ltd (TEG) announced the formation of a joint venture with Czech Republic scientists to develop a 60MW pilot plant in Prague. The nascent partnership would have 50 scientists and engineers working on the pilot plant project, expected to cost over $300 million dollars.

Canada ventures are examining thorium MSR opportunities.

Thorium Power Canada states that it has a design for a molten salt thorium reactor. The company says it is in the preliminary licensing process to build a 10 MW unit in Chile and a 25 MW unit in Indonesia.

Thorium One has tried to market thorium solid fuels for LWRs and CANDU (Canadian deuterium uranium) reactors. The fuel rods would incorporate plutonium fissile material with thorium, with technology similar to the MOX fuel manufactured in France. Instead of solid fuels, Thorium One is now considering the liquid fuel technology of MSRs.

Canada's nuclear engineers are not tied to LWR technology; they developed the CANDU reactor technology, which employs easily fabricated tubes instead of a massive reactor vessel and heavy water moderator. With Canada's divesting of AECL, that talent is available for new ventures in MSRs.

Canada is USA's largest oil supplier. The tar sands of Alberta are mined and retorted to supply much of this, raising environmental concerns about the additional CO_2 releases from this extraction method. Process heat from a DMSR or LFTR can supply steam for in situ recovery of the 175 billion barrels of reserves. Multiple, small, distributed, modular reactors are appropriate because the nuclear heat can only be efficiently transported about 10 km. An MSR configured to supply process heat does not need the expensive Brayton or SCO2 cycle power conversion turbine, saving 30-40% of the cost of an electric power plant. David LeBlanc, Ottawa Valley Research Associates, and Penumbra Energy of Calgary are raising the interest of engineering and oil sands firms.

The Canadian Nuclear Safety Commission may be more conducive to such advanced technology than the US NRC. For example, tritium release standards are more lenient, enabling CANDU reactors in Canada but not the US.

Dr. Kazuo Furukawa founded IThEMS to build the FUJI MSR.

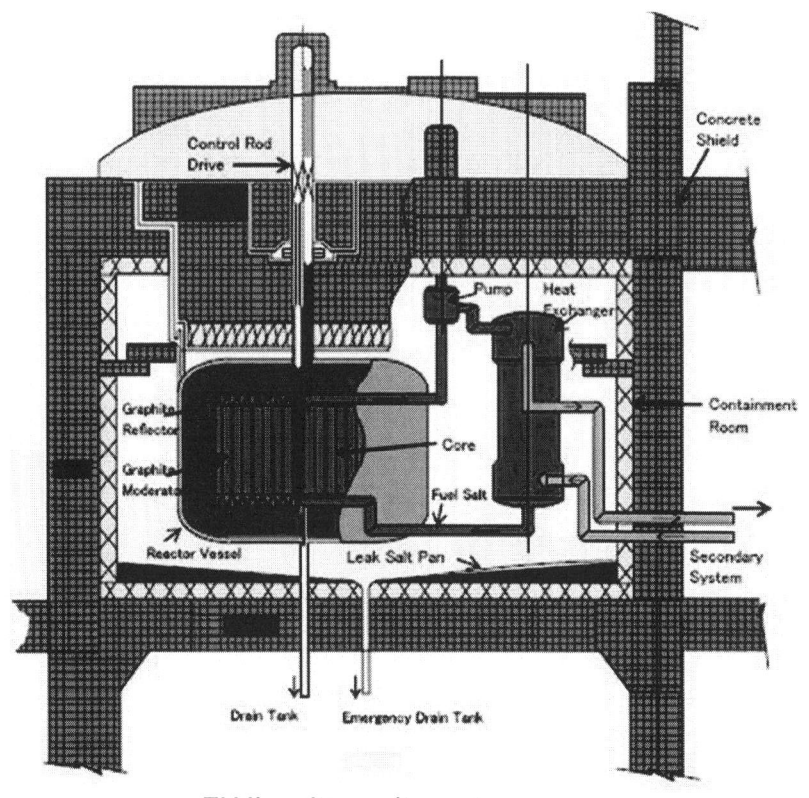

FUJI molten salt reactor

Dr. Kazuo Furukawa led Japanese research in molten salt reactor technology up until his death in 2011. In 2010 he founded IThEMS, an company seeking to build thorium MSRs in Japan.

The business plan was to raise $300 million to develop a 10 MW(e) MiniFUJI MSR within 6 years. Costs for production versions were estimated at $30 million ($3/W). The follow-on 200 MW(e) FUJI design was for a single fluid molten salt reactor, with thorium in the fuel salt. The projected power production cost was 6.1 cents/kWh. The venture had difficulty raising money and is not now proceeding following the death of Dr. Furukawa.

Graphite occupies 90% of the volume of the reactor vessel. Fuel salt temperature is about 600°C. The Th/U cycle reactor is a near breeder, requiring regular addition of supplemental fissile material. The reactor design accommodated any fissile material, including U-233, U-235, Pu-239, etc. The fuel salt is 7LiF-BeF2-ThF4-UF4.

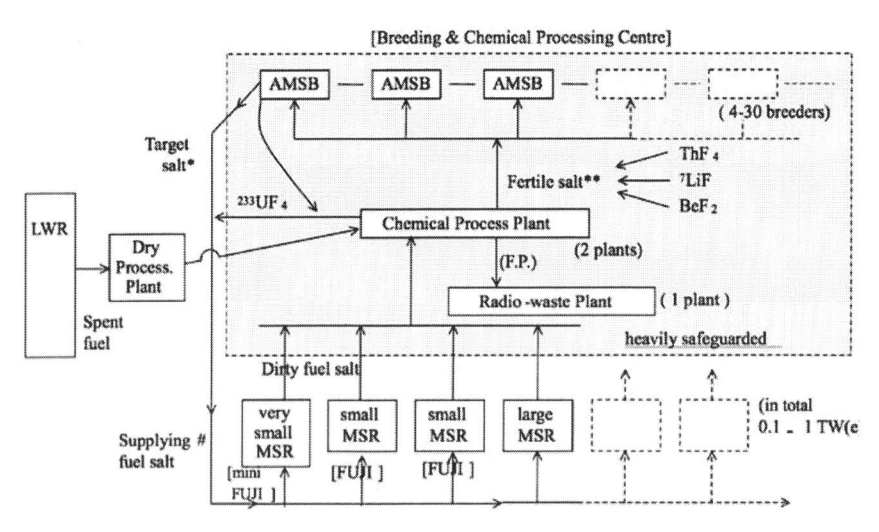

FUJI centralized fuel breeding and distributed MSR system

To provide necessary makeup U-233, the concept included a highly safeguarded U-233 central breeding facility utilizing accelerator molten salt breeder (AMSB) reactors. AMSB's conceptual 1,000 MeV, 300-milliamp proton accelerator would produce 400 kg/year of U-233 from fertile Th-232 in a molten salt target. The startup inventory of the 200 MW(e) FUJI) is 800 kg, thus each AMSB could support commissioning of one FUJI reactor every two years.

Takashi Kamei is continuing research and development for the FUJI MSR concepts. One design uses plutonium from spent LWR fuel rather than U-233 for fissile startup material, summarized in this table from his December 2011 Nuclear Safety and Simulation article. Fuel reprocessing takes place on a 7.5 year cycle over a reactor lifetime of 30 years.

FUJI 200 MW(e) MSR		
	FUJI-PU2	FUJI-U3
Th starting inventory	31.3 t	56.3 t
Pu starting inventory	5.78 t	
Pu 30-year additions	1.16 t	
U-233 starting inventory		1.132 t
U-233 30-year additions		0.344 t
U-233 ending inventory	0.295 t	1.505 t
Transuranics at end	.285 t	0.005 t
Conversion ratio	0.92	1.01

Kamei, who has written Japanese articles and books on this subject, says that Chubu Electric Power Company is considering thorium reactors for future nuclear power. Chubu now operates three LWR nuclear power plants in central Japan.

CONTENDERS

This book promotes thorium-fueled molten salt reactors such as LFTR and DMSR because they have the potential to produce energy cheaper than coal. Coal-burning power plants are the largest single, fixed sources of CO_2 and soot emissions. The thesis of the book is that LFTR can provide a market-based solution to our global environmental crises by underselling coal-burning power stations. LFTR-produced power will be low cost because the fuel is cheap and the capital cost is relatively low, because of its compact, high temperature, low pressure, and inherently safe design.

However several other advanced nuclear power projects may be contenders for this goal of energy cheaper than coal. A sustainable world needs such a solution, whether or not it is LFTR. Many contending technology advocates do not put such a high priority on energy cheaper than coal. Such a goal should be considered at every step of design and development for all technologies, including LFTR. This section will discuss other, contending advanced nuclear power technologies that are being investigated worldwide.

Following are other technologies that might be considered to have the potential for energy cheaper than coal.

NGNP

NGNP is US DOE's choice for next generation nuclear power.

NGNP (next generation nuclear power) is the US Department of Energy's chosen technology for advanced nuclear power, budgeted at $50 million for 2012. The objective is to develop a high-temperature heat source not only for efficient electric power generation, but also for efficient hydrogen dissociation and industrial process heat. The technology is based on TRISO fuel particle embedded in graphite spheres in a pebble bed reactor or in fixed prismatic graphite compacts. The reactor core is cooled by high-pressure helium gas using an external heat exchanger to transfer thermal energy to the steam.

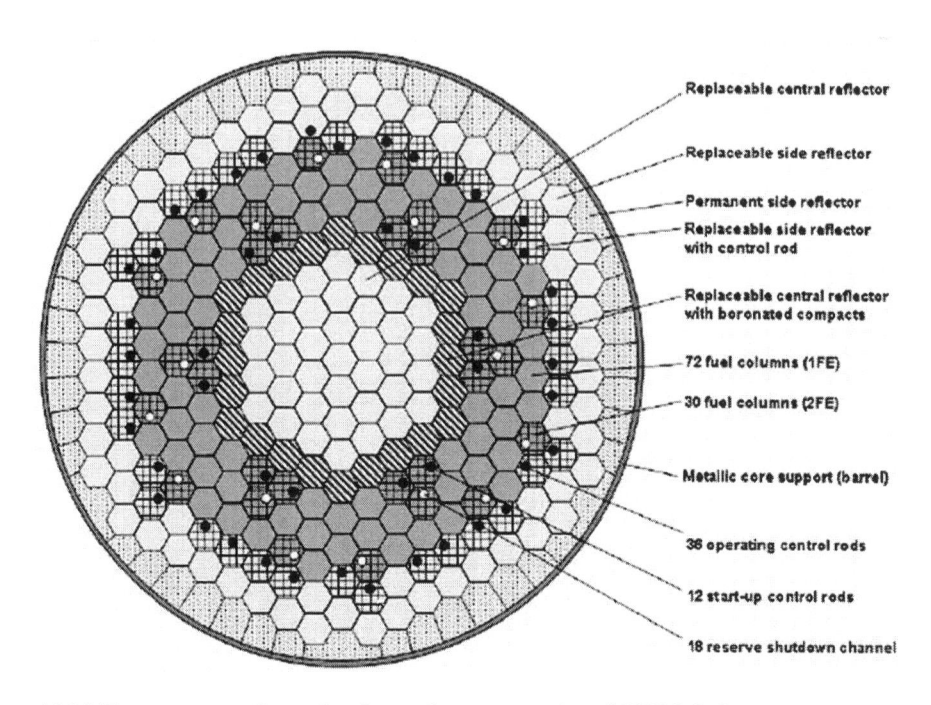

NGNP cross section of prismatic compacts of TRISO fuel particles

This cross section of a prismatic core shows a ring of hexagonal fuel compacts surrounding central neutron reflectors, with more outside reflectors protecting the reactor vessel.

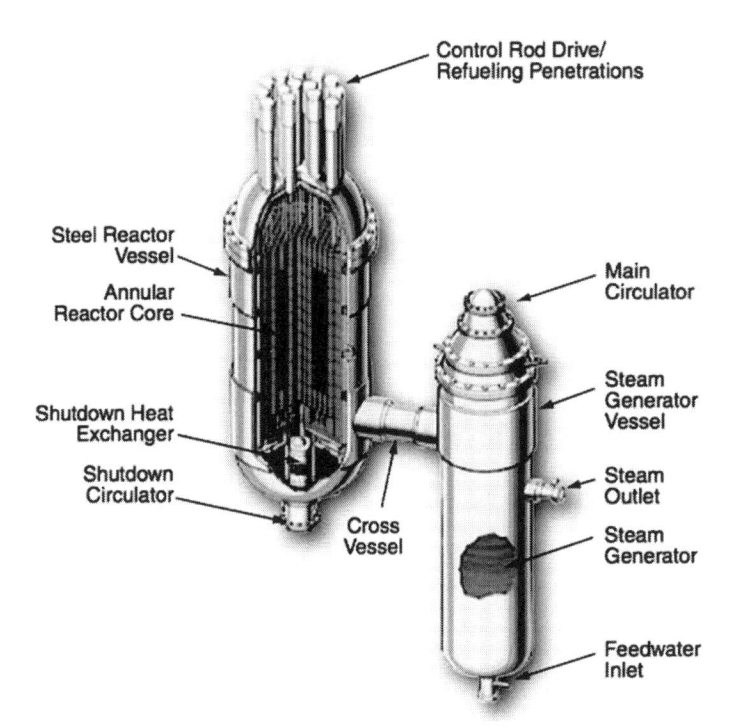

NGNP pressure vessel and steam generator

The 2005 Energy Policy Act authorized $1.5 billion towards NGNP, estimated to cost about $4 billion. The law funds Idaho National Labs work on NGNP and requires industry cost sharing, so the NGNP Alliance was established comprising a dozen companies such as Areva, Westinghouse, Dow Chemical, and Entergy. The Alliance has selected the Areva Antares design, with prismatic TRISO fuel and a conventional steam generation power conversion cycle.

In 2012 INL published a 59-page project plan with about 2000 steps with initial operation in 2021. INL has estimated NGNP capital costs in the range of $2/W(t) for power plants over 600 MW(t) capacity. The NGNP Alliance expects plant costs to be competitive with natural gas heat at $6-9/MBTU, 2-3 cents per kWh(t). Adding a 33% efficient steam turbine and generator at $1/watt might conceivably generate electricity at a cost near 7-10 cents/kWh(e), not quite lower than coal power electricity at 5.6 cents per kWh.

High-temperature, TRISO fuel, helium gas-cooled, pebble bed reactors have been successfully operated in Germany and China. These used TRISO fuel in recirculating pebbles rather than the fixed prismatic graphite compacts selected for the NGNP. South Africa attempted to build a pebble bed reactor, but the project was mothballed in 2010 when it ran out of funds.

Westinghouse AP1000

AP1000 design evolved from Westinghouse PWR experience.

The Toshiba Westinghouse AP-1000 1.1 GW PWR is a possible contender for energy cheaper than coal. While designs such as LFTR are disruptive to the industry, the AP1000 results from the evolution of decades of experience in construction and operation of pressurized water reactors.

Evolutionary change is especially important in an industry such as nuclear power with very long lead times and regulatory delays. Many successful industries have succeeded in the face of disruptive competitors and technologies. Two examples follow.

Magnetic disk drives, threatened by optical disk and solid-state memories, have been predicted to become obsolete since 1956. Yet magnetic disk drives have survived and improved performance dramatically. Consumer prices are under $0.10 per gigabyte; even lower industrial costs make services such as Google possible.

Piston engines successfully evolved over an even longer period, in the face of turbines, Wankel engines, and electric motors.

Advances in computing and engineering enable new designs.

Modern nuclear reactor design development makes extensive use of computer aided design (CAD) and engineering techniques not available to 1970s reactor designers. Computers are a million times more powerful, interconnected by optical fibers and the internet. Database management systems and search engines now store and retrieve information more reliably and quickly.

Engineering advancements include 3-D CAD coupled to static and dynamic finite element analysis, with software such as Fluent, MATLAB, AutoCAD, Catia, and Pro/E. This allows virtual exploration of thermal stresses and strains, viscous fluid flow, thermal conductivity, electrical conductivity, and neutron flux. New engineered system management techniques include systems such as total quality management, GE's 6-sigma concept, ISO 9000 design and production process control, manufacturing resource planning (MRP), and enterprise-scale management systems such as SAP. Probabilistic risk assessment (PRA), used extensively by NASA, now quantifies and manages safety of nuclear power plant designs, where PRA is beginning to replace vague "as reasonably achievable" regulatory safety guidance.

Westinghouse built the first PWRs.

At the 1893 Worlds Fair George Westinghouse demonstrated Nikola Tesla's new invention – alternating current electric power distribution – sowing the seeds for a new company. Westinghouse Electric Company, located in near Pittsburg, is now owned by Toshiba and its minority partners, including Shaw Group. Westinghouse built the first atomic engine, which powered the Nautilus submarine in 1953. Westinghouse built the first US nuclear power plant at Shippingport, PA, in 1957. Half the world's nuclear power plants use Westinghouse pressurized water reactor technology. In 2011 Westinghouse received US Nuclear Regulatory Commission AP1000 design certification, and in 2012 the NRC approved construction and operating licenses for four AP1000 reactors in the US.

Westinghouse's new AP1000 has fewer costly components.

Compared to previous PWR designs, the AP1000 uses many fewer components, reducing costs and improving reliability. Canned coolant pumps have no seals to leak; the hydraulic impeller and motor armature are totally within the pumped coolant, which is also the lubricant; all electric circuits are outside the can. The portion of the building that must withstand earthquakes is relatively small, reducing costs and site footprint.

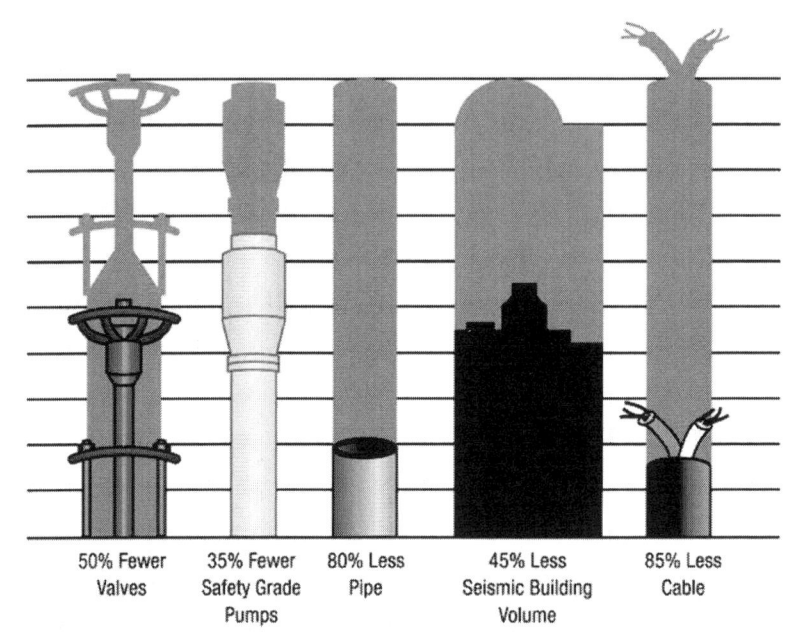

| 50% Fewer Valves | 35% Fewer Safety Grade Pumps | 80% Less Pipe | 45% Less Seismic Building Volume | 85% Less Cable |

AP1000 improvements form Westinghouse brochure

This simplified design philosophy encompasses all instrumentation, operating and control systems, safety systems, control room, and construction techniques, leading to a power plant that is less expensive to build, operate, and maintain. The design includes a 36-month construction schedule.

The new designs reduce the chance of core damage accidents to a tenth of NRC specifications, or 100 times better than that of today's operating nuclear power plants.

The AP1000 uses new modular construction techniques.

Utilizing CAD-expressed designs, modern computer aided manufacturing enables modular construction techniques, where components can be fabricated in factories, transported, then assembled reliably on site. A new construction technique utilizes steel plates to sandwich poured concrete, replacing iron reinforcing bar and temporary plywood forms, along with the time to set up and tear them down. Unlike conventional reinforced concrete, the resulting structure can continue to support substantial loads even in beyond-design-basis accidents that fracture the concrete, which would spall off rebar structures. The AP-1000 is designed to withstand the impact of a commercial airliner. The AP1000 uses under a fifth of the concrete and reinforcing bars of previous designs.

Westinghouse has built a module fabrication factory in China, and its partner Shaw Group has built such a factory in Lake Charles, LA, in the US.

AP1000 shutdown decay heat is passively removed.

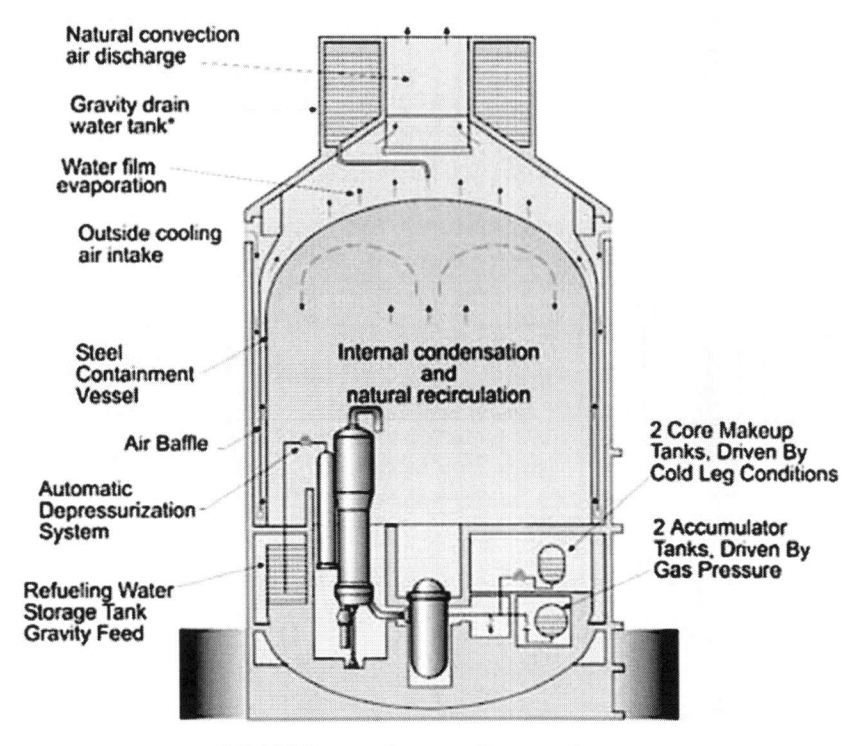

AP1000 passive cooling systems

When a nuclear reactor shuts down and fission stops, fuel rods still contain unstable fission products that decay and produce heat. One minute after shutdown the reactor still generates 4% of its full power heat, and one day later 0.5%. Current LWRs such as at Fukushima continue to cool the fuel with circulating coolant pumps powered by electricity, which became unavailable after the tsunami swamped the backup diesel generators. Like all new reactors, the AP1000 has passive cooling that can operate without electric power. Cooling comes from natural convection, compressed gas coolant transfer, evaporation, and gravity-fed water from a high tank. No operator actions are required for 7 days, when external power must be supplied.

The AP1000 may generate electricity cheaper than coal.

China's objective is to build Westinghouse AP1000 reactors for less than $2/watt capital cost – the same as our LFTR cost objective. Four are already under construction, 8 more are firmly planned, and 30 more are proposed. The first 4 are expected to cost under $2/watt with later reactors costing $1.60/watt.

Will US AP-1000 construction costs drop from $5/watt to $2/watt? Probably not in the short term; labor costs are more expensive in the US. Also, Westinghouse has the most advanced technology, a full order book, and little competition. Areva is experiencing cost overruns and delays building its EPR in Finland. GE seems not to be marketing its BWR design.

China will compete with Westinghouse in the future. Their contract provides China with full rights to AP1000 technology. China is starting to build a derivative design, the CAP1400, a 1.4 GW unit, which could be exported. This could lead to competition between China and Westinghouse that could bring down capital costs for utilities.

SMALL MODULAR REACTORS

New entrants in the nuclear power industry can not face up to the Westinghouse AP1000 and the risks of multi-billion dollar investments in the large scale power generation market. The small modular reactor (SMR) sector attracts new ventures because:

1. Smaller reactors reduce venture capital at risk.
2. New technologies can be exploited.
3. Utilities may only risk smaller capital investments, now that multi-billion dollar US loan guarantees are unavailable.
4. Utilities can add modules as electric power demands rise.
5. Costs will drop with factory production of more, smaller units.

The newly announced SMRs have not yet been demonstrated. They share common characteristics. They are in the 25-300 MW power range. They are almost all PWRs with an integral steam generator heat exchanger within the pressure vessel. Modular components are rail-shippable for assembly on site. The reactor vessels are located underground to improve defense against airliner impacts or other terrorist attacks. The earth can also provide radiation shielding and some heat transfer for passive decay heat removal. Passive heat removal is more easily accomplished with smaller reactors because of their higher surface-to-volume ratios.

The US DOE is encouraging development of American-made small modular reactors with $450 million to support engineering, design certification, and licensing for up to two SMR designs. To obtain an award, the SMRs must have potential for NRC licensing and operation by 2022, and industry must provide at least 50% of the funds.

Babcock & Wilcox applies naval reactor expertise to SMRs.

Babcock & Wilcox has years of experience providing services and materiel to US nuclear powered submarines and aircraft carriers. B&W operates a uranium fuel production facility for the fleet, and it also converts excess weapons-grade uranium into fuel for commercial nuclear power plants. With this experience B&W has designed the mPower 180 MW PWR. Where cooling water is not available, mPower can provide 155 MW using direct air cooling. Refueling intervals are 4 years; 40 years of spent fuel storage is incorporated. The emergency core cooling system operates with passive heat transfer without the need for AC power. B&W has partnered with contractor Bechtel to build the units.

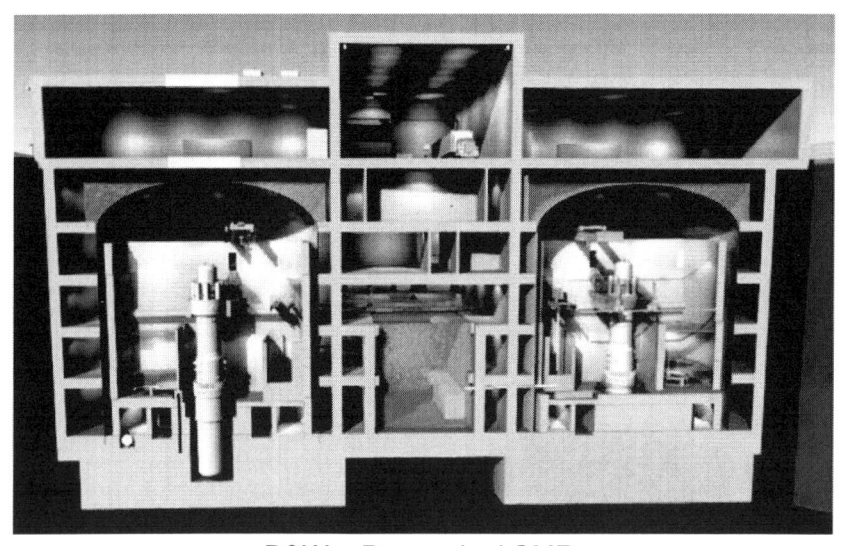

B&W mPower dual SMR

B&W has established an integrated system test facility in Virginia to test all technical features of a scale model of mPower, using electric rather than nuclear heat, for testing and to support NRC licensing activities. Tennessee Valley Authority has notified the NRC of its intent to construct up to six mPower modules at the TVA Clinch River site. B&W has responded to DOE's funding opportunity that offers $450 million for up to two SMR developers. Capital costs are projected to be under $6/watt.

NuScale's SMR evolved from INL and Oregon State R&D.

Idaho National Laboratory and Oregon State University conducted research in small nuclear power plants beginning in 2000, concentrating on passive safety systems that use natural air convection to cool the reactor. OSU built a third-scale model electrically heated test facility to provide data for possible NRC licensing. OSU continued the design work and granted NuScale a technology license and use of the test facility.

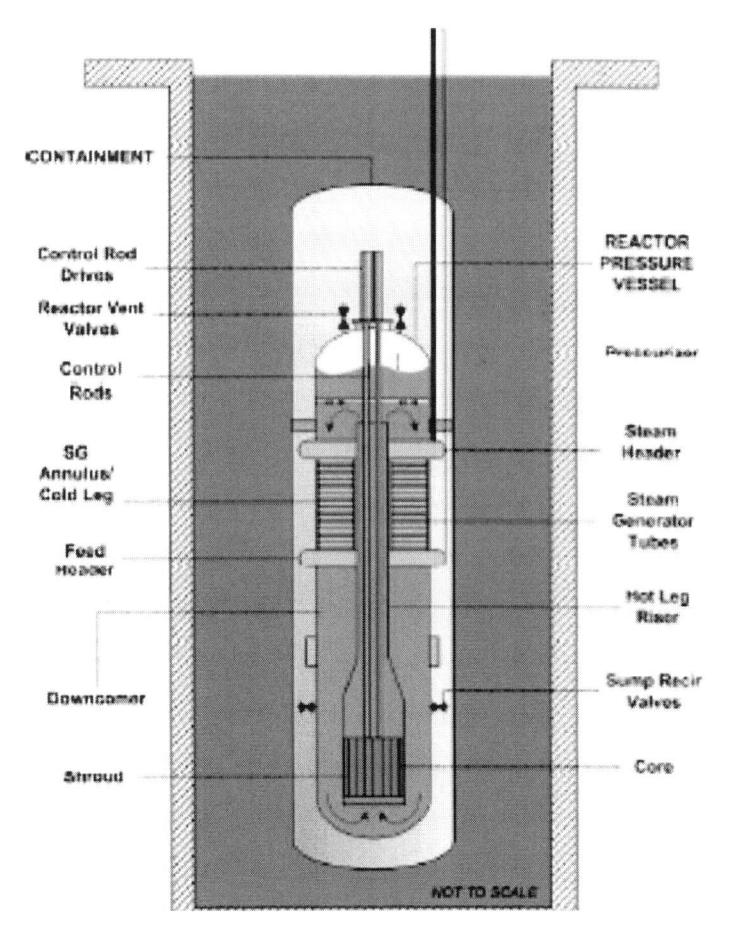

NuScale SMR immersed in emergency cooling water

The previous diagram shows the core at the bottom of the pressure vessel, within the containment. Water surrounds the containment to cool it after shutdown and power blackout. The cooling water boils away over a period of about a month, when the decay heat has reduced sufficiently that convective air cooling replaces the water evaporative cooling.

NuScale's reactor modules are 45 MW PWRs. Uranium oxide fuel enriched to < 5% is in standard 17 x 17 fuel rod assemblies, but only 6 feet long. It is a natural circulation reactor with passive cooling safety.

NuScale is now 55% owned by Fluor, a large engineering and construction firm. In May, 2012, South Carolina Electric & Gas Company and NuScale have submitted their proposal to deploy a small, modular reactor at the DOE Savannah River Site.

Holtec plans its first 140 MW SMR at Savannah River.

Holtec supplies fuel handling and management systems to the nuclear power industry, working closely with utilities at over a dozen US power plants.

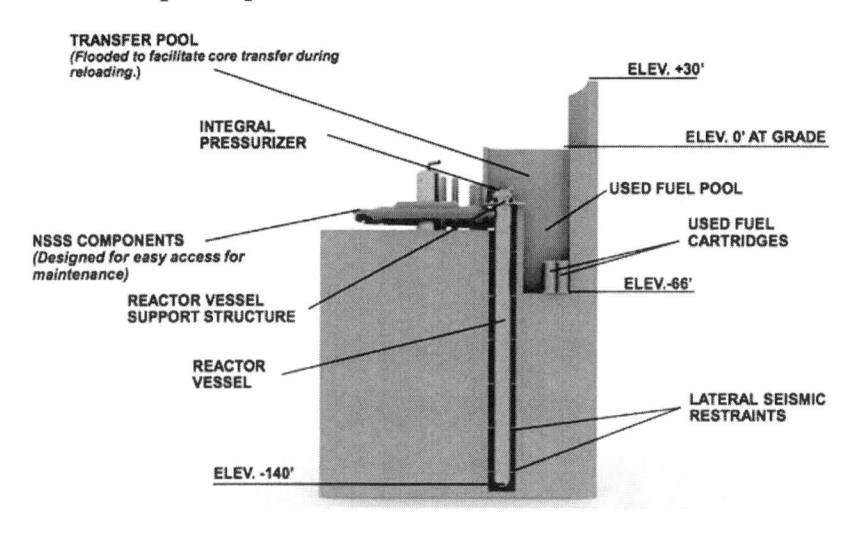

Holtec SMR

Holtec has a preliminary PWR design and has engaged nuclear-industry-contractor Shaw Group to design power conversion and support systems.

Holtec's design features gravity convection to circulate coolant in normal operation and also under accident conditions. There are no pumps requiring electrical power. Refueling is accomplished every three years by removing and replacing the entire core cartridge. Direct air cooling is an option where water use is restricted.

DOE operates a 300 square mile nuclear research complex at its Savannah River site in South Carolina, where Holtec will build the prototype reactor. DOE will provide the project site, transmission lines, roads, and security, and will buy power from the completed reactor.

Westinghouse is designing a 225 MW SMR.

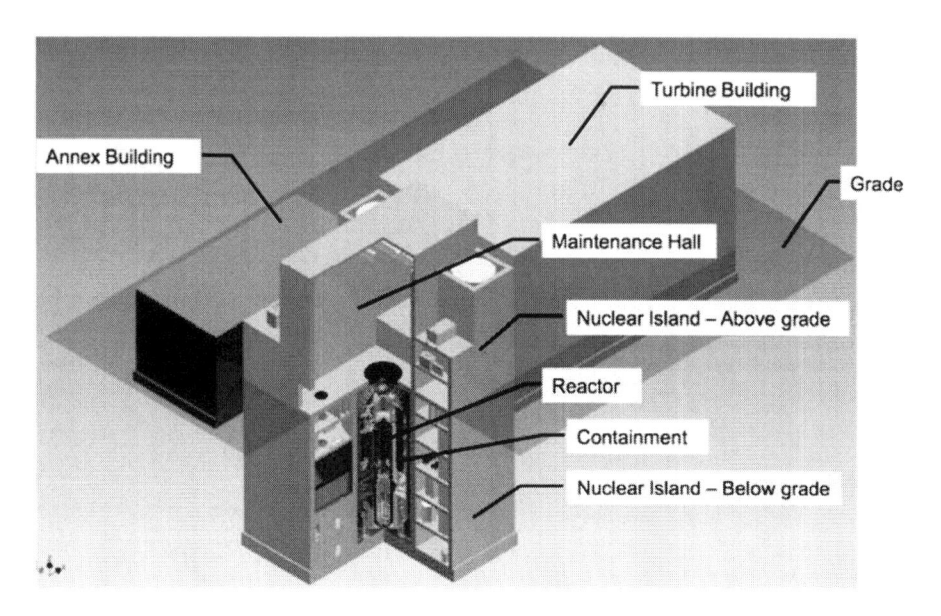

Westinghouse SMR

Westinghouse is designing a small modular reactor based on the technology used for its AP1000. Fuel assemblies are industry standard 17 x 17 fuel rods containing < 5% enriched UO2. Eight redundant canned pumps circulate cooling water.

Passive cooling is similar to the AP1000, using gravity and pressurized gases operating with no external power; after 7 days makeup water or electric power must be supplied externally. The refueling cycle is 24 months.

Westinghouse is seeking NRC design certification for its SMR in 2013. Missouri utility Ameren is working with Westinghouse to apply to the NRC for a construction and operating license for 5 SMRs at the Callaway site, replacing the previously proposed Areva EPR. Together they are also seeking part of DOE's $450 million cost-sharing funding for SMR development and licensing.

Gen4 Energy, née Hyperion, is designing a 25 MW SMR.

Gen4 Energy is the new name for Hyperion Power Generation, which changed management in 2012. The market for such small SMRs includes remote communities, industrial process heat applications, and grid-independent military bases.

The unusual Gen4 technology is not common water cooled uranium oxide in fuel rods. Los Alamos National Laboratories licensed the new technology to Hyperion, now Gen4 Energy. The fuel is uranium nitride, a high-temperature ceramic, enclosed in stainless steel fuel rods. The coolant is a liquid metal mixture of lead and bismuth. This allows operation at 500°C, higher than water-cooled PWRs, which is helpful for process heat applications and more efficient for electricity generation. Similar technology was used in Soviet Alpha-class submarines. This fast reactor can operate for 10 years before refueling.

Gen4 Energy has an agreement with the Savannah River Site, where the first reactor could be built.

Liquid Metal Fast Breeder Reactors

As the name suggests, liquid metal fast breeder reactors (LMFBs)

- are cooled by molten metal.
- use fast neutrons to fission the Pu-239.
- breed Pu-239 fuel from fertile U-238.

In the 1950s the motivations for developing the LMFBR were the worry of running out of fissile uranium-235 and the appealing 99.3% abundance of fertile U-238 in natural uranium, compared to the meager 0.7% of fissile U-235. Newly discovered uranium reserves promise sufficient U-235 for the near future, so LMFBR research slowed in the US. Current concerns for the climate and the search for plentiful, sustainable, zero-carbon energy have renewed interest in LMFBR technology.

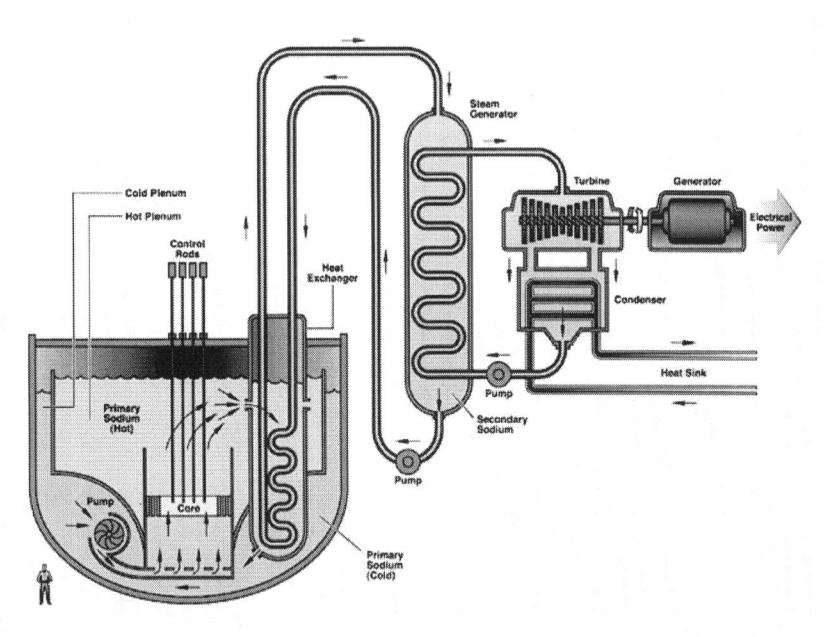

LMFBR in pool of sodium metal

In the diagram above, the LMFBR reactive core is contained in a large pool of molten sodium metal. The heated metal coolant is pumped through a heat exchanger also within the pool. The coolant then circulates back to the pool to be reheated. The sccondary loop contains nonradioactive sodium that transfers thermal energy to an external heat exchanger to make steam or other hot gas for a turbine generator.

The molten metal coolant may be sodium, or lead, or a lead-bismuth mixture. LMFBRs were operated in the US, UK, Russia, India, Japan, and France. Several suffered accidents including sodium fires and core melts. In 2012 only Russia's BN-600 operated in commercial power generation.

LFTR and PWRs use slow, moderated neutrons for efficient fissioning of U-233 or U-235. For Pu-239 fuel, such slow neutrons are too often absorbed rather than causing fission, so plutonium-fueled reactors use fast, unmoderated neutrons because the fission is so more likely. Hence the term "fast reactor".

The US developed three LMFBRs for power generation. In 1951 in Idaho the experimental breeder reactor I (EBR-I) became the world's first electric power generation station, producing 200 kW.

Experimental Breeder Reactor II used metal uranium fuel.

In 1965 the experimental breeder reactor II (EBR-II) came into operation at today's Idaho National Laboratory.

The sodium metal coolant did not react with steel or the metal fuel. It allowed a breeding ratio greater than one, so the sodium cooled fast reactor could create more fissile atoms than it consumed. Sodium cooled reactors have had problems with fires, because sodium spontaneously burns on contact with air or water. Leaks have occurred in the secondary loops causing fires. A leak in the primary loop would be more dangerous because the radioactive sodium would be burned and dispersed.

EBR-II was demonstrated to be passively safe even with shutdown control rods disabled. Two tests were loss of coolant flow, and loss of heat sink normally provided by electric power generation.

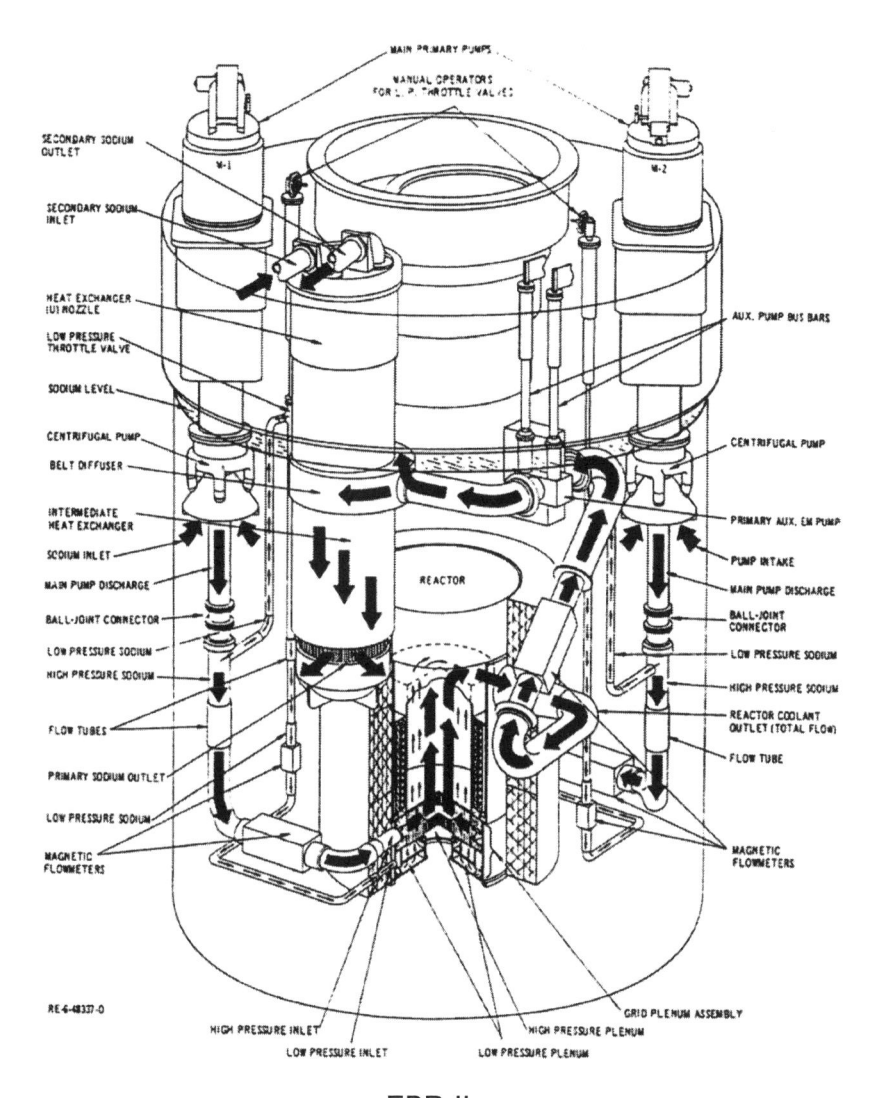

EBR-II

The 20 MW EBR-II ran for 30 years. It demonstrated two novel technologies: metal fuel, and onsite reprocessing.

Metals conduct heat better than ceramics, but the fuel form for LWRs and previous LMFBRs was ceramic uranium oxide, UO_2. EBR–II solved the problem of swelling of irradiated metal fuel. It used uranium/plutonium metal fuel alloyed with 10% zirconium, encased in steel clad fuel pins, with space inside to expand. The superior heat transfer enables high power density and a more compact reactor. The metal fuel bypasses the ceramic fuel overheating that caused fuel melts in other, earlier LMFBRs. In operation neutron irradiation weakens the steel cladding, so the valuable fuel must be periodically reprocessed.

The integral fast reactor is based on EBR-II.

The integral fast reactor (IFR) plan is to conduct reprocessing on site, hence the word "integral". No plutonium is transported in or out of the plant site.

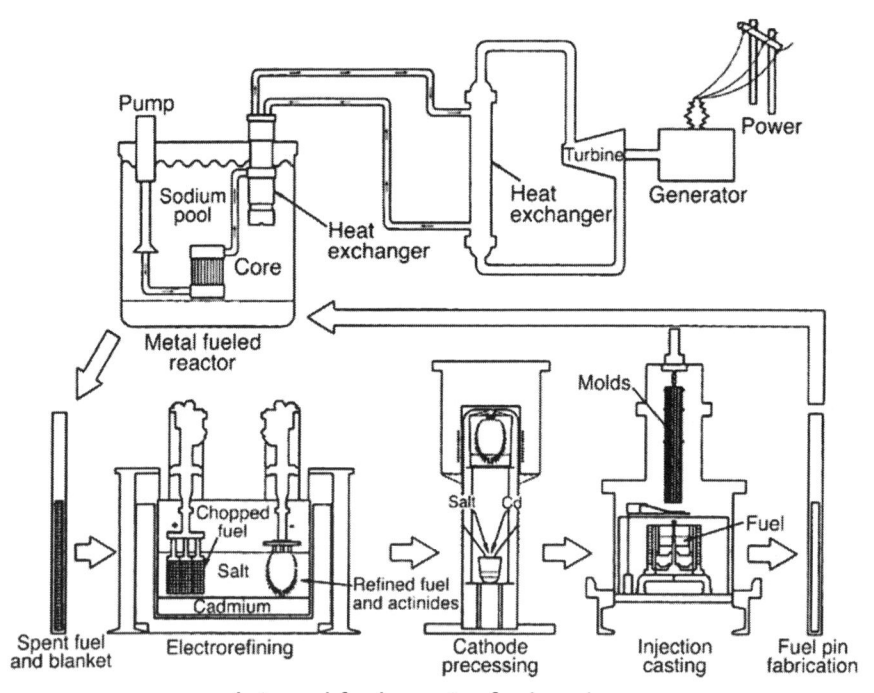

Integral fuel reactor fuel cycle

The reprocessing involves chopping up the fuel rods, placing them in a steel basket in molten chloride salt electrolyte, then passing electric current between the anode basket and two cathodes – of cadmium and of steel. The process is able to separate the uranium and the plutonium, which is always mixed with highly radioactive isotopes of the heavier actinides neptunium, americium, curium, etc. The metals are recast and placed into new steel fuel pins. All these processes take place in an argon-atmosphere radiation-shielded hot cell, using robotic and remote handling equipment.

The reprocessing of spent fuel is proliferation resistant for several reasons. The potentially weapons-useful Pu-239 is too contaminated with other plutonium isotopes that fission spontaneously. Also, the plutonium is always mixed with highly radioactive actinides that can not be separated with the electro-refining equipment. Handling such plutonium would be fatal to unshielded workers or military personnel.

The IFR design has a blanket of uranium in which Pu-239 is bred and processed independently of the spent fuel reprocessing. The IFR electro-refining can not feasibly separate the plutonium, but this could be accomplished with other known processes: PUREX or fluoride volatility, which are not within the IFR plant.

Argonne National Labs had advanced the IFR design enough to begin construction, but in 1994 the US Congress shut down the program, at the urging of the Clinton administration, which claimed that weapons proliferation risks were too high, and released near disinformation about a weapons test with plutonium from a power reactor that turned out to be a dual-purpose UK Magnox reactor. In the 1994 State of the Union address President Clinton said "We are eliminating programs that are no longer needed, such as nuclear power research and development".

A nation intent on developing nuclear weapons would use demonstrated technology – either centrifuge uranium enrichment or plutonium production with frequent fuel exchange in purpose-built reactors. Attempting to modify the IFR would be much more difficult, risky, and expensive.

GE-Hitachi's S-PRISM based on EBR-II and IFR development.

GE-Hitachi (GEH) derived its 311 MW S-PRISM design from the work by Argonne National Labs on IBR-II and IFR. S-PRISM has a breeding ratio of only 0.8, so it requires fissile makeup fuel. This is intended to be Pu-239 and U-235 remaining in spent LWR fuel, providing a way to reduce nuclear waste and generate power.

Differing from the IFR, the GE concept has a separate advanced recycling center (ARC) serving six S-PRISM reactors. GEH estimates that a first-of-a-kind S-PRISM and ARC could be designed and built for $3.2 billion. The initial fuel could be excess weapons grade plutonium that the US is committed to destroy, currently by fabricating mixed oxide (MOX) plutonium-uranium fuel in a Savannah River plant still under construction in 2012.

GEH has not built such a reactor, but has a 2010 agreement with the US DOE Savannah River National Lab site that would allow construction of an S-PRISM there without an NRC license. In 2011 GEH entered discussions with the UK about using the S-PRISM to generate power and consume 100 tonnes of plutonium stored at Sellafield in Cumbria.

Russia's SVBR-100 is based on Alfa submarine experience.

The Russian Navy powered its 40-knot interceptor submarine with a fast reactor cooled by a lead-bismuth eutectic. Apparently capable of speeds to 52 mph, the last Alfa class submarine was decommissioned in 1981. With 80 reactor-years of operation, this technology is resurfacing as a 100 MW small modular reactor. The steam generator and reactor core are in the same pool of lead-bismuth coolant. A demonstration unit is planned for 2017.

Russia's BN-600 LMFBR has operated since 1980.

Currently (2012) the BN-600 is the only liquid metal cooled fast breeder reactor in commercial power production. The 540 MW reactor has operated with uranium oxide fuel enriched up to 26% and also with MOX. Russia plans to replace its fertile breeding blanket with steel reflectors, turning it into a net consumer of fissile material, to consume excess plutonium from military stockpiles.

The BN-800 is an improved version under construction in Russia, with power production planned for 2014. China has agreed to buy two BN-800 units from Russia, with construction to start in 2013 at Sanming in Fujian province.

Bill Gates backs TerraPower's traveling wave reactor.

TerraPower is a spin-off of Intellectual Ventures, an intellectual property licensing and investment firm founded by Nathan Myhrvold, formerly the chief technology officer of Microsoft. TerraPower was created in 2008 and led by John Gilleland, with the goal of developing a traveling wave reactor.

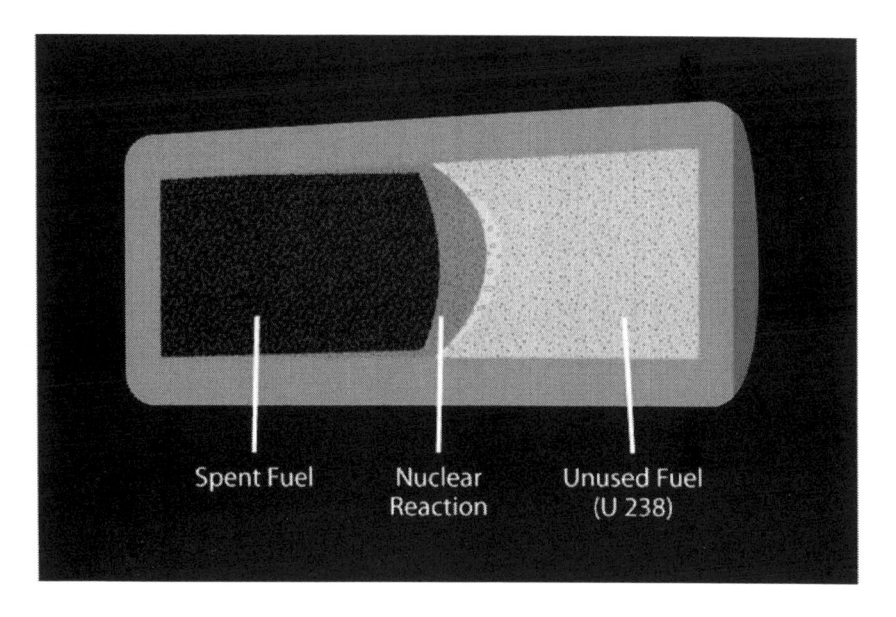

TerraPower traveling wave reactor concept

The original idea was sometimes termed the burning cigar. In the center of the diagram is an area containing fissile Pu-239 fuel participating in a critical nuclear chain reaction. Excess neutrons penetrate into the fertile U-238 fuel, breeding it to more Pu-239 fuel. The boundary area proceeds from left to right over a period of decades, leaving behind the spent fuel, depleted of about 20% of the uranium, along with fission products. After refueling, TerraPower says the spent fuel could be recast into new fuel pins with no chemical reprocessing. New U-238 is relatively cheap, though.

TerraPower has received funding from Bill Gates, Khosla Ventures, Charles River Ventures, and Reliance Industries. TerraPower has assembled a team of over 50 nuclear engineering professionals and nine research partners. The design has evolved considerably and drawn on the experience of EBR-II and the IFR design work.

TerraPower's TWR-D shuffles its fuel pins internally.

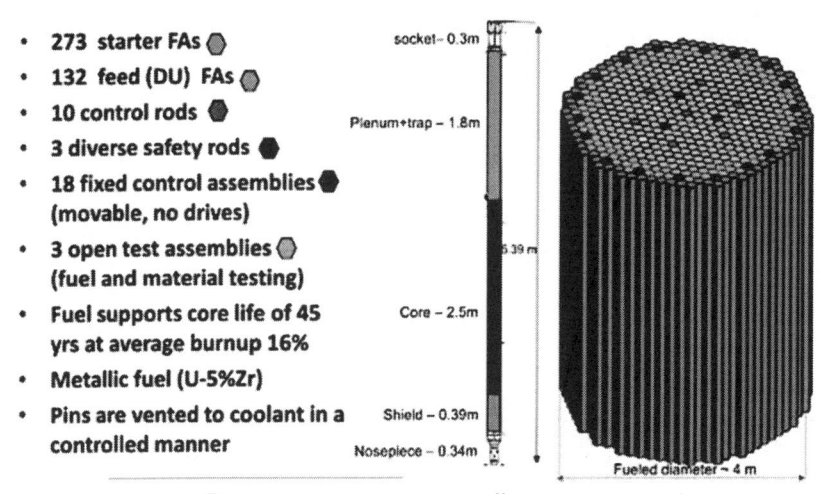

- 273 starter FAs
- 132 feed (DU) FAs
- 10 control rods
- 3 diverse safety rods
- 18 fixed control assemblies (movable, no drives)
- 3 open test assemblies (fuel and material testing)
- Fuel supports core life of 45 yrs at average burnup 16%
- Metallic fuel (U-5%Zr)
- Pins are vented to coolant in a controlled manner

socket - 0.3m
Plenum+trap - 1.8m
5.39 m
Core - 2.5m
Shield - 0.39m
Nosepiece - 0.34m
Fueled diameter - 4 m

TerraPower center-out traveling wave reactor

In the current design the fissile fuel is in fuel pins near the center of the core where the critical reaction takes place. The fertile fuel pins surround the center pins. Neutrons from the critical reaction in the center are absorbed by the fertile fuel in the surrounding pins, causing conversion of U-238 to Pu-239. The traveling wave moves out radially from the center.

Every 18-24 months a machine within the reactor vessel shuffles the pins, replacing spent fuel pins from the center with fresh fuel pins from the surrounding region.

Like other pool LMFBRs, the 500 MW TWR-D reactor is cooled with liquid sodium exiting the core at 510°C, then passing through in-vessel heat exchangers. Secondary sodium loops transfer the thermal energy to a steam generating heat exchanger to power a conventional steam turbine, for an electric/thermal efficiency of 42%.

As in EBR-II, the uranium-zirconium metal fuel is within steel cylinder pins, with an expansion plenum to accommodate fuel swelling and fission product gases. To relieve pressure, some gases and volatiles are vented from the pins, so the resulting radioactive cesium and krypton are continuously removed from the circulating sodium. Unlike EBR-II or IFR, these fuel pins remain in the reactor core for the full 40-year core life. There is no fuel reprocessing until, optionally, end of core life.

Fuel for the TWR-D is readily available. The US government already owns over 500,000 tonnes of U-238 orphaned by enrichment plants making LWR fuel. That alone is enough fuel to supply all US electric power demand for 500 years. Known uranium reserves are ten time this, and uranium in seawater is 10,000 times more. The fuel supply is inexhaustible.

Control rods regulate the reactivity, with backup shutdown control rods. Even if these fail, the TWR-D design includes passive safety cooling for loss-of-coolant-flow and for loss-of-heat-sink (from loss of generator load), with no electric power.

Proliferation resistance is similar to that of LWRs. No plutonium exists outside of the reactor core; U-238 is converted to Pu-239 and consumed within the core. The Pu-239 is mixed with other isotopes making it undesirable for weapons. The spent fuel pins contain highly radioactive fission products that would kill unshielded weapons workers.

Befitting its founders, the design is based on computer-intensive modeling and simulation, taking into account cross-sections and decay rates for 3400 different involved isotopes including about 1300 decay products. The TerraPower team can run a Monte Carlo simulation of 110,000 zones, out 60 years and provide results in one day.

The economic goal for TWR-D is to be competitive with LWRs. Contributing to lower costs are lack of need for enriched uranium (except for startup), and higher temperature leading to 20% more electricity generation than LWRs.

Start-up fissile U-235 inventory is not published, but if TWR-D operates like other fast reactors, would probably would be 5-10 times higher than for LFTR reactors. The fuel will be enriched to "below 20%", about 5 times that of LWR fuel.

TerraPower has completed the conceptual design of TWR-D in conformance with IAEA safety requirements. Construction could begin by 2015 with operation by 2020, however there are no specific plans for this.

LFTR advantages require a longer R&D path than LMFBRs.

Compared to LMFBRs, the LFTR has several advantages.

1. LFTR can operate at higher temperatures (700°C rather than 510°C) enabling more efficient Brayton power conversion cycles.
2. LFTR's higher temperature enables more efficient hydrogen generation and industrial process heat.
3. Fluoride salts have 4.5 times more volumetric heat capacity than liquid sodium, so the reactor is 2-4 times smaller than IFR.
4. LFTR fissile start-up fuel is 5-10 times less.
5. LFTR liquid fuel technology is simpler; all US LMFBR reactors were shut down; worldwide only one is still in commercial service, in Russia.

LMFBR may be farther along the path to commercial development.

1. The US government invested over $16 billion (2012 dollars) in IFR development.
2. Prototype EBR-II, the world's only metal-fueled LMFBR, operated successfully for 30 years.
3. Fuel recycling processes have been designed and tested.
4. LMFBR technology is supported by a strong commercial company, GE-Hitachi, which has prepared initial materials for NRC licensing applications.
5. Traveling wave reactor technology is being developed by a skilled, well-funded TerraPower team that has completed a conceptual design, with construction completion possible by 2020.

ACCELERATOR-DRIVEN SUBCRITICAL REACTOR

An accelerator-driven reactor is subcritical.

In today's nuclear power plants power is sustained by a chain reaction. Fissioning of uranium atoms by neutrons produces more neutrons, which fission yet more uranium atoms, creating the chain reaction. Each fission typically releases 2 or 3 neutrons -- some may be absorbed, and some may cause fission. The average number of neutrons from one fission causing another fission is termed the effective neutron multiplication factor, or criticality, k. For a stable chain reaction $k = 1$. If $k > 1$ the chain reaction rate increases and the increased heat causes the reactor to decrease the criticality. For $k < 1$ the chain reaction dies out, because the reactivity is subcritical.

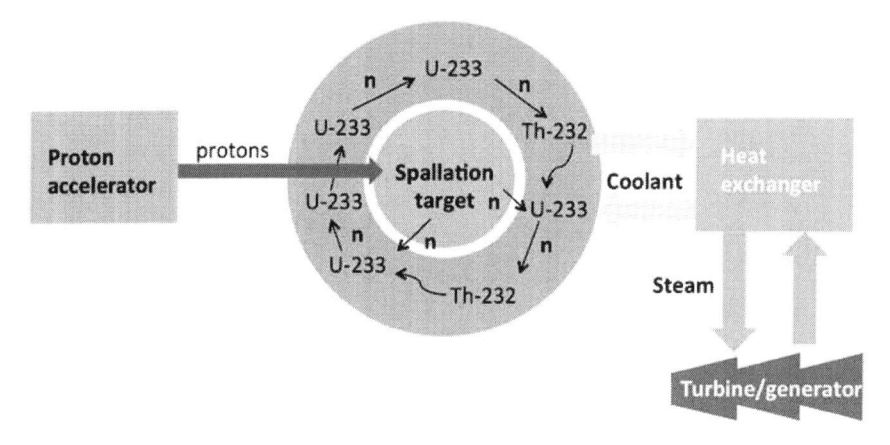

Accelerator-driven reactor

An accelerator-driven reactor is normally subcritical. It can not sustain a chain reaction unless additional neutrons are injected. In the ring in the diagram above the fissioning U-233 atoms emit neutrons (n), k of which fission more U-233 atoms. If $k = 0.95$, then on average the chain reaction will die out after about 20 more fissions. To sustain it would require supplying one external neutron for each 20 generated by fission.

The source of the external neutrons is high energy protons impinging on a heavy metal target such as lead. Advanced multistage accelerators can increase the kinetic energy of charged protons by 1 GeV (a large amount of energy – about equal to the energy of the proton mass itself). Such an energetic proton impinging on a lead target creates a cascade of particles, including about 24 neutrons on average. Most of these neutrons can fission a U-233 atom, each creating a chain of about 19 more fissions in a subcritical reactor with $k = 0.95$.

In an accelerator-driven subcritical reactor (ADSR) the core contains Th-232 as well as U-233. Some of the neutrons are absorbed by Th-232, leading to two-stage decay to U-233, represented by the wiggly line. Thus the fuel for ADSR is fertile thorium.

The accelerator uses considerable power to generate the stream of energetic protons. Depending on design, a 600 MW ADSR might require 15 MW of power to drive the accelerator. Sometimes the ADSR is termed "energy amplifier" because the output power is a multiple of the input power.

E. O. Lawrence described the idea in 1948. Another Nobel prize winning physicist Carlo Rubbia reintroduced and patented the ADSR in 1995. It has some of the advantages characteristic of LFTR, including plentiful fertile thorium fuel and reduced long-lived radiotoxic transuranic waste.

The ADSR can be switched off.

The public-perceived safety advantage of the ADSR is that it is subcritical, unable to sustain a nuclear reaction unless the accelerator is operating. The operator could just "switch it off".

However, a manual switch might not be operated fast enough. Automated systems for turning a reactor off have failure potential too. LWRs are inherently protected from runaway explosions because the moderating water would be evaporated to steam. LFTR and DMSR fuel salts expand with heat, expressing fuel, and reducing criticality. These physical safety factors require no active control systems.

All Fukushima reactors turned off properly using their control rods; the damage to flooded ones was caused by fission product decay, which can't be turn off in an ADSR, either.

Large proton accelerators are expensive and unreliable.

Proton accelerators adequate for ADSRs have not yet been demonstrated. They are not yet capable of continuous operation required for electrical power generation.

The largest spallation neutron source (in 2012) is at Oak Ridge National Laboratory, capable of producing 1.4 MW of 1 GeV protons at 1.5×10^{14} protons per sec. It is used for neutron science research; its operating schedule is not continuous. Construction was completed in 2006 at a cost of $1.4 billion. Because of accelerator unreliability, most ADSR designs incorporate multiple accelerators. Just the accelerators could cost more than a traditional nuclear power plant.

ADSRs need reliable control rods.

Because the accelerator cost is so high, it is attractive to improve the neutron production from fission. Rubbia's presentation calculation shows $k = 0.997$, which is close to stable criticality of 1.000. The mixtures of U-233, fission products, and fissile transuranics change during the fuel life cycle, so that at some time $k > 1$, except that control rods are used to absorb excess neutrons and are adjusted during the fuel burn cycle. So a stuck control rod could cause runaway overheating that could not be "switched off" by shutting down the accelerator. An ADSR must have high-reliability shutdown control rods as today's LWRs have.

The best safety systems are those that are inherent to the physics. For example, LFTR fuel salt expands if it heats up, diluting the fissionable uranium, so it becomes less critical. For another example, the high-temperature gas-cooled NGNP reactor can not run away because at high temperatures the Doppler broadening of the U-238 neutron absorption steals too many neutrons to allow criticality to continue, so the NGNP reaction stops and it idles at a high temperature. For another example, traditional LWRs will stop if their cooling water boils, because this reduces moderator density.

Starting up an ADSR with no fissile material is impractical.

The ADSR description above uses U-233 as the fissile material. An ADSR could be started up with U-233, U-235, Pu-239, or the mixture of fissile materials in spent LWR fuel. As it operates, these are consumed and the fissile material becomes only U-233 generated from neutron absorption by thorium.

A suggested advantage of the ADSR is that it might be started up without the need to transport radioactive fissile material to the site. The accelerator would simply be turned on to start the process of converting Th-232 to U-233. However to generate enough U-233 this way would take 40-400 years of accelerator operation, for a thermal reactor.

Most ADSR designs are for liquid metal cooled fast reactors, where the coolant is lead or lead-bismuth. These would require about five times more startup fissile material that thermal reactors.

Ralph Moir's estimated $500 per gram cost of accelerator-produced fissile U-233 is ten times that of U-233 from fission. Just the electricity to generate enough U-233 for a 1 GW ADSR would cost $240 million.

Britain's ThorEA promotes ADSR research.

The Thorium Energy Association (ThorEA) is a UK not-for-profit association to promote thorium nuclear fuel, organizing workshops and meetings related to ADSR research.

ThorEA, formerly the "thorium energy amplifier association" published a 2010 report proposing a $500 million initial public investment in ADSR research and development over five years, followed by a $3 billion private investment over ten years to develop a 600 MW prototype electric power station in operation in 2025. ThorEA encourages Britain to lead in this industry. ThorEA's blog links to press coverage of ADSR technology.

ADSRs have been studied by several other start-ups.

In 2010 Norwegian oil services company Aker Solutions collaborated with Carlo Rubbia and undertook a feasibility study to develop a commercial 600 MW ADSR (termed ADTR™) power station. It would be a sub-critical, thorium-fueled, lead-cooled, fast reactor with a proton accelerator. Aker purchased Rubbia's patents and invested $3 million in studies.

In 2010 with ADNA (Accelerator Driven Neutron Applications) Corporation in Virginia proposed a $160 million research project into a molten salt version of ADSR.

There have been two *International Workshop on ADS and Thorium Utilization* meetings, in Virginia and India.

Much of the interest in ADSR technology is initiated by physicists who use accelerator technology for fundamental particle research and discover this possible further use.

Much of the press appeal is due to the involvement of Carlo Rubbia, the Nobel prize winning Italian physicist who patented the idea of the energy amplifier. It's an interesting idea, but even Rubbia says he is a physicist, not an engineer.

ADSR has no advantage over LFTR.

Although the novel use for particle accelerators is interesting, ADSR certainly does not contend for energy cheaper than coal. It offers no advantage over other contending designs. The "switch it off" characteristic does not increase safety. It requires all the safety systems and other components of other designs. It's really a nuclear reactor with an expensive accelerator added on.

For LFTR the liquid molten salt is the key to low costs. It is the vehicle for fissile and fertile material. It has high heat capacity enabling excellent cooling and high power density, leading to low costs – energy cheaper than coal. The safety is inherent. No fissile materials need by transported to or from the LFTR after startup. The fissile inventory is low (about the same as ADSR). MSRs have been demonstrated to work. The fluid form allows removal of fission products; the noble gases bubble out; the noble metals plate out; the fission product fluorides dissolved in the molten salt can be removed chemically. Removal of FPs increases safety. The LFTR has no FP waste tied up in solid fuel rods within the reactor.

LFTR ADVANTAGES

Nuclear engineer Ed Phiel compiled this summary of LFTR advantages.

1. Low Pu-239 production.
2. Compared to LWR, 1% of radioactive waste volume.
3. Waste contains virtually no fissile materials, so criticality concerns during waste handling are eliminated.
4. No xenon neutron absorption causing LWR startup instability.
5. Inherently safe by thermal expansion shutting down fission with increasing temperatures.
6. Core is already molten so can't melt down.
7. No risk of coolant flashing to steam sending fission products airborne and losing cooling.
8. Entire core can be automatically dumped to sub-critical configuration, with indefinite air cooling decay heat removal if it overheats or for any other reason.
9. Core is load-following and self-controlling based on temperature changes with no control rods needed for load changes.
10. Thorium is available almost for free from rare earth mining activities waste streams.
11. Thorium is less radioactive than uranium.
12. LFTR has on-line refueling, so no periodic shut downs.
13. LFTR does not need excess reactivity in its core requiring suppression with control rods and neutron-absorbing poisons.
14. LFTR breeds its own fissile fuel and which can be removed from the blanket by fluorine gas purging, then hydrogen gas treatment, then transferred to the core
15. Single fluid proof-of-concept molten salt reactors been built and operated by ORNL.
16. Fissile U-233 contains U-232, a decay precursor to thallium emitting 2.6 MeV gamma radiation, making U-233 unsuitable for military weapons production.
17. No core infrastructure cost since the core is molten salt, greatly reduces operational costs versus PWR/BWR
18. LFTR can be 44-50% efficient compared to 33% for PWR/BWR.

19 LFTR efficiency allows possibility of having purely air-cooled reactors for arid or cold regions.
20 Fluorine salt is less corrosive than hot water.
21 PWR/BWR zirconium plus water to hydrogen production is eliminated, so Fukushima-like hydrogen explosions are eliminated.
22 Fluorine forms ionic salts that are extremely stable in a radiation field, even compared to water.
23 Fluorine salt's boiling point is 400-700°C above the operational temperature, so boiling is not a possibility, the reactor self shuts down far below those temperatures.
24 Fission gases are continuously removed and stored safely, so breach of containment would not release them.
25 Other fission products can also be removed on-line and safe-stored away from the reactor core.
26 Thorium fuel is 4x as common as Uranium.
27 Thorium is 560x as common as U-235.
28 Thorium is found in higher concentrations than uranium due to higher chemical stability.
29 Much less mining is required compared to LWRs.
30 LFTR is highly scalable from small plants to large plants
31 LFTR reactors are very compact.
32 No CO2 emissions.
33 LFTR only requires a small containment structure, because there is no steam or hydrogen explosion to contain.
34 Because LFTR is small it can be built underground for further protection.

6 SAFETY

ACCIDENTS

Accidents happen. In the energy sector accidents can have large effects because there are large amounts of potential energy stored in fuel tanks, hydroelectric reservoirs, or nuclear fuel rods, for example. Although the risk of accidents can never be zero, engineers and regulators work to keep the number of accidents small and compatible with public experience and expectations.

It is helpful to compare the frequency and severity of accidents in nuclear power generation with those that generate power from other energy sources.

22 energy disasters killed 608 people in 2010.

Alexis Madrigal of *The Atlantic* published an illustrated list of energy disasters of 2010, summarized below.

2010 energy disasters			
Accident	Place	Date	Deaths
Natural gas power plant explosion	Middletown CT, USA	Feb 7	6
Refinery explosion	Artesia NM, USA	Mar 2	2
Coal mine fire	Zhengzhou, China	Mar 15	25
Coal mine collapse	Quetta, Pakistan	Mar 20	45

Coal mine flood	Shanxi, China	Mar 28	28
Refinery explosion	Anacortes WA, USA	Apr 2	5
Big Branch coal mine explosion	Raleigh County WV, USA	April 5	29
Deepwater Horizon drilling rig explosion	Gulf of Mexico	Apr 20	11
Coal mine explosion	Mezhdurechensk, Russia	May 8	91
Gas explosion	Anshun City, Chizhou, China	May 14	21
Coal mine explosion	Zonguldak, Turkey	May 18	28
Coal mine explosion	Shanxi Province, China	May 19	10
Coal mine dynamite explosion	Chenzhou City, China	May 30	17
Coal mine gas explosion	Amaga, Colombia	Jun 17	73
Coal mine carbon monoxide poisoning	Pingdingshawn City, China	Jun 21	46
Natural gas explosion	Los Angeles CA, USA	July 30	1
Natural gas pipeline explosion	San Bruno CA, USA	Aug 10	5
Coal mine explosion	Yuzhou, China	Oct 16	20
Coal mine gas explosion	Greymouth, New Zealand	Nov 19	29
Coal mine explosion	Heilongjiang, China	Nov 21	87
Oil pipeline explosion	San Martin Texmelucan, Mexico	Dec 19	27
Natural gas explosion	Wayne IN, USA	Dec 29	2

None of these accidents involved nuclear power plants. What about Fukushima, Chernobyl, and Three Mile Island? No one was killed or injured at Fukushima or Three Mile Island. The Chernobyl event is included below in the Paul Scherrer Institut comprehensive study of severe accidents in the energy chain for electrical power generation. Severe accidents are defined to have five or more fatalities. The energy chain involves all the chain of activities, from drilling, transporting, refining, and distributing oil, for example.

Severe accidents in the energy chain, 1969-1996			
Energy chain	Accidents with ≥ 5 fatalities	Fatalities	Fatalities per GW-year
Coal	185	8,100	0.35
Oil	330	14,000	0.38
Natural gas	85	1,500	0.08
LPG	75	2,500	2.9
Hydro	10	5,100	0.9
Nuclear	1	28	0.0085

To compare accident rates for different energy intensities, the last column divides the fatalities by the amount of electricity produced for that energy source. Nuclear power is the safest electricity power source by far -- 9 times safer than natural gas – 41 times safer than coal.

US NRC studied consequences of severe nuclear accidents.

In 2012 the US Nuclear Regulatory Commission published the results of its five-year State-of-the-Art Reactor Consequence Analysis (SORCA) research study. The study collected detailed information about two different plants' layouts and operations and modeled severe accident consequences using state-of-the-art computer codes that incorporate decades of research into severe reactor accidents. The report says:

> "SOARCA's analyses show essentially zero risk of early fatalities." (pg. xix)

> "The calculated cancer fatality risks from the selected, important scenarios analyzed in SOARCA are thousands of times lower than the NRC Safety Goal and millions of times lower than the general U.S. cancer fatality risk." (pg. xxiii)

This report was based on detailed analysis of US BWR and PWR power plants. LFTR and DMSR plants are expected to be even safer, because of the unpressurized fissile materials, the impossibility of melt-down, and the passive decay heat removal.

IONIZING RADIATION

Radiation radiates; energy travels outward from a source along a radius. Common radiation is sunlight radiating from the sun, radio waves radiating from a cell phone, infrared light radiating from a TV remote control, and television signals radiating from a satellite in the sky. Ionizing radiation has higher energy that these examples and can disrupt chemical bonds in molecules in cells.

0.18% of ionizing radiation comes from nuclear power.

Half of natural background radiation comes from radon gas, which decays from radium in the earth's crust. The rest comes from cosmic rays and naturally occurring radioactive elements. Most man-made radiation comes from medical imaging and therapy. Only 1% of man-made radiation results from the nuclear power.

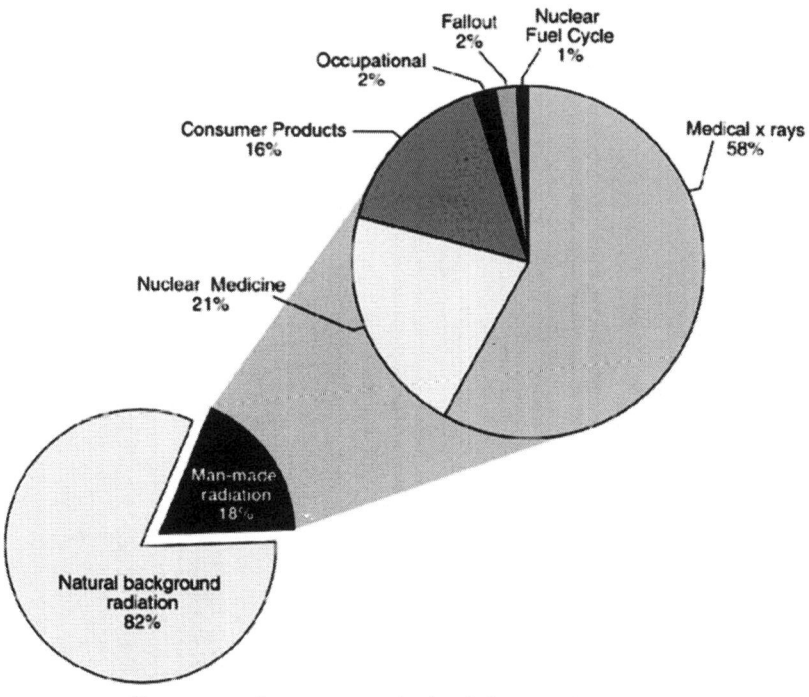

Sources of exposure to ionizing radiation

Four ionizing particles come from four sources.

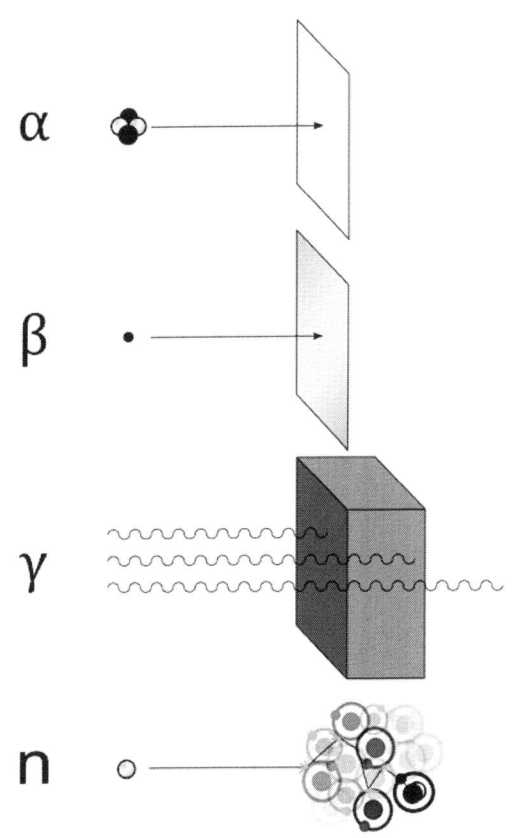

α — Energetic, heavy alpha particles (He nuclei) from actinide decay can not penetrate epidermis.

β — Beta particles, electrons ejected as neutron-rich isotopes become stable, do not penetrate metal foil.

γ — Gamma radiation, photons from nuclei energy level changes, are absorbed by dense material such as bone to make X-ray images.

n — Neutrons from nuclear reactor fission are slowed by collisions with light elements like H in H2O.

Although alpha radiation is easily stopped by the skin's outer layer of dead epidermis, alpha particles released on the surface of internal lung tissue ionize molecules in living cells and can damage them. Radon gas from decay of radium in granite can collect in homes. Its half life is 4 days, and in your lungs it may alpha decay to polonium and then lead. Less known is that tobacco phosphate fertilizers contain some radium, decaying to radon, polonium, then lead. A pack-a-day cigarette smoker is exposed to an extra 12 mSv per year, compared to normal background radiation of about 2 mSv/year.

Ionizing radiation can damage cells.

Molecular binding energies are of the order of 1 eV, and ionizing energy over 10 eV can strip an electron from a molecule making an ion or free radical, which is chemically reactive. Normal cellular metabolism is the principal source of reactive oxygen molecules as such as hydrogen peroxide. Enzymes normally convert these reactive oxygen molecules back to oxygen and water, but excess amounts of residual peroxides can damage DNA, RNA, and proteins.

DNA damage and repair occurs frequently, about once per cell per second, for each of the 100 million million cells in the human body. The overwhelming damage source is normal metabolism, not ionizing radiation.

Radioactivity is measured by counting decays.

Each familiar click of a radiation counter results from one nuclear decay that releases an alpha, beta, or gamma particle (although most detectors only count gammas). A Becquerel (Bq) is the count per second of decays of a source. For example, the potassium-40 in a banana beta-decays at about 20 Bq – 20 counts per second. Becquerels count the activity of a source. A bunch of ten bananas might have an activity of 200 Bq. Hardly any of the electrons from the potassium-40 beta decay exit the banana, but when you eat the banana its potassium may end up in a molecule in a cell in your body. It's natural and normal.

The table on the next page gives some examples of activity, measured in Becquerels.

Radioactivity examples, nuclear decays per second (Bq)	
Bq	Radioactivity source
20	one banana
100	Japan max iodine-131/liter for babies drinking water
300	Japan maximum iodine-131/liter for drinking water
650	one cubic meter of typical soil (500 Bq from K-40)
740	EPA maximum tritium/liter for drinking water
1,000	one kg of coffee
1,000	one kg of granite (such as a kitchen countertop)
2,000	one kg of coal ash
2,000	Japan maximum iodine-131/kg of fish and vegetables
3,000	radon in a 100 square meter Australian home
3,000	IAEA maximum iodine-131/liter for drinking water
5,000	one kg superphosphate fertilizer
7,000	human adult (70 kg)
7,000	Canada (Ontario) max tritium/liter for drinking water
10,000	Switzerland maximum tritium/liter for drinking water
30,000	household smoke detector with americium
30,000	radon in a 100 square meter European home
500,000	one kg uranium ore (Australian, 0.3%)
925,000	tritium in one wristwatch
1 million	one kg of low-level radioactive waste
25 million	one kg of uranium ore (Canadian, 15%)
70 million	radioisotope for medical diagnostic purposes
4 billion	Iodine-131 source for thyroid cancer treatment
1,000 billion	one luminous EXIT sign with tritium (1970s)
10,000 billion	one kg of 50-yr-old vitrified reactor nuclear waste
100,000 billion	radioisotope source for medical treatment
.4 billion billion	Fukushima
3 billion billion	500 historical atmospheric nuclear weapons tests
4 billion billion	Chernobyl

Radiation dose is measured in energy units.

A radiation dose is a measure of the energy deposited in biomass by ionizing radiation. The energy is measured in joules. (One joule is one watt-second.) A joule is 6×10^{18} eV, which is a lot of energy compared to the typical 1 eV chemical bond.

Because more flesh absorbs more radiation, the dose is stated in joules per kilogram of biological mass, J/kg. That dose unit is named a gray; one Gy = 1 J/kg. A kilogram of biomass has on the order of 10^{22} atoms, but 1 Gy = 6 x 10^{18} eV is still a lot of energy to spread around 10^{22} atoms. 1 Gy is a big dose.

The Bq measures the activity of a _source_. The Gy measures the energy _absorbed_ by a biomass.

Heavy alpha particles have twenty times the effect of beta or gamma particles of the same energy. The "effective dose" is twenty times the absorbed dose for alpha particles; for beta and gamma particles it is the same. The effective dose unit is the sievert, in the same units as gray, J/kg.

So for gamma and beta radiation, 1 Sv = 1 Gy. For alpha radiation 1 Gy = 20 Sv. Doses are most often reported in sieverts, usually the same as grays. Alpha radiation does not penetrate skin epidermis to living tissue, so the distinction is only important for inhalation or ingestion of radioactive alpha-decaying materials.

Since 1 Sv is a large dose, the examples in the table on the following page are in millisieverts, mSv. The number in the last row is very large because the radiation is concentrated on and absorbed by a very small biomass.

Absorbed energy dose examples, millisieverts	
Dose, mSv	Cause
0.001	one 10-sec airport backscatter wave scan
0.007	one bitewing X-ray, F-speed film
0.010	living near a nuclear plant for one year
0.014	one dental X-ray, Panorex, digital
0.02	sleeping next to another person for one year
0.03	one 6-hour airplane cross-country flight
0.04	eating one banana per day for a year
0.05	living at nuclear plant perimeter fence one year
0.1	living in a brick house for one year
0.1	skull X-ray
0.2	chest X-ray
0.3	mammogram
1	abdomen X-ray
1.5	EPA maximum for an uninformed adult for one year
2	airline crew member, short flights for one year
3	one head CT scan
4	cross-country airline crew member, 900 hours/year
6	one barium X-ray
10	cooking with natural gas (radon) for a year
10	one full-body CT scan
9	airline crew member on polar flights, 900 hours/year
13	smoking one pack of cigarettes per day for a year
20	nuclear plant worker, max 5-year annual average
50	cardiac catheterization, coronary angiogram
50	nuclear plant worker, max total exposure in 1 year
100	lowest clearly carcinogenic level
150	whole body dose from I-131 cancer therapy
250	temporary sterility in men
500	nausea, fatigue within hours
750	vomiting and hair loss in 2-3 weeks
1500	in one hour, 0 to 5% fatal
4000	In one hour, immediate severe skin burns, 50% fatal
20,000	in one hour, 100% fatal
400,000	thyroid dose from I-131 therapy

LNT theory warns that any radiation is dangerous.

A very protective model of the health effects of ionizing radiation is the Linear No Threshold (LNT) theory. It states that halving the effective dose halves the risk of cancer, and that no matter how much radiation is reduced there is no safe threshold below which ionizing radiation is safe. In 2005 the National Academy of Sciences published this controversial report, Biological Effects of Ionizing Radiation VII (BEIR).

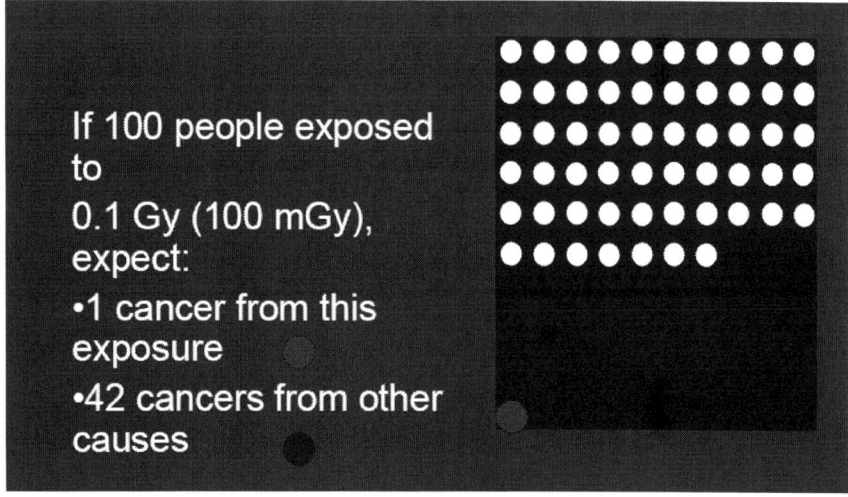

If 100 people exposed to 0.1 Gy (100 mGy), expect:
- 1 cancer from this exposure
- 42 cancers from other causes

LNT, linear no threshold theory

This statement says 100 mSv of radiation creates a 1% lifetime risk of cancer. The death rate from cancer is about 50%, so this also says that 100 mSv creates a 0.5% death risk for one person, or that 1 mSv creates an 0.005% death risk. By LNT logic the 815 billion airline-passenger-miles traveled annually in the US increases cosmic ray exposures resulting in 10,000 passenger-sieverts that kill 500 people per year. Nuclear power opponents similarly multiply small exposures by large populations, claiming that an accident that exposed 100 million people to 1 mSv (half of background radiation) would kill 5,000 people.

I find the 400 page BEIR report difficult to read and not really persuasive. There is little observational data in it. It is largely a discussion of many other published research papers and radiation effects models.

However, LNT is the theory that guides NRC and EPA. Under LNT cancer risk is linearly proportional to exposure. Typical exposures and LNT-derived risks are plotted on the following graph. Note it is a log-log scale covering 6 orders of magnitude.

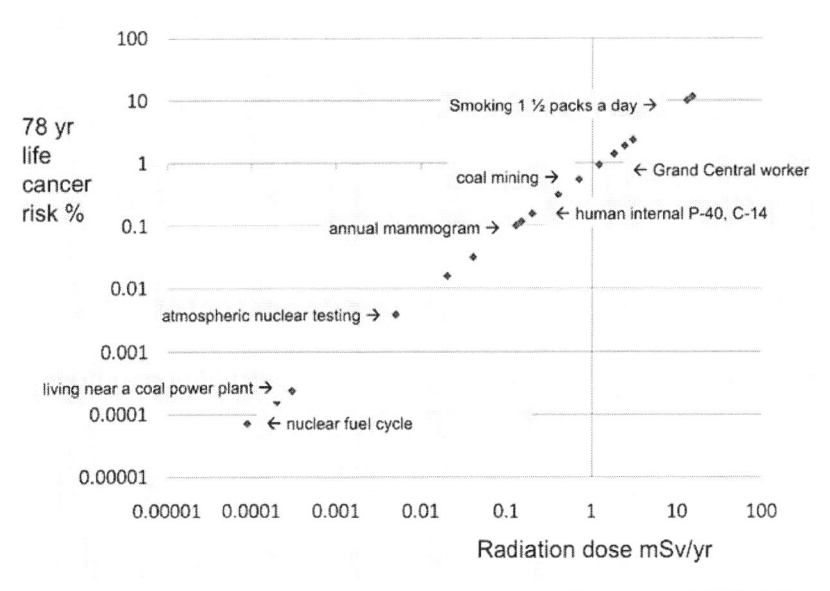

Cancer risk from radiation exposure predicted by BEIR VII

For example, a worker in Grand Central Station is exposed to an extra 1.2 mSv per year from the radiation from the granite structure and has an extra 1% lifetime chance of getting cancer.

An issue with LNT is that it is difficult to verify, as illustrated on the following graph. Most of the dose examples are well below the 2.4 mSv/year background radiation dose – in the noise. Similarly most of the cancer risks are well below the 42% lifetime cancer incidence – in the noise. The shading represents the area of health interest, precisely where the experimental noise from background radiation and normal cancer incidence masks the relationship.

SAFETY

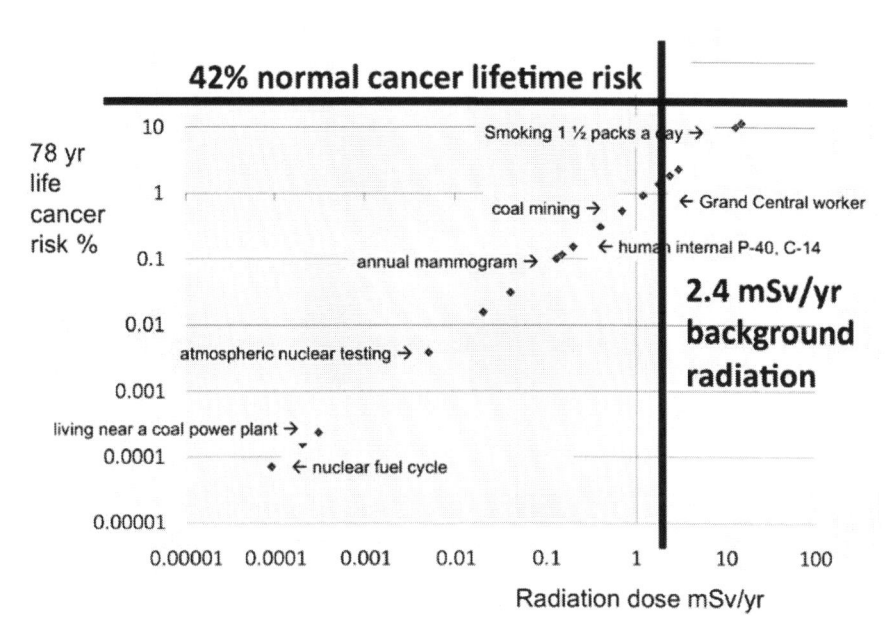

42% normal cancer lifetime risk

(Chart: 78 yr life cancer risk % vs. Radiation dose mSv/yr)

- Smoking 1 ½ packs a day →
- ← Grand Central worker
- coal mining →
- ← human internal P-40, C-14
- annual mammogram →
- **2.4 mSv/yr background radiation**
- atmospheric nuclear testing →
- living near a coal power plant →
- ← nuclear fuel cycle

Background radiation and cancer risk compared to exposures

Everyday life activities bring a risk of death.

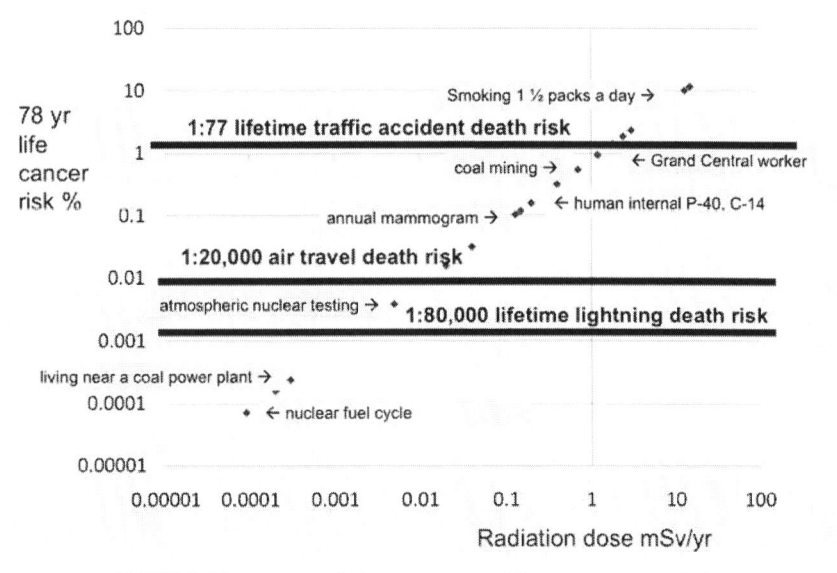

(Chart: 78 yr life cancer risk % vs. Radiation dose mSv/yr)

- Smoking 1 ½ packs a day →
- **1:77 lifetime traffic accident death risk**
- coal mining →
- ← Grand Central worker
- annual mammogram →
- ← human internal P-40, C-14
- **1:20,000 air travel death risk**
- atmospheric nuclear testing → **1:80,000 lifetime lightning death risk**
- living near a coal power plant →
- ← nuclear fuel cycle

BEIR VII cancer risk compared to everyday risks

One in 77 of people in the US end their lives in a traffic accident, yet no one suggests banning driving because transportation is critical to our civilization. What level of risk should be permit for generation of electric power that is critical to our civilization?

Life is fatal. Everyone dies. Every activity runs a risk of injury or death. Every breath brings a risk of DNA breaks by reactive oxygen molecules. Every moment runs a risk of spontaneous cancer. Every heartbeat runs a risk of stroke. What is reasonable risk?

EPA tries to balance regulatory costs and values.

To compare a regulatory cost to the benefit EPA assigns a value to a life saved. The EPA used $7.9 million as economic value of a statistical life. For example, if a proposed new rules for automobile crash resistance would save 1,000 lives per year then society should be willing to require the auto industry to spend $7.9 billion a year for improved crash resistance. What should society be wiling to spend for nuclear power safety?

Paul Slovic in the Bulletin of the Atomic Scientists reports research by Tengs that we spend $69 per life-year saved by seatbelts, but $100 million per life-year saved by some nuclear power radiation emission controls. Our society's approach to balancing risk and cost is wildly inconsistent – by a factor of a million.

David Ropeik and Stephen Levitt differentiate statistical risk and perceived risk, which unduly influences regulators and lawmakers. We perceive a high threat to children from keeping a gun at home, but in homes with both a gun and a swimming pool a child is a hundred times more likely to drown in the pool than die of a gunshot.

Fear saves lives.

In the debate about fear versus reason, we need to understand and value the life-protecting value of fear. The human animal survives because of its quick, unthinking response to a perceived threat, such as a moving snake or an oncoming truck. Fear triggers the fight-or-flight-or-freeze response. Reason evolved much later than did the brain's amygdala with its control over adrenalin, heart rate, visual awareness, and blood flow.

Fear is not so helpful for making decisions about complex issues such as nuclear waste storage. These require rational thought of a calm, clear mind. But when emotions of fear are triggered the subconscious mind's fast signaling pathways flood consciousness and reasoning ability.

Frank Furedi, at the University of Kent, cites fear of genetically modified crops, mobile phones, global warming, and foot-and-mouth disease. He argues that perceptions of risk, ideas about safety, and controversies over health, the environment and technology have little to do with science or empirical evidence.

Fear sells.

Rational marketers know how to make good use of emotional fear. Fear for child safety lets Verizon Wireless sell tracking of kids' cellphone locations, though child abductions have not increased. Fear of household invasion is raising sales of household alarms and security devices, though crime rates have fallen. Fear of disease sells unnecessary full body CT scans, reporting pseudo diseases and instigating unnecessary tests that raise health care costs.

Politicians, too, use fear. Bush used fear of Hussein's atomic bomb development to rally support to invade Iraq. Former National Security Advisor Zbigniew Brzezinski argued that the use of the term *War on Terror* was intended to deliberately generate a culture of fear, because it "obscures reason, intensifies emotions and makes it easier for demagogic politicians to mobilize the public on behalf of the policies they want to pursue." Furedi continues, "Politics has internalized the culture of fear. So political disagreements are often over which risk the public should worry about the most."

In the debate about the Vermont Yankee nuclear power plant politicians have used fear this way. Arousing fears of Vermont

Yankee enabled politicians to posture themselves as public saviors, focus on a single topic, and get elected. Yet rational thought concludes this power plant is the safest, carbon-free, least expensive source of electricity Vermonters can have.

Fear causes flight to the perceived safety of nature.

Fear causes flight -- back to nature -- back to the forest and the Tree of Souls in Pandora in the movie *Avatar*. Producer David Cameron knew his audiences' emotions when he designed this all time best-selling film grossing over $2.7 billion.

Frightened people think that natural power – from windmills, waterfalls, sunlight, and wood-burning -- will save civilization. The arousal of radiation fear and the flight to renewable, natural energy sources obscures the facts about their risks, costs, and environmental impacts. Even if green, renewable, natural energy is the ultimate goal of Vermonters and Germans, their fearful flight from nuclear radiation makes them thoughtlessly accept burning more natural gas and coal, temporarily, even if it does emit more CO_2 and pollution.

Here is an emotional conflict -- fear vs nature. Einstein said we can't solve problems by using the same kind of thinking we used when we created them. We need to encourage use of a higher level of consciousness -- rationality.

We need to encourage people to identify fear-arousing messages, and to consider their sources and content. Do they come from scientists, engineers, regulators, or radiation oncologists? Do they have numbers? What are the costs? What risks are acceptable? One in 77 people end their lives in traffic accidents, yet we accept this death rate as the price for driving. Can we be rational about the risks, costs, and environmental impacts of nuclear power?

The Linear No Threshold theory is disputed.

Many people dispute the validity of the LNT model. Here are short forms of some arguments against LNT.

- People living at high altitudes absorb about 1 mSv/year more radiation but exhibit no more cancer.
- People living in places with 5 times normal background radiation exhibit no more cancer.
- Radiation therapy to destroy cancer is not given as a single acute dose but multiple smaller doses to allow nearby exposed healthy tissue time to recover.
- The observed death rates for highly-exposed Chernobyl heroes are not linear: 2% @ 2.5 Sv and 33% @ 5 Sv.
- Workers in the nuclear industry have less cancer.
- Residents of a Taiwan building with steel contaminated by radioactive cobalt-60 had fewer cancers.

The United Nations Scientific Committee on the Effects of Atomic Radiation generally supports LNT but also says "a strictly linear dose response should not be expected in all circumstances." The French Academy of Sciences "doubts on the validity of using LNT for evaluating the carcinogenic risk of low doses (< 100 mSv) and even more for very low doses (< 10 mSv)". In the US the Health Physics Society "recommends against quantitative estimation of health risks below an individual dose of 50 mSv in one year or a lifetime dose of 100 mSv above that received from natural sources". The American Nuclear Society says "Below 100 mSv (which includes occupational and environmental exposures) risks of health effects are either too small to be observed or are non-existent."

The late Bernard Cohen, Professor of Physics at the University of Pittsburgh, has actively opposed LNT in many of his writings, which have had wide audiences.

Another physicist, Wade Allison, writes clearly about this in his book, *Radiation and Reason*, which includes raw data about the Hiroshima and Nagasaki atomic bomb survivors' cancer incidences.

Leukemia deaths were not affected by radiation < 200 mSv.

mSv step to..	Survivors	Survivor deaths	Control deaths
<5	37,407	92	84.9
100	30,387	60	72.1
200	5,841	14	14.5
500	6,304	27	15.6
1,000	3,963	20	9.5
2,000	1,972	39	4.9
>2,000	737	25	1.6

For Hiroshima and Nagasaki atomic bomb survivors, the absorbed radiation range steps in the left column are 0-4, 5-99, 100-199, etc. Besides survivors, the researchers selected a control group of 25,580 people who lived in Japan outside the bombed cities. Their leukemia death rates were normalized to the same number in the Survivors column, so the right column exhibits the expected number of deaths.

Solid cancer deaths were not affected by radiation < 100 mSv.

mSv step to..	Survivors	Survivor deaths	Control deaths
<5	38,507	4,270	4,282
100	29,960	3,387	3,313
200	5,949	732	691
500	6,380	815	736
1,000	3,426	483	378
2,000	1,764	326	191
>2,000	625	114	56

This table exhibits a similar result. Absorbed radiation exposures under 100 mSv did not lead to more solid cancer deaths.

These radiation exposures < 100 mSv took place in an instant, acutely. In contrast, chronic radiation exposures over months allow time for DNA repair mechanisms to act. Chronic exposure should lead to less damage and fewer cancers than acute exposure.

Residents of Fukushima will not exhibit extra cancers.

As many as 20,000 people died from the 2011 earthquake and tsunami near Fukushima, though none of these deaths are attributable to radiation from the damaged nuclear reactors. Prof. Robert Gale of Imperial College estimated radiation exposure to nearby residents. The workers gaining control of the damaged Fukushima power plants were exposed to an average of 9 mSv, with 37 workers receiving doses over 100 mSv, increasing their lifetime cancer risks by 1-2%.

mSv step to ...	People exposed
< 1	5800
10	4100
20	71
23	2

John Boice, president of the National Council on Radiation Protection and Measurements, said "The exposures to the population are very, very low, As such, there is no opportunity to conduct epidemiological studies that have any chance of detecting excess risk. The doses are just too low." Despite this, the Japanese government will undertake large studies to "reduce anxiety and provide assurance to the population" [?!]. These studies include:

- A 10-page questionnaire for all 2 million residents in the Fukushima prefecture, with a 30-year follow-up study.
- 360,000 children under the age of 18 having their thyroid glands scanned.
- A health exam of people in the proximal area, including blood exams.
- A special survey of 20,000 pregnant and nursing mothers.

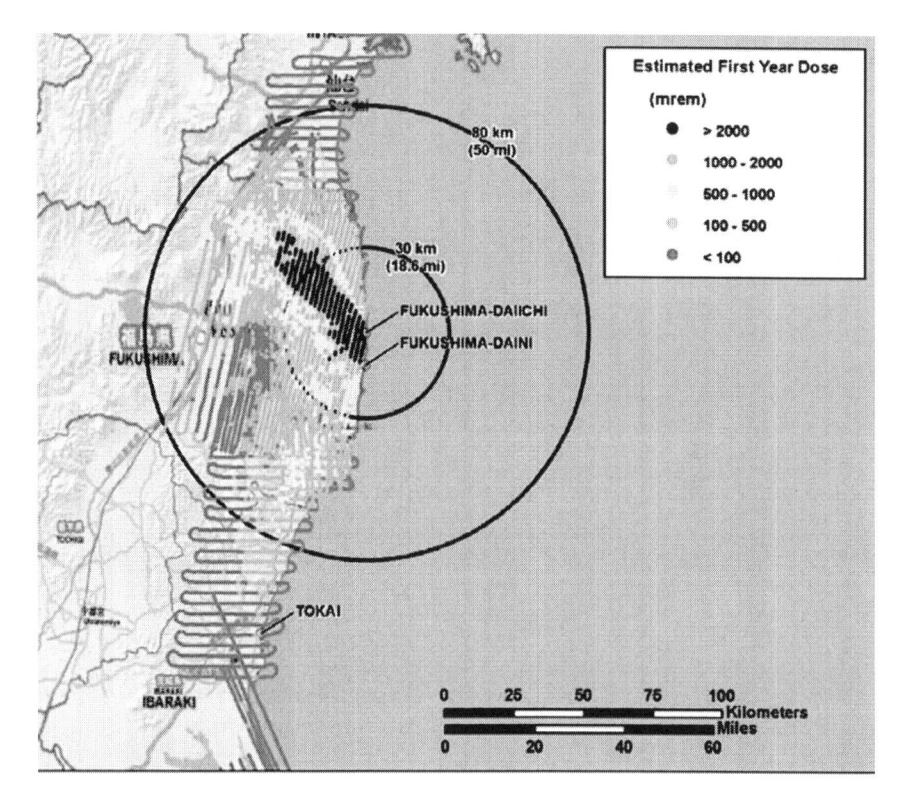

The US National Nuclear Security Administration monitored radiation at Fukushima and produced this map of total first year absorbed radiation. The darkest band of lines northwest of the Fukushima-Daichi plant indicate areas of radiation exceeding 20 mSv (2000 mrem in US units). The map is consistent with the table above. Jerry Cuttler documents the mistake of evacuations from the entire area, where radiation levels nowhere exceeded 680 mSv/yr, a dose rate once considered safe for medical X-rays.

Low-dose radiation research programs contradict LNT theory.

With the controversy about global warming, nuclear power, and the LNT model of health effects of low-level radiation, it would seem important to understand this better. For example, excessive radiation protection standards may unnecessarily be costing extra billions of dollars for site cleanup at the World War II Hanford weapons plutonium production facility. We are also in an age where we may need to respond sensibly to a future terrorist's dirty bomb, rather than abandoning whole cities.

Until 2012 the US had funded research for the DOE Office of Science Low Dose Radiation Research Program. Previous research and educational materials are still available on their website lowdose.energy.gov and links to new research by others are currently maintained there. Here's an experiment illustrating nonlinear response to radiation.

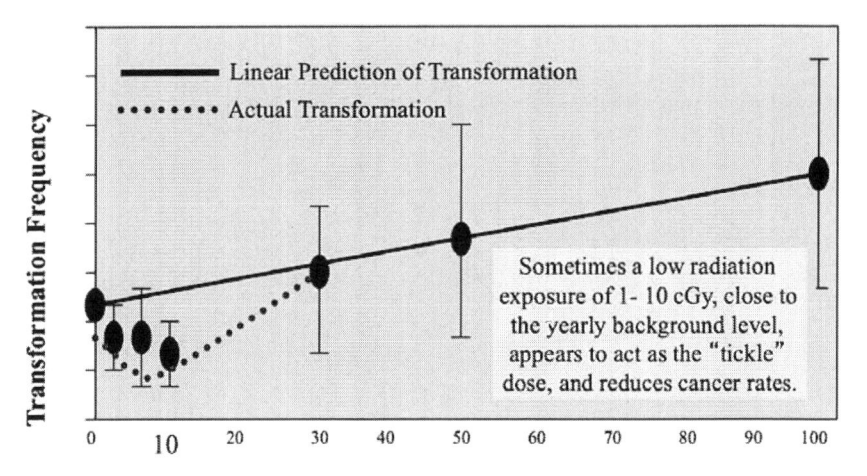

[Chart: Transformation Frequency vs dose (0-100 cGy), showing Linear Prediction of Transformation (solid line) and Actual Transformation (dotted line). Annotation: "Sometimes a low radiation exposure of 1- 10 cGy, close to the yearly background level, appears to act as the 'tickle' dose, and reduces cancer rates."]

Observed cancers at low doses (cGy) vs linear extrapolation

New Mexico State University's Low Background Radiation Experiment took place 2000 feet underground, excluding almost all radiation from cosmic sources, the sun, and radioactive minerals. Preliminary results showed that bacteria growth is inhibited by the lack of radiation.

In December 2011 Lawrence Berkeley National Laboratory reported research that actually observed more rapid repair of DNA strand breaks at lower levels of radiation. Exposure to 100 mSv created 4 times as many repair sites as exposure to 1000 mSv. "Our data show that at lower doses of ionizing radiation, DNA repair mechanisms work much better than at higher doses," says Mina Bissell, a world-renowned breast cancer researcher with Berkeley Lab's Life Sciences Division. "This non-linear DNA damage response casts doubt on the general assumption that any amount of ionizing radiation is harmful and additive."

A 2012 study by Engelward and Yanch at MIT exposed mice to prolonged radiation exposure at a rate of 100 mSv/year. After 5 weeks (about 10 mSv) of exposure the researchers tested for several types of DNA damage, including DNA strand breaks and base lesions. No significant increases were found. DNA damage and repair occurs spontaneously and naturally at a rate of about 10,000 per cell per day, increased by only 12 per day from the 100 mSv/year exposure. Previous studies have shown some DNA damage from a single, acute 10 mSv dose, stressing the cells' inherent DNA repair mechanisms. But the chronic 10 mSv dose showed no damage.

Radiation protection guidelines based on acute doses are far too conservative, needlessly displacing thousands of people from areas near Fukushima where radiation levels are well below the safe level of 100 mSv/yr.

Cobalt-60 radiation reduced cancer in Taiwan.

Recycled steel, accidentally contaminated with radioactive cobalt-60, was used in construction of apartment buildings in Taiwan. Over 20 years 8,000 people were exposed to an average of 400 mSv of radiation. The health effects were positive! Is chronic radiation is an effective prophylaxis against cancer?

Natural, predicted, and observed cancers for 8,000 people		
Normal	LNT predicted	Observed
186	242	5

Low dose radiation hormesis may protect against high doses.

The Taiwan experience and Berkeley Lab's research may even support a phenomenon called hormesis, a cellular defensive response to damage, sometimes termed adaptive response. The International Dose-Response Society at the University of Massachusetts School of Public Health runs an annual conference and publishes a journal on biological response to small doses of chemicals, drugs, and radiation. For example, Krzysztof Fornalski's article *The Healthy Worker Effect and Nuclear Industry Workers* analyzes whether the workers are healthier from radiation exposure or from better health care.

The adaptive response to low-level ionizing radiation may have developed during evolution. Life first appeared on earth 3 billion years ago, when natural background radiation was about 10 mSv/year -- 4 times what it is today.

Radiophobia is harmful.

Japan's government may be exaggerating radiophobia by shutting all nuclear reactors and undertaking massive health surveys. 80,000 people within 30 kilometers of the plant were evacuated, regardless of radiation intensity. Only 16,000 were allowed to return by March 2012, a year after the accident. In the Fukushima evacuation area 10 people died in vehicles evacuating them from hospitals. Officials certified 573 deaths as disaster-related, defined as not directly caused by a tragedy, but by fatigue or the aggravation of a chronic disease due to the disaster.

Zbigniew Jaworowski, former chair of the United Nations Scientific Committee on the Effects of Atomic Radiation, wrote that current standards for radiation protection are unethical because they needlessly caused psychosomatic disorders in 15 million people in Belarus, Ukraine, and Russia after the Chernobyl accident. They lead to hundreds of billions of dollars wasted on unnecessary radiation protection from nuclear power. He proposed reasons for the radiophobia.

1. The psychological reaction to the devastation and loss of life caused by the atomic bombs dropped on Hiroshima and Nagasaki at the end of World War II.
2. Psychological warfare during the cold war that played on the public's fear of nuclear weapons.
3. Lobbying by fossil fuel industries.
4. The interests of radiation researchers striving for recognition and budget.
5. The interests of politicians for whom radiophobia has been a handy weapon in their power games (in the 1970s in the US, and in the 1980s and 1990s in eastern and western Europe and in the former Soviet Union).
6. The interests of news media that profit by inducing public fear.
7. The assumption of a linear, no-threshold relationship between radiation and biological effects.

Jaworowski's proposal is simple. Raise limits on public radiation exposure from 1 mSv/yr to 10 mSv/yr. This is a tenth of the level at which any health issues have been observed. Note that 10 mSv/yr is also the level of radiation at the start of life's evolution over 3 billion years ago -- good evidence of 10 mSv/yr harmlessness to life and genetic evolution.

Unreasonably low radiation limits injure people.

Over 80,000 people who lived within 20 km of the Fukushima power plants were evacuated and not permitted to return to their homes. There is a human cost to creating so many displaced persons. People moving to polluted cities breath unhealthy air. Suicide rates increase. There should be some balance between the risks of displacement and the risks of radiation to human health.

Standards of the ICRP, the International Commission on Radiological Protection, are guided by "as low as reasonably achievable" (ALARA). ICRP standards guide national standards. ALARA is not a reasoned safety level determined by observation. In our world facing global warming, air pollution, and resource contention, excluding clean, safe, economical nuclear energy will cause tremendous injury. Wade Allison proposes a limit set "as high as relatively safe". He suggests 100 mSv/month to initiate discussion of studies of safety-guided radiation limits.

American Nuclear Society documents LNT fallacies.

In June 2012 the American Nuclear Society held a special session on the effects of low-level ionizing radiation and published a compendium of papers and references that disprove the linear no threshold (LNT) theory of health effects of ionizing radiation. The publication includes internet links and also reprints of papers. A link to this large file is in the reference section of this book.

No mainstream media published an account of this report or meeting. Journalists publish articles that arouse fear to gain attention to sell more advertising. Safety is boring. Fear sells.

Waste

Nature safely buried its own nuclear reactor waste in Gabon.

Uranium-235 has a half life of 700 million years, so 1,700 million years ago Earth's concentration of U-235 in natural uranium was close to 3%, rather than 0.7% as it is today. By chance, enough uranium ore was then concentrated in sandstone in Oklo, Gabon, Africa, to create a natural nuclear fission reactor.

Natural nuclear fission reactor site

Groundwater H2O provided the hydrogen moderator that slowed neutrons so they fissioned the U-235. The heat changed the water to less dense steam, so the moderation decreased and the reactor stopped until the water cooled and returned. The three-hour on-off cycling persisted for more than 100,000 years, while the natural reactor generated 100 kW of power. A total of 16 such natural fission reactor sites were identified. The fission products have remained localized at the sites of the reaction zones for over a billion years.

US military nuclear waste is safely buried underground.

The US Waste Isolation Pilot Plant (WIPP) near Carlsbad, New Mexico has been in operation since 1999. It has accepted over 10,000 shipments of transuranic materials left over from US research and production of nuclear weapons.

The radioactive materials are stored 600 meters underground, within a 1000 meter thick salt formation.

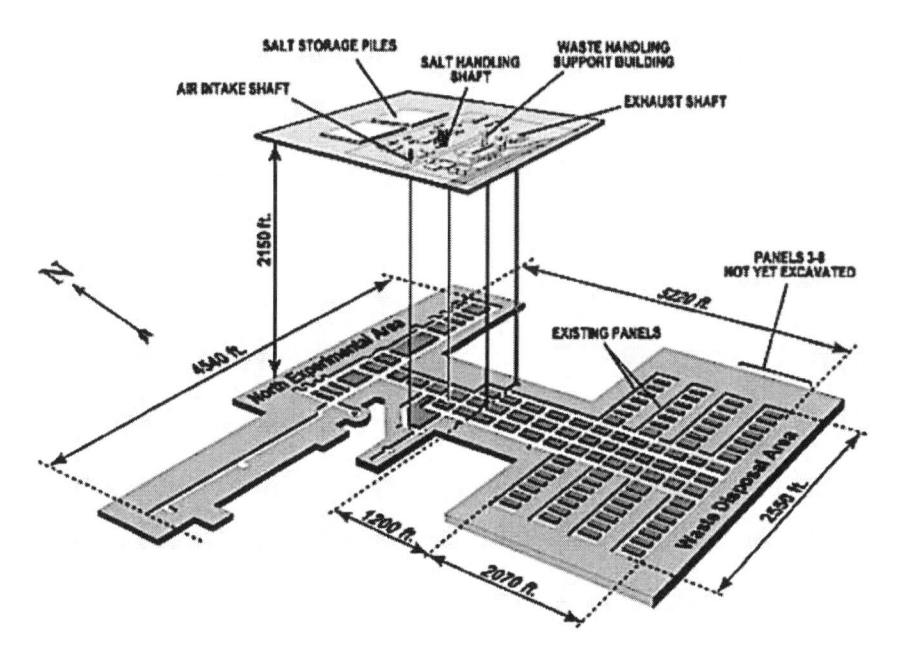

New Mexico waste isolation pilot plant

The NaCl salt is somewhat plastic; cracks and holes close up, stopping water flow. Eventually the hollowed out caverns containing the waste will be engulfed by the salt, permanently sequestering the stored materials.

Other operational underground radioactive materials storage sites exist in Finland, Germany, and Sweden. In planning are underground sites in Argentina, Belgium, Canada, China, France, Germany, Japan, Korea (under construction), Switzerland, United Kingdom, and USA (Yucca Mountain).

Long-lived waste may be sequestered in deep boreholes.

LFTR will produce less than 1% of the long-lived radioactive waste of current LWRs. Molten salt reactors can be used to reduce existing radioactive materials in stored spent LWR fuel rods. But in all scenarios some long-lived radiotoxic materials must be sequestered from the environment.

Per Peterson, a member of the President's Blue Ribbon Commission of the Future of Nuclear Power, corresponded with me on this subject. Land-based deep geologic disposal appears to be the most practical approach. Deep seabed disposal could work from a technical perspective, but the legal issues are sufficiently difficult that land-based approaches are more attractive. Multiple options for geologic disposal meet highly protective long-term safety standards at affordable costs.

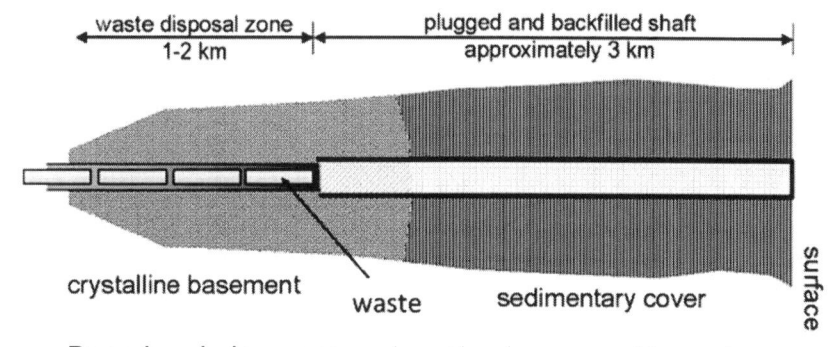

Deep borehole permanent waste storage, not to scale

Sandia National Laboratory reports that 70% of the U.S. has geologic conditions appropriate for deep borehole disposal -- crystalline basement rock over one billion years old within 2 kilometers of the ground surface. Borehole disposal has low fixed costs because it does not require the development, staffing, and maintenance of underground facilities. Sandia estimated that deep borehole disposal of spent fuel may cost $2.1 billion dollars per 10,000 tonnes, requiring approximately 3.2 square kilometers for 85 boreholes spaced 0.2 km apart.

LFTRs will generate about 1 tonne of waste per GW-year. Prorating Sandia's storage cost results in $210,000 per GW-year, or 0.002 cents/kWh. Today utility companies pay 0.1 cents/kWh into a government-managed nuclear waste disposal fund. If a fleet of LFTRs powered all of the 500 GW US electricity demand for a whole century, they would generate 50,000 tonnes of waste stored in deep boreholes beneath 16 square kilometers of land area.

Salt caverns are another option for geologic disposal. The local community in Carlsbad, New Mexico has extensive experience with salt gained at the WIPP facility, and has expressed strong interest in becoming involved in a wider variety of fuel-cycle related activities including disposal. Besides extensive salt deposits, the U.S. has extensive resources for granite and clay shale that also could be used to host geologic disposal facilities. In 2011 the US lacked a legal framework to develop such resources, but this will happen once Congress amends the Nuclear Waste Policy Act to implement the BRC's recommendations.

An alternative to disposing of existing LWR spent fuel is to reprocess it in molten salt reactors such as DMSR. The production of fuel for DMSRs from LWR spent fuel will be less expensive than fuel production and fabrication of TRISO fuel or LWR fuel. This could create a market for LWR spent fuel. The reprocessing might also recover platinum-group noble metal fission products, which would also be marketable.

Less geological storage is needed for LFTR waste.

LFTR reduces nuclear waste storage issues from millions of years to a few hundred years. The radiotoxicity of nuclear waste arises from two sources: the highly radioactive fission products from fission and the long-lived actinides from neutron absorption. Thorium and uranium fueled reactors produce essentially the same fission products, whose radiotoxicity in 500 years drops below even that of the natural uranium ore to power a LWR. Compared to LWRs, LFTRs create far fewer transuranics because Th-232 requires seven neutron absorptions to make Pu-239, whereas U-238 requires just one. After 300 years the LFTR waste radiation would be 10,000 times less. In practice, about 0.1% of the LFTR transuranics might pass through the chemical waste separator escaping into the waste stresam, so the LFTR waste radiotoxicity would be 1/1000 that from PWRs. Geological repositories smaller than Yucca mountain would suffice.

SAFETY

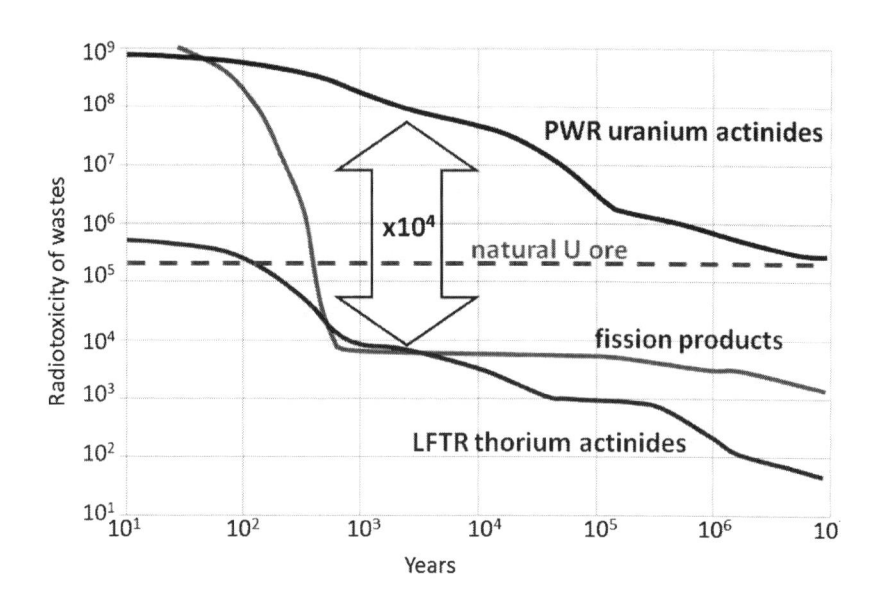

Radiotoxicity of waste from 1 GW(t) reactor

WEAPONS PROLIFERATION

The safety of world civilization may depend on limiting the spread of nuclear weapons, which can destroy whole cities. Many advocates of nuclear power are environmentalists who wish to protect civilization from global warming, pollution, and wars over resource contention. Such advocates would not take steps that might increase the spread of nuclear weapons to untrustworthy regimes that might initiate nuclear warfare.

Weapons arose from political ambitions, not nuclear power.

Even poor, developing nations such as India, Pakistan, and North Korea have obtained nuclear weapons. The acquisition process was not technology development seeded by commercial nuclear power; it was international politics. Weapons technologies have been transferred for the perceived political gain of the provider and/or the receiver.

Many people remain misinformed of the role of nuclear power in weapons proliferation. For example, former US Vice-President Al Gore said

> "During my eight years in the White House, every nuclear weapons proliferation issue we dealt with was connected to a nuclear reactor program. Today, the dangerous weapons programs in both Iran and North Korea are linked to their civilian reactor programs."

This statement is not correct. North Korea has no civilian reactor program. Iran's newly operational electric power generation reactor is fuelled by Russia, not by uranium from Iran's weapons-threatening centrifuge enrichment program.

The role of international politics is revealed in a book by Thomas C. Reed, of the Livermore weapons laboratory and former secretary of the Air Force, and Danny B. Stillman, former director of intelligence at Los Alamos. *"The Nuclear Express: A Political History of the Bomb and its Proliferation"* states that since the birth of the nuclear age no nation has developed a nuclear weapon on its own.

Weapons technology has been transferred from weapons-owning-nations to weapons-seeking-nations by international political processes.

1. The US cooperated with Canada and the United Kingdom in the Manhattan Project, helping the UK build weapons.
2. France gained atomic bomb knowledge through Manhattan Project veterans.
3. Russian spies from the Manhattan project enabled Stalin to build and explode an exact replica of the Nagasaki bomb.
4. China obtained information freely from Russia. Manhattan Project spy Klaus Fuchs provided Mao's program with details of the US weapon after Fuchs' release from prison in 1959.
5. China provided Algeria, Pakistan and North Korea with technical information about the atomic bomb.
6. Pakistan provided the weapon design to Libya and Iran.
7. India obtained an experimental nuclear reactor from Canada, heavy water from the US, and technical advice from France; India promised only peaceful use, then built the bomb.
8. Dozens of Israeli scientists participated in the French weapons program.
9. Israel provided cooperation and nuclear know-how to South Africa, which subsequently dismantled its nuclear weapons.

No nation other than the US has independently invented nuclear weapons. Nuclear power generation has never been a source for nuclear weapons. Nuclear power technologists will continue to make such a route difficult. LFTR is proliferation resistant. DMSR is even more so.

Advanced nuclear power must be proliferation resistant.

Nuclear weapons can cause terrible destruction of whole cities and contaminate entire regions, so expansion of nuclear power must come with assurances that the risk of proliferation of nuclear weapons is not increased. The technology for making such weapons is widely known, although the process is difficult and expensive. Building commercial nuclear power plants has not led to weapons development; nations that have nuclear weapons have developed them with purposeful programs and facilities. However dual-use technologies such as centrifuge enrichment of U-235 that can make fuel for PWRs can be adapted to make highly enriched uranium for weapons.

After President Eisenhower's *Atoms for Peace* speech the US helped nations to acquire the knowledge and materials to use nuclear technology for peaceful purposes. Unexpectedly this knowledge led India to develop nuclear weapons instead.

Selling advanced nuclear power plants worldwide does not require providing each nation with the technical skills and materials to build nuclear power plants or nuclear weapons. Consider the airplane and jet engine industry: nations want prestigious national airlines. Fully 83 countries, from Algeria to Yemen, operate airlines using the Boeing 747 airliner, yet these nations do not have their own airframe or engine production or maintenance capabilities. General Electric makes a business of maintaining and overhauling engines at GE's own service centers. This is a technology-transfer-resistant model suitable for LFTR installation and maintenance.

The liquid fluoride thorium reactor is proliferation resistant.

LFTR requires fissile material to be transported to the site for startup, but not thereafter. LFTR then creates and burns fissile U-233 that conceivably could be used instead for a nuclear weapon. Would this ever happen?

China, USA, Russia, India, UK, France, Pakistan, and Israel, which account for 57% of global CO_2 emissions, already have nuclear weapons and no incentive to subvert LFTR technology. So just implementing LFTRs in these nations would be a big step in addressing global warming. Many additional nations, such as Canada, Japan, and South Africa, have the capability to build nuclear weapons but have chosen not to, so there is no incentive for them to subvert LFTR technology for this purpose.

Should LFTRs be implemented in other non-weapons states? Certainly terrorists could not steal this uranium dissolved in a molten salt solution along with even more radioactive fission products inside a sealed reactor. IAEA safeguards include physical security, accounting and control of all nuclear materials, surveillance to detect tampering, and intrusive inspections.

LFTR's neutron economy contributes to securing its inventory of nuclear materials. Neutron absorption by uranium-233 produces about 2.4 neutrons per fission--one to drive a subsequent fission and another to drive the conversion of Th-232 to U-233 in the blanket molten salt. Taking into account neutron losses from capture by protactinium and other nuclei, a well-designed LFTR reactor will direct just about 1.00 neutrons per fission to thorium transmutation. This delicate balance doesn't create excess U-233, just enough to generate fuel indefinitely. If this conversion ratio could be increased to 1.01, a 100 MW LFTR might generate 1 kilogram of excess U-233 per year. If meaningful quantities of uranium-233 are misdirected for non-peaceful purposes, the reactor will report the diversion by stopping because of insufficient U-233 to maintain a chain reaction.

Yet a sovereign nation or revolutionary group might expel IAEA observers, stop the LFTR, and attempt to remove the U-233 for weapons. Accomplishing this would require that skilled engineers, working in a radioactive environment, modify the reactor's fluorination equipment to separate uranium from the fuel salt instead of the thorium blanket salt. What would happen to them?

The neutrons that produce U-233 also produce contaminating U-232, whose decay products emit 2.6 MeV penetrating gamma radiation, hazardous to weapons builders and obvious to detection monitors. The U-232 decays via a cascade of elements to thallium-208, which builds up and emits the radiation.

^{232}U (α, 72 years) → ^{228}Th (α, 1.9 year)
→ ^{224}Ra (α, 3.6 day, 0.24 MeV) → ^{220}Rn (α, 55 s, 0.54 MeV)
→ ^{216}Po (α, 0.15 s) → ^{212}Pb (β−, 10.64 h)
→ ^{212}Bi (α, 61 s, 0.78 MeV) → ^{208}Tl (β−, 3 m, 2.6 MeV)
→^{208}Pb (stable)

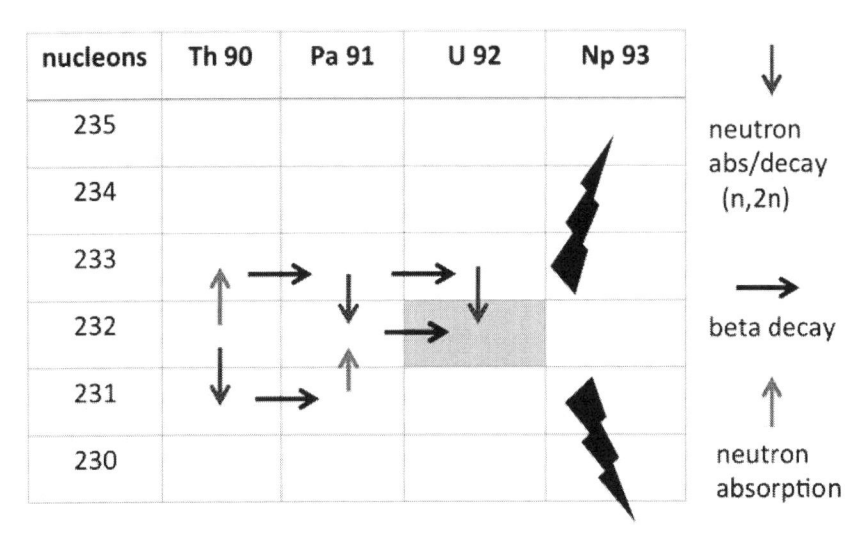

nucleons	Th 90	Pa 91	U 92	Np 93
235				
234				
233				
232				
231				
230				

↓ neutron abs/decay (n,2n)

→ beta decay

↑ neutron absorption

U-232 production in a thorium fueled reactor

Depending on design specifics, the proportion of U-232 would be about 0.13% for a commercial power reactor. A year after separation, a weapons worker one meter from a subcritical 5 kg sphere of such U-233 would receive a radiation dose of 43 mSv/hr, compared to 0.003 mSv/hr from plutonium, even less from U-235. Death becomes probable after 72 hours exposure. After ten years this radiation triples.

A resulting weapons would be highly radioactive and therefore dangerous to military workers nearby. The penetrating 2.6 MeV gamma radiation is an easily detected marker revealing the presence of such U-233, possibly even from a satellite.

U-232 can not be removed chemically, and centrifuge separation from U-233 would make the centrifuges too radioactive to maintain. Conceivably, nuclear experts might try to stop the reactor, chemically extract the uranium, and devise chemistry to remove the intermediate elements of the U-232 decay chain before the thallium is formed, except that the isotopes are continually replaced by U-232 decay. They might try to quickly separate the small amount of Pa-233 from the uranium and let it decay to pure U-233, but they would have to design and build a special chemical plant within the radioactive reactor. Bomb-makers might attempt quickly fabricate a weapon from newly separated U-233 before radiation hazards become lethal; even so there will be sufficient U-232 contamination that penetrating 2.6 MeV gamma rays will be readily detected. The challenge of developing and perfecting such new processes will be more difficult and expensive than creating a purpose-built weapons factory with known technology, such as centrifuge enrichment of U-235 conducted in Iran or PUREX for extracting plutonium from solid fuel irradiated in LWRs.

Bruce Hoglund wrote a fuller report of the challenges to would-be bomb makers, and there is a discussion in the comments of the energy from thorium blog, both linked in the references section.

A LFTR operating under IAEA safeguards might additionally be protected by injecting U-238 from a remotely controlled tank of U-238. The U-238 would dilute (denature) the U-233 to make it useless for weapons, but it would also stop the reactor and ruin the fuel salt for further use.

For personnel safety, any U-233 material operations must be accomplished by remote handling equipment within a radioactively shielded hot cell. This can be designed to make it very hard for any insiders or outsiders to remove material from the hot cell.

Another hurdle for the would-be pilferer of uranium from 700° C molten salt is the retained radioactive fission products. Even with a 1-hour cooling period to allow decay of the short-lived isotopes, the salt still releases ~350 W/liter of heat. That heat comes from deadly ionizing radiation that would kill a nearby pilferer in minutes unless shielded by meters of concrete or water or heavy lead. This fission product radiation is the same self protection that protects spent LWR fuel from theft.

The single-fluid DMSR is highly proliferation resistant.

The DMSR contains enough U-238 mixed with fissile U-233 and U-235 that the uranium can not sustain the rapid fission reaction necessary for a nuclear weapon. Uranium enriched to less than 20% U-235 is termed LEU, low-enriched uranium. The LEU fuel is not suitable for a nuclear weapon, which typically requires over 90% U-235. The DMSR with at least 80% U-238 is said to be denatured with it.

The DMSR has less chemical processing equipment than the two-fluid LFTR, which uses fluorine chemistry to direct U-233 generated in the thorium blanket to the core. The DMSR has no chemical processing equipment in the reactor plant that might somehow be modified to divert U-233 for a weapons program.

Because of the substantial amount of U-238 in the DMSR, it does breed plutonium from neutron capture, just as does a standard LWR. Some Pu-239 fissions. However the fissile Pu-239 isotope that might be desired for a weapon is only 31% of the plutonium, mixed with other isotopes (Pu-238, 240, 241, 242) that make the plutonium unsuitable for a weapon. Because the plutonium is dissolved in the fuel salt, there is no opportunity to remove it early to obtain weapons grade Pu-239 before neutrons convert it to other isotopes, as in a LWR, CANDU, RBMK, or military plutonium production reactor. Further, plutonium's chemistry makes it difficult to remove from the salt. Also, the salt contains highly radioactive fission products as well as U-232, whose decay daughters emit a penetrating 2.6 MeV gamma ray. DMSR is the most proliferation-resistant nuclear reactor.

There are easier paths than U-233 to make nuclear weapons.

Pakistan has illustrated how a developing nation can make uranium weapons using centrifuge enrichment; in a dual path it simultaneously developed the methods to extract weapons grade plutonium from uranium reactors. India and North Korea developed plutonium weapons from heavy water or graphite moderated reactors with online fuel exchange capability. Iran has built centrifuge enrichment plants capable of making highly enriched U-235 for nuclear weapons. These proven weapons paths eliminate the incentive for nations to try to develop nuclear weapons via the technically challenging and expensive U-233 path.

Only a determined, well-funded effort on the scale of a national program could overcome the obstacles to illicit use of uranium-232/233 produced in a LFTR reactor. Such an effort would certainly find that it was less problematic to pursue the enrichment of natural uranium or the breeding of plutonium.

LFTR reduces existing weapons proliferation risks.

Deploying LFTRs on a global scale will not increase the risk of nuclear weapons proliferation, but rather decrease it.

Starting up LFTRs with existing plutonium can consume inventories of this weapons-capable material.

The thorium-uranium fuel cycle reduces demand for U-235 enrichment plants, which can make weapons material nearly as easily as power reactor fuel.

Abundant energy cheaper than coal can increase prosperity and enable lifestyles that lead to sustainable populations, reducing the potential for wars over resources.

7 A Sustainable World

We can create a sustainable world by capitalizing on the benefits of safe, plentiful, inexpensive energy from the liquid fluoride thorium reactor. There are many ways that LFTR energy can substitute for energy we now derive from burning fossil fuels. This section covers the improvements that will lead to a sustainable world.

Electricity is the most valuable, useful energy source for advancing the health, safety, and prosperity of the world's civilizations. LFTR "energy cheaper than coal" not only dissuades nations from burning fossil fuels in fixed station power plants, it enables affordable electricity to developing nations who sorely need it for water processing, sanitation, food processing, communications, commerce, industry, and many other activities.

Oil is today essential to world transportation and commerce, because it is a liquid, energy-dense fuel that can be carried on board cars, trucks, trains, and airplanes. LFTR-provided cheap heat and power can help fabricate synthetic fuel substitutes – synfuels – that are affordable, with no net CO_2 emissions to the atmosphere.

Fresh, clean water is essential for improving the health of over a billion people lacking it and for growing food. Pumping, distributing, and processing water and waste water consumes 8% of the world's energy. LFTR can provide this energy for the rest of the world's people, and LFTR heat and electricity can replace and expand desalinization, now accomplished from burning fossil fuels.

COAL POWER REPLACEMENT

Taiwan's Taichung 4.4 GW coal power plant is the world's largest. The world's 1200 largest coal plants are together responsible for 30% of all global warming CO2 emissions.

LFTR can zero world coal power plant emissions.

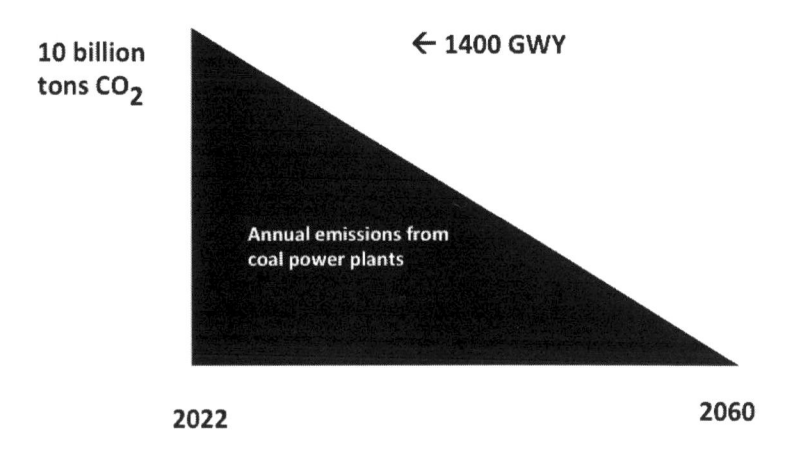

10 billion tons CO_2

← 1400 GWY

Annual emissions from coal power plants

2022   2060

LFTR-reduced CO2 emissions from coal power plants

Each year's coal-fired electricity production adds approximately 10 Gt of CO2 to the atmosphere. Coal-to-LFTR replacement can eliminate the single largest global source of this gas that drives global warming. Daily production of 100 MW LFTR power plants can replace all the world's coal power plants by 2060.

Ending these emissions will slow and eventually stop ocean CO2 absorption and acidification, harmful to sea life and food supplies.

The US Environmental Protection Agency estimates that 34,000 US lives could be saved annually by stopping particulate air pollution from coal power plants. China could save hundreds of thousands of lives by stopping its coal burning.

Thorium: energy cheaper than coal is the economic key to dissuading 9 billion people in 250 nations from burning coal.

SHIPPING

LFTR can power commercial ships.

Powering ocean cargo vessels with LFTR electric power will eliminate global oil demand of 7 million barrels per day and eliminate 4% of man-made greenhouse gas emissions. Nuclear power is successfully used today to power navy submarines, ice breakers, and aircraft carriers. The first ever use of nuclear power was to power the submarine USS Nautilus on and in the ocean. Since 1955 the US Navy has accumulated 5,400 reactor years of accident-free experience with its nuclear power plants. Nuclear-powered commercial shipping is a low-hanging-fruit opportunity.

Reducing the cargo space occupied by tanks for 380 tons of fuel for every day at sea will increase paying cargo. LFTR energy cheaper than coal is also cheaper than from the asphalt-like, refinery residues burned for fuel, reducing operational costs. The elimination of frequent refueling not only ends refueling delays but also allows ships to plan shipping routes without refueling port constraints.

The largest container ship in operation in 2012 has a 90 MW power plant, close to the 100 MW size of the small modular LFTR example. The largest, Nimitz-class super-carrier has a 200 MW nuclear power plant.

Just as the shipping industry changed from coal power to oil power, it can change from oil power to LFTR power.

Oil

The world is running out of cheap, pumpable oil. The extractive industries now turn to unconventional oil sources that require more energy for extraction and refining.

Postponing peak oil lowers EROI and raises CO2 emissions.

Peak oil is the posited time when oil consumption exceeds new discoveries, predicting the impending time when the world runs out of oil. Peak oil is postponed each time a new technology uncovers a new or unconventional petroleum source, but extraction requires increasing amounts of energy and increases emissions of CO2.

Hydraulic fracturing has been successfully used to extract natural gas from "tight" shale that was impervious to methane flow. Fracking is also beginning to be used similarly to extract tight oil and this might replace a quarter of US imported oil.

Canadian oil sands contain bitumen, which is mined and then heated and upgraded with natural gas. This process requires much more energy than simply pumping oil; the energy comes from natural gas that is burned, emitting more CO2 into the atmosphere. The resulting low grade oil is refined more expensively, with an EROI of just 4. Already 10% of US imported oil comes from these oil sands. Increased demand is the impetus for constructing the controversial Keystone XL pipeline from Canada to the US.

South Africa's SASOL coal-to-liquids plants already produce 150,000 barrels of oil per day, about 35% of national consumption. However the energy required comes from burning more coal, so that the total CO2 emissions from burning gasoline produced this way are about 50% more than with traditional pumped oil. South Africa's now-defunct pebble bed modular reactor (PBMR) project was to supply heat to enable the Bergius coal-to-liquid transformation process without producing the additional CO2. China is tripling the capacity of its coal-to-liquids plant in Shenhua.

In Qatar, Shell has completed its Pearl gas-to-liquids plant, capable of producing 260,000 barrels of oil-equivalent per day. Powered by burning natural gas, it converts natural gas methane into liquids such as ultraclean diesel fuel.

Although US demand for gasoline is slowly dropping, world demand is increasing as developing countries such as China buy more vehicles.

Peak oil may never come, but we are past peak cheap oil. In 2008 Shell's chief Jeroen van der Veer spoke,

> "After 2015, easily accessible supplies of oil and gas probably will no longer keep up with demand. As a result, we will have no choice but to add other sources of energy – renewables, yes, but also more nuclear power and unconventional fossil fuels such as oil sands."

The US has more oil than mankind has ever pumped.

The Green River Basin in Wyoming, Utah, and Colorado has vast underground resources of oil shale. Most is on federal land. This shale contains kerogen, which when heated in situ becomes pumpable liquid oil, leaving behind carbon char and gases.

Kerogens are composed of high molecular weight (over 1,000) molecules containing both carbon and hydrogen atoms in about a 2:3 ratio. Like oil and coal, kerogens are formed from the demise of living matter. When heated kerogen can release crude oil and natural gas. Most of the world's 7,000 billion barrels of kerogen deposits are in the Americas – Canada, USA, and Venezuela.

In the US the kerogen in the Green River Basin represent 1,500 billion barrels of oil, of which 1,000 billion barrels might be recoverable. US oil consumption is about 7 billion barrels per year, so the Green River Basin represents a century-scale petroleum resource.

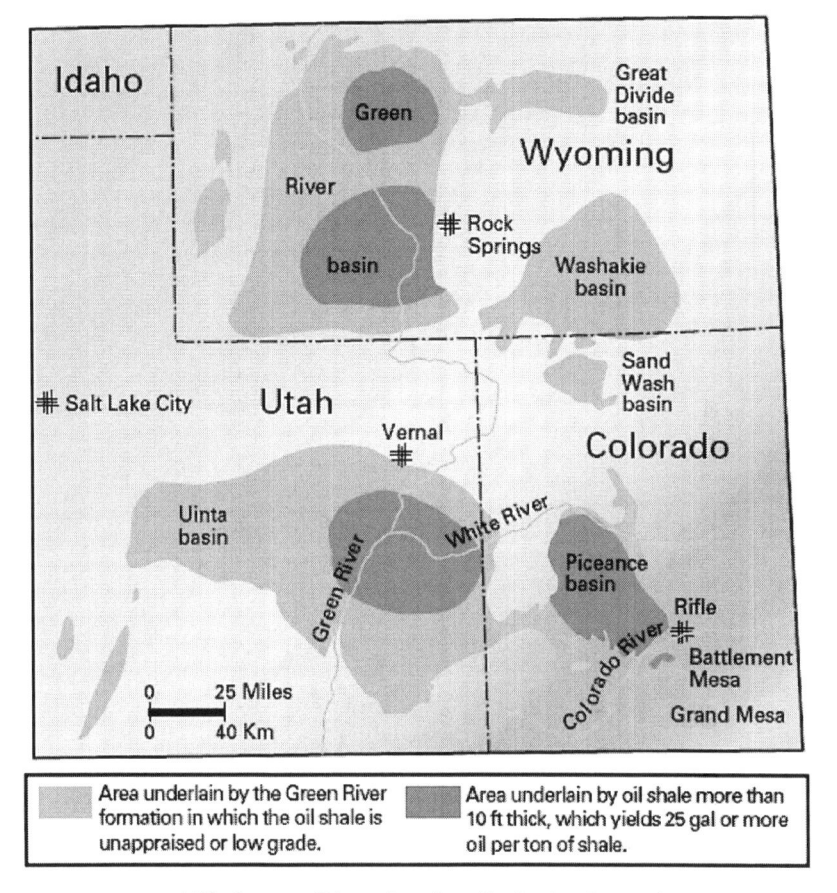

US Green River basin oil shale deposits

The EROI (energy return on investment) for extracting oil from underground kerogen is estimated to be less than 4, meaning that over a quarter of the harvested energy must be used to power extraction. If this is supplied by fossil fuels such as natural gas or kerogen, then over 25% more CO2 is released in comparison to conventional pumped oil. Alternatively, this energy might instead be supplied by nuclear heat and electricity from a LFTR.

Surface mining of oil shale is environmentally harsh.

Converting kerogen in oil shale to oil is not yet a commercial practice in the US because of the costs, new technology, and environmental concerns. Elsewhere 18,000 barrels per day are produced this way. In surface mining heavy equipment is used to remove the overburden earth, exposing the oil shale, which is dug by large machines and transported by large trucks to a processing center. This mining is similar to the processes employed for the Alberta tar sands and West Virginia coal mining by mountaintop removal.

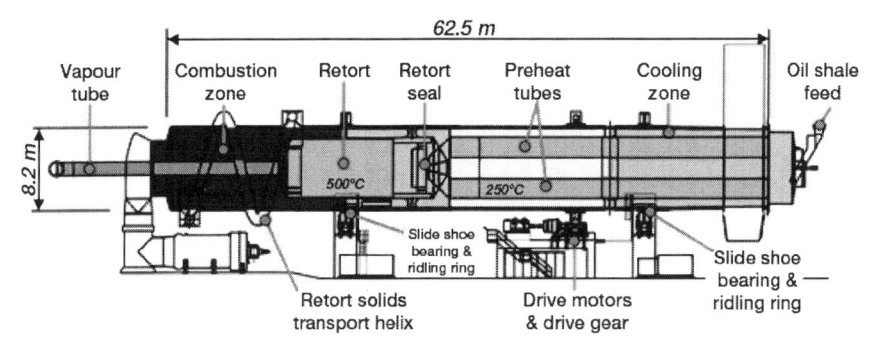

Retort for heating oil shale to 750°C to vaporize kerogen

The oil shale is then crushed to increase the surface area through which oil flows out, then heated in a retort heated by burning gas or oil. At temperatures near 500°C the kerogen dissociates into oil and gas flowing out of the shale. These oil and gas products are captured and further refined. The process creates up to 10 gallons of waste water per ton of shale, and the spent shale may contain pollutants including sulfates, heavy metals, and polycyclic aromatic hydrocarbons, some of which are toxic and carcinogenic.

In situ shale oil extraction has less environmental impact.

Rather than mining oil shale, the in situ techniques heat the oil shale in place, decomposing the kerogen. The oil and gas are removed and the char and waste are left deep underground. The cost is the substantial heat that must be employed to heat the earth for months or years to reach the kerogen decomposition temperature, because the thermal conductivity of the rock is low.

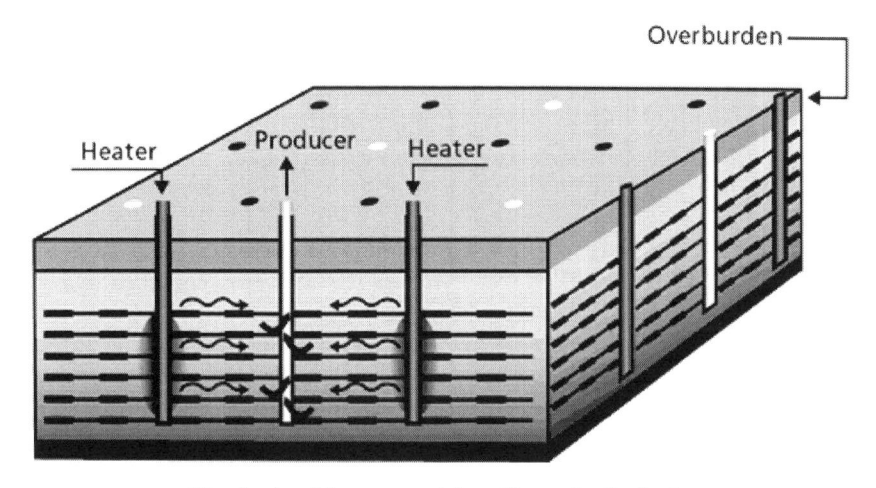

Shell electric current heating of oil shale

Shell's experimental process uses electrical heating elements in heater wells to raise the oil shale layer temperature to abut 350°C over a period of approximately 4 years. The gas and oil are taken from the producer wells. The processing area is isolated from surrounding groundwater by a freeze wall consisting of wells filled with a circulating super-chilled fluid. Disadvantages are large electrical power consumption, extensive water use, and the risk of groundwater pollution. Shell estimated an EROI of 3-4 for this process.

In the process proposed by American Shale Oil, superheated steam is circulated through a series of horizontal pipes placed below the oil shale layer to be extracted. Vertical wells provide vertical heat transfer through refluxing of converted shale oil and a means to collect the produced hydrocarbons. Heat is supplied by fossil fuel gases.

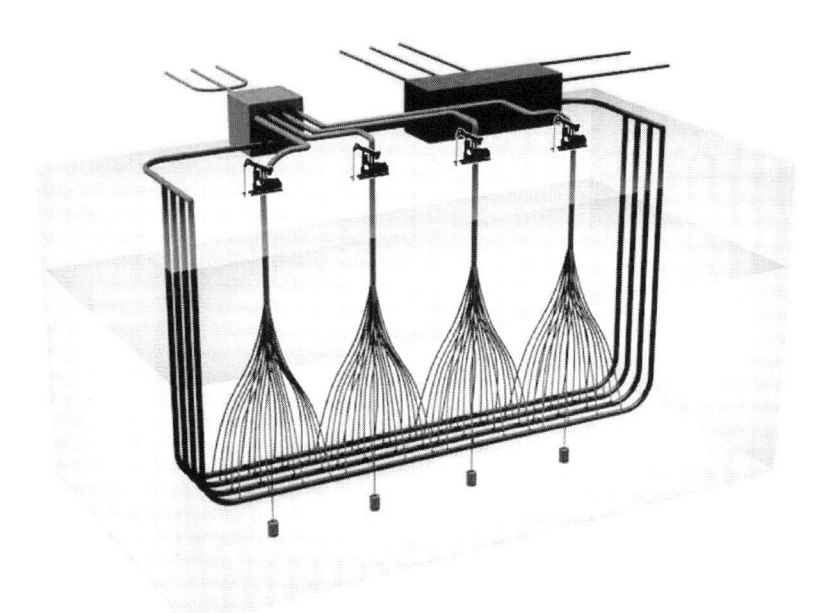

Superheated steam heating and converted kerogen extraction

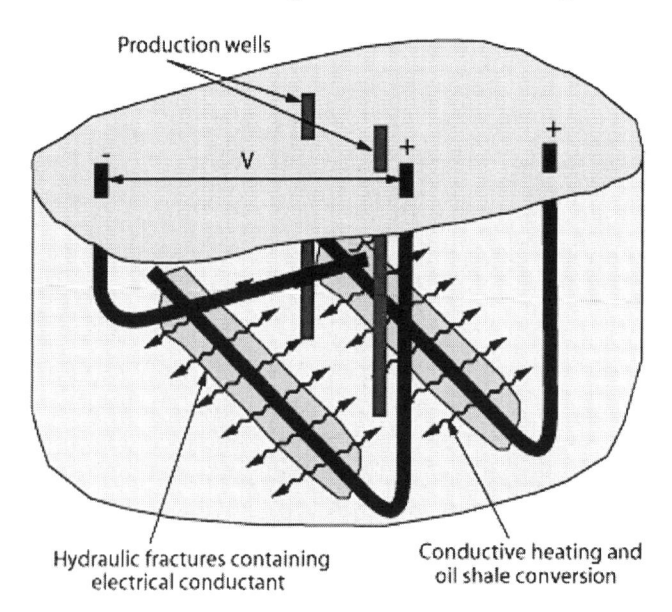

Resistance heating of conductant in fractured oil shale

ElectroFrac technology conducts electrical current through the earth, so that its resistance heats the oil shale. Hydraulic fracturing permits injection of an electrical conductant to make a low resistance connection between the electric power lines and the shale rock. Production wells are separate.

Chevron's process injects hot CO_2 gas to liquefy the kerogen.

General Synfuels proposes superheated air.

None of these technologies are in commercial operation.

LFTR can supply cheap energy for shale oil extraction.

Techniques for shale oil extraction are extremely energy intensive, with estimated EROI between 2 and 4. An EROI of 2 would double lifecycle CO_2 emissions for liquid fuel production and use if the needed energy comes from fossil fuels such as oil or gas.

What might be the LFTR energy cost of extracting shale oil? For the ElectroFrac process assume an onsite LFTR can produce electric power at $0.03/kWh for use with electric conductive heating. Because the resistive heating is released throughout the conductive shale, not just at the connection points, the efficiency of uniform heating might lead to an EROI as high as 4. That means the electricity cost for extracting the oil would be a quarter of the energy extracted, or $0.0075 per kWh(t) of oil, or $13/barrel. [If EROI is only 2, LFTR electricity costs $26/barrel.]

Using a simpler process-heat-generating onsite LFTR to produce steam heat to melt the kerogen should cost about $0.01/kWh(t) for the heating. Assuming the poor EROI of 2, this gives $0.005/kWh(t) or about $8/barrel for energy for the extracted oil.

This $8-26/bbl is a rough estimate of just the energy cost for extracting the oil from kerogen, not the cost of the refinery, transport, labor, drilling, etc. Still, this would seem an affordable cost with imported oil at $100/bbl.

A large 2 GW(t) LFTR power plant could provide heat to produce about 10 million barrels of oil per year. The US consumes 7 billion barrels of petroleum per year. If the US would scale back annual imported oil demand by 3 billion barrels, heating and extracting shale oil from the Green River Basin could supply that demand with 300 2 GW(t) reactors. That would be a lot of power! For comparison, the US now has only about 100 such size reactors dedicated to electricity production. But petroleum is a big business; one Exxon-Mobil refinery in Texas produces 40 GW(t) of petroleum product power.

LFTR heat can extract crude from Canadian tar sands.

Canada is the largest US supplier of petroleum products. Canada's largest source are the tar sands in Alberta. The tar sands are excavated by surface mining. Bitumen, a very heavy crude oil, is extracted by heating. The bitumen is then converted to a more liquid petroleum than can be transported and refined. The substantial energy and hydrogen for this process comes from natural gas. EROI is about 5. Total CO_2 emissions for fuels produced this way are about 15% higher than for sweet crude petroleum. Increased CO_2 emissions is the reason the Keystone XL pipeline is opposed.

Canada's House of Commons, Energy Alberta, Shell, and Idaho National Labs are exploring the use of nuclear power for tar sands oil extraction. One report estimates a 600 MW(e) nuclear power plant could provide energy to extract 60,000 barrels/day, however it is not practical to distriubute steam much over 10 km, so smaller, distributed reactors such as 100 MW(e) LFTRs could supply energy to distributed tar sands projects. At 3 cents/kWh electricity costs would be $7/barrel.

Using LFTR for CO_2-free extraction of oil from oil shale or tar sands does not change the fact that burning the extracted fossil fuels releases CO_2 into the atmosphere.

Synthetic Liquid Vehicle Fuels

Carbonaceous fuels have valuable high energy density.

How would we power trucks and airplanes in a post-fossil-fuel era? The advantage of carbonaceous (carbon based) fuels such as gasoline, diesel, and jet fuel is their high energy density. It permits vehicles to carry their own fuel supplies on board at reasonable costs. The highest cost of an airline is for jet fuel. Even with today's highly engineered and optimized aircraft and turbine engines, a long distance Boeing 747 airplane weighs half as much at landing as at takeoff. Over half the takeoff weight is fuel; the rest is the aircraft, passengers, and payload. Airline operation would be impractical with heavier fuel. There is no good substitute for carbonaceous fuels, so civilization will need a carbon-neutral fuel cycle if airplanes are to fly with no net CO_2 emissions.

Iso-octane, butane, aromatic

Gasoline, refined from petroleum, is a carbonaceous fuel mixture of compounds such as iso-octane, butane, and aromatics. These are hydrocarbons that burn to form water and CO_2, for example:

$$2\ C_8H_{18} + 25\ O_2 \rightarrow 16\ CO_2 + 18\ H_2O + \text{heat}$$

Diesel fuel and jet fuel are similar, with different mixtures of hydrocarbon molecules containing 8 to 21 carbon atoms.

Hydrocarbon fuels can be produced from natural gas, especially methanol (CH_3OH) to substitute for gasoline and dimethyl ether (CH_3OCH_3) to substitute for diesel. However these have a third less energy density than gasoline, requiring a 50% larger tank in a vehicle for the same range.

The world economy depends on petroleum for transport.

The world gets 37% of its energy from petroleum, vs 21% from coal. A typical nuclear reactor power plant generates about 1 GW (1000 MW) of electric power. A large refinery produces 40 GW of power in the form of gasoline, diesel, and jet fuel.

Petroleum's high energy density and a century of engineering experience in its use have made it essential to the world economy. The US thirst for it runs to 19 million barrels per day, of which we import 45% at a cost near $1 billion per day. Our protective presence in the Persian Gulf is estimated to have cost over $7 trillion.

An alternative way to benefit from LFTR's inexpensive power is to synthesize liquid fuels to replace petroleum. We can certainly use LFTR-produced electric power for more high-speed electric trains and for more small short-range battery-powered automobiles. But we can't electrify commercial airliners and big trucks because they cannot carry heavy, bulky batteries with them.

Hydrocarbon fuels can use LFTR-produced hydrogen.

Synthesizing hydrocarbon fuels requires a source of hydrogen and a source of carbon. Today commercial hydrogen comes from natural gas methane, CH_4, but this process is not carbon-neutral because the carbon is removed and becomes CO_2 in the atmosphere. To be carbon-neutral, hydrogen can be obtained from water, H_2O by high-temperature dissociation.

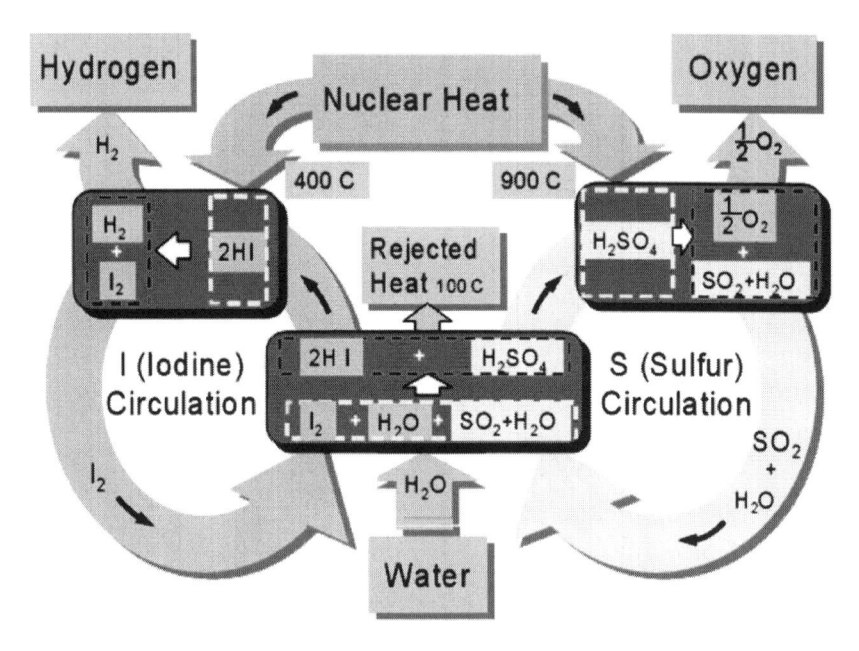

Sulfur-iodine cycle

Nuclear heat can power the production of hydrogen. At a temperature of 950°C, the sulfur-iodine process works at a chemical/thermal conversion efficiency approaching 50%. The 43% efficient copper-chloride process can operate at 530°C, a temperature compatible with currently certified nuclear structural material.

At low temperatures electrolysis can dissociate hydrogen from water at chemical/electrical efficiencies up to 60%. If the electricity was generated with an electric/thermal of efficiency of 40%, the comparable chemical/thermal efficiency is 24%. Although it is a possible vehicle fuel, the most practical use for hydrogen is as a feedstock for liquid fuels.

Hydrogen can be combined with coal to make synfuels.

To fabricate hydrocarbon fuels we also need carbon. Carbon can be obtained from coal. Rather than using nuclear hydrogen, existing chemical processes such as Fischer-Tropsch can be used to manufacture gasoline from coal. This F-T process also emits considerable CO2, so that the total CO2 when the gasoline is burned is 50% more than gasoline refined from petroleum. This process was perfected in South Africa during the Apartheid embargos and now produces 160,000 barrels of synfuels per day. It is occasionally proposed in the US a means to increase energy independence.

Our objective is a carbon-neutral synfuel manufacturing process that does not emit CO2 during manufacturing. We want to combine hydrogen and carbon in a process something like

$$8\ C + 9\ H_2 + energy \rightarrow C_8H_{18}$$

which is octane, a form of gasoline. We can make nuclear hydrogen, but we need a carbon source. Coal can be that carbon source. Locke Bogart has described a process that makes good use of both hydrogen and oxygen from water dissociation, and full use of water from the Fischer-Tropsch reactor. The "-CH$_2$-" stands for chains of hydrocarbons such as C_8H_{18}.

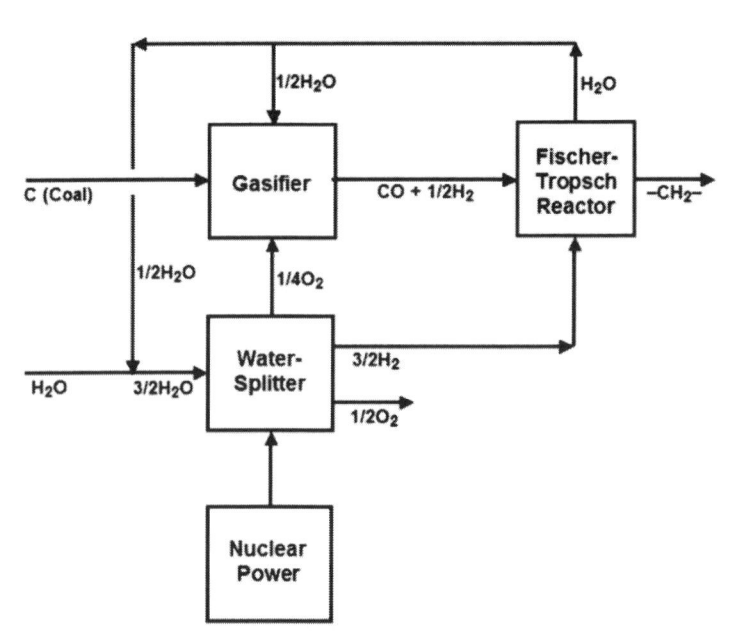

Idealized water splitter, coal gasifier, and F-T synfuel generator

LFTR energy can combine coal and natural gas for synfuels.

Combining coal and natural gas with in a nuclear heated chemical plant is a possible way to make liquid fuels such as gasoline or diesel. Compared to octane (C8H18), methane (CH4) has too much hydrogen, and coal (approximately H,2C) has too little. We can mix them in the required proportions of H and C to make liquid petroleum substitutes.

$$4 \times (CH_4) + 2 \times (2C,H) + energy \rightarrow C_8H_{18}$$

The heats of combustion of the feedstocks and the product are:

Methane	56 kJ/g
Coal	27 kJ/g
Gasoline	47 kJ/g

Equal masses of methane and coal would have a heat of combustion of only 41 kJ/g so we would need to add at least 6 kJ/g of energy, from carbon-neutral nuclear power. This possible chemical process would not emit CO_2 during the synfuel manufacturing so it would be an improvement over commercial coal-to-liquids and gas-to-liquids processes. Coal-to-liquids plants are powered by burning coal, so they emit more CO_2 than drilling for oil. Gas-to-liquids plants are powered by burning natural gas, emitting CO_2. The US has ample coal and natural gas for raw materials to make gasoline and diesel without emitting any additional CO_2 compared to drilling for petroleum.

No such coal-and-gas-to-liquids refinery has been constructed, and its development would be a multibillion chemical engineering project on the scale of coal-to-liquids in South Africa or gas-to-liquids plants in Qatar.

However this synfuel would emit just as much CO_2 as petroleum-derived gasoline when burned. The principal advantage for the US would be improved energy independence from reducing petroleum imports. It would not address CO_2 emissions and the impending global warming catastrophe.

Synfuels could be carbon-neutral by recycling CO2.

Coal plants burn coal and emit CO2 into the atmosphere. With advanced coal technology such as integrated-gasification combined-cycle (IGCC) it would be possible to add CO2 capture to the coal plant. The CO2 could be used a source of carbon to manufacture the synfuel. The carbon-neutrality argument is that the coal plant emissions were going into the atmosphere anyhow, so the synfuel combustion process only releases the CO2 the coal plant generated and causes no net increase in CO2 emissions. If all 1.9 Gt/year of US coal plant CO2 emissions were captured this way, using nuclear power to make synfuels, it would provide all the transportation fuel for the US economy. Production costs are estimated at roughly $3/gallon with carbon captured from coal plant stack CO2 emission, or roughly $2/gallon using carbon directly from coal. One problem with this carbon neutrality argument is that coal plants will be replaced with natural gas and LFTR nuclear power.

More carbon sources are air, vegetation, and cement.

We will examine three other sources of carbon for synfuels. We might extract CO2 from the air, but it's difficult because the CO2 density is low, about 0.04%. Another option is biomass, principally harvesting the carbon the vegetation captured, rather than just the energy stored via photosynthesis. A surprising third option is using CO2 created by making cement.

Green Freedom proposes extraction of CO2 from air.

Project Green Freedom is conceived by Jeffrey Martin and William Kubic of Los Alamos National Laboratory. Their idea is to use a nuclear power plant to provide the energy to synthesize fuel, and use the air flow of the cooling towers as a source for carbon from CO2 that makes up about 0.037% of the atmosphere.

A SUSTAINABLE WORLD

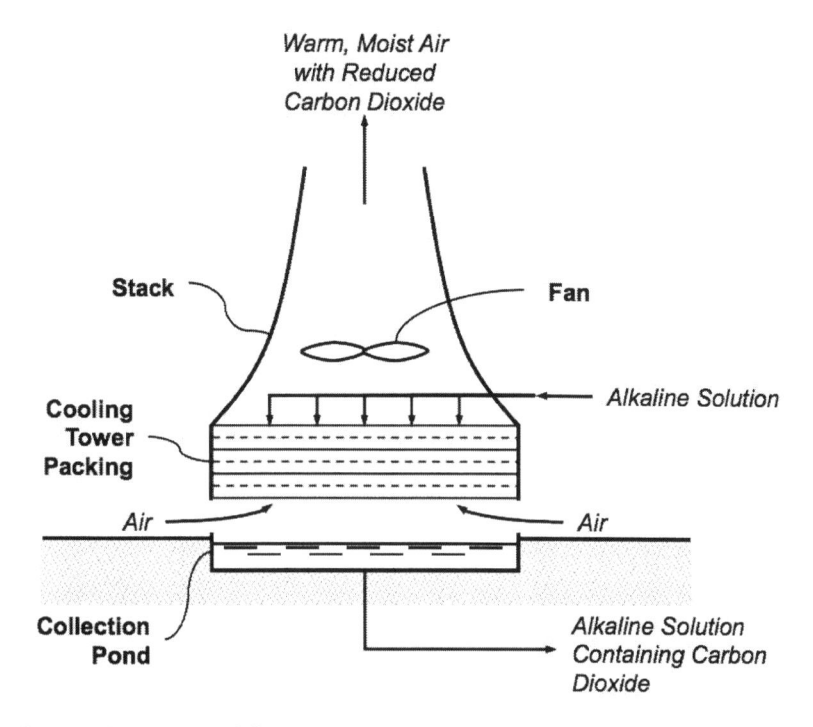

Green Freedom CO2 extraction in airflow of cooling tower

They observe that alkaline lakes absorb about 30 times the CO2 of similar size fields of switchgrass, and they conceived of trays of potassium carbonate solution exposed to the airflow within the nuclear plant cooling towers. The potassium carbonate readily absorbs CO2 by

$$CO_2 + K_2CO_3 + H_2O \rightarrow 2\ KHCO_3$$

creating potassium bicarbonate. The CO2 would be electro-chemically removed from the bicarbonate solution, requiring ~410 kJ/mole-CO2 of electric energy and 100 kJ/mole-CO2 of thermal energy. This power would be supplied by a nuclear power plant such as LFTR. Because the CO2 is so dilute in the air, Martin and Kubic envision a large air processing facility with nuclear reactors providing 1000 MW(e) and 470 MW(t) power with six CO2-capturing towers for cooling the reactors and chemical plant cooling.

The chemical manufacturing processes for conversion of CO_2 and hydrogen to methanol are proven and commercialized. Mobil has developed a process for converting methanol to gasoline. The complete facility could produce 17,000 barrels per day of gasoline at an estimated consumer cost of $5/gallon (2007), requiring an investment of approximately $5 billion. Martin and Kubic anticipate cost reductions of about 20% from improved technologies. Less expensive nuclear power from LFTR will also reduce the synthetic gasoline fuel costs. The process benefits from use of commercialized technologies, so the only technical risk is scaling up CO_2 capture from air.

Methanol has been used for decades to power race cars at the Indianapolis-500. Although it has about half the energy density of gasoline, methanol can readily be used in flex-fuel vehicles or modified engines in ordinary vehicles. Today methanol (CH_3OH) is produced from natural gas (CH_4), so natural-gas-produced methanol could be a transition fuel until carbon-neutral sources such as Green Freedom are perfected.

The complete Green Freedom fuel cycle would be carbon neutral, because just as much CO_2 would be put into the atmosphere by burning gasoline as removed by Green Freedom.

Nuclear heat and hydrogen can make biomass into synfuel.

Plants absorb carbon from air. Biomass and hydrogen can be combined with nuclear heat to manufacture synfuels such as diesel more efficiently than does cellulosic ethanol technology.

Many kinds of biomass can be processed in a heated, entrained-flow chemical reactor to create liquid fuels. The required energy can be supplied by burning biomass. To reduce the biomass feed to the synfuel production process we can supply the energy externally. This can be accomplished with LFTR-produced hydrogen and raising the temperature of the oxygen-free production process to approximately 1000-1200°C. LFTR heat is below this temperature, but the high temperature can be achieved instead by an electricity-powered plasma arc.

The role of the biomass is not so much to provide energy but to contribute the carbon that is combined with hydrogen and LFTR energy to synthesize the biofuel. This table illustrates the contribution of the additional, LFTR-supplied energy for diesel synthesis.

Biomass reductions from LFTR-supplied energy		
	Biomass / diesel mass ratio	kWh(e) in per kWh(t) synfuel out
Biomass gasification	5.6	0
Biomass gasification with LFTR energy	1.7	1.08

By avoiding oxidation of the biomass, the synfuel mass yield of the process can be 5.6/1.7 = 3.3 times that of anticipated cellulosic ethanol processes such as enzymatic fermentation or gasification. This means that land use requirements for biomass production are reduced by 70%, reducing competition with land for food crops.

Estimated costs for diesel fuel production in this manner are 0.89 euros/liter, or $4 per gallon. No such biomass refineries are in production, and there is considerable chemical engineering development to be accomplished before constructing such billion-dollar plants. The major oil companies have the expertise to develop them.

LFTR-energized biomass fuels might supply US needs.

A 2005 DOE study projected that the US could produce 3 billion barrels of synfuel from 1366 million tons of dry biomass. With LFTR-energy conversion, synfuel production could be tripled. The US consumes about 7 billion barrels of petroleum products per year. Dry biomass growth is about 6 tonnes/ha/yr, so to supply all US petroleum substitutes this way would require 160 million hectares for biomass crops. Forestland and farmland area in the

US totals about 670 million hectares, so meeting US fuels needs this way is barely conceivable. Cutting liquid fuel use would help.

Gasoline motor fuel accounts for 44% of US petroleum consumption, and this could be substantially reduced through more efficient cars and electric cars. Nearly half of all rail freight is moving coal from mines to power plants, so this use of diesel fuel will diminish as coal-fired power plants are retired. Railroad electrification can further reduce diesel fuel consumption, and high speed rail service can diminish demand for air travel. Trucks and airplanes will be the principal consumers of carbonaceous liquid biofuels.

AMMONIA

What if the previous examples of production of carbonaceous fuels turn out to be too difficult or expensive to be a practical source of liquid fuels?

Ammonia can transport much of hydrogen's energy.

Hydrogen is difficult to use as a vehicle fuel. To contain it requires either costly refrigeration at -253°C to liquefy it or costly compression to 5000 psi, requiring 30% more energy. The small molecules of H2 leak and can embrittle metals.

Like hydrocarbons, nitrogen can also transport the chemical potential energy of hydrogen. The liquid forms of such hydrogenated fuels can be readily contained in tanks at standard temperatures and modest pressures.

Building molecules with nitrogen instead of carbon can create another fuel – ammonia, NH3. Nitrogen is abundant, comprising 78% of the atmosphere. It can be obtained at much lower cost than carbon. This opens up an additional way to benefit from LFTR's inexpensive power -- manufacturing ammonia.

Ammonia energy density is higher than that of hydrogen.

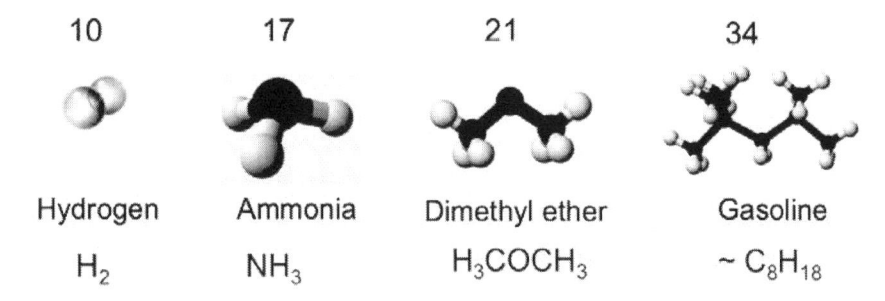

10	17	21	34
Hydrogen H_2	Ammonia NH_3	Dimethyl ether H_3COCH_3	Gasoline ~ C_8H_{18}

Energy densities of liquid fuels, MJ/L

The energy densities above, in megajoules per liter, show that even compressed to 5000 psi a hydrogen tank would take up almost twice the volume of a tank of ammonia. Dimethyl ether is an example carbonaceous synfuel that can substitute for diesel fuel.

The higher energy density and portability of carbonaceous fuels do make them the most attractive energy transport liquid fuels for vehicles. Ammonia only has half the energy density of gasoline or diesel, so would require a bigger tank for on-vehicle fuel storage. But there are many opportunities to use zero-carbon ammonia.

Ammonia is a common industrial chemical.

The US uses 20 million tons of ammonia and ammonia fertilizer products annually. Energy for production of ammonia uses 1-2% of all world energy. Over 80% of ammonia is used for fertilizers that are responsible for food production sustaining 1/3 of the world population. Ammonia fertilizers were a component of the 20th century Green Revolution credited with saving over one billion people from starvation. Today ammonia is principally produced from natural gas, releasing CO_2. World food production is highly dependent on fossil fuels.

Ammonia being injected into soil as fertilizer

Ammonia can fuel internal combustion engines.

In Belgium during World War II a fleet of ammonia-fueled motor busses carried passengers thousands of miles.

Ammonia fueled bus in Belgium

Today engineers are improving spark-ignited internal-combusition engines and diesel engines fueled with ammonia or ammonia with additives such as biodiesel, ethanol, hydrogen, cetane, or gasoline. Sturman Industries is developing an ammonia fueled hydraulic engine – no crank, no cam, no carbon.

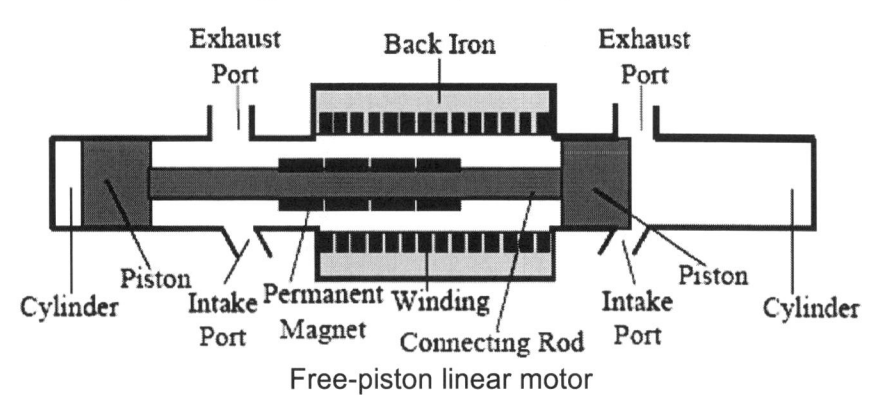

Free-piston linear motor

Development continues on the electricity-generating free-piston linear motor said to achieve an efficiency of 50% on a lean mixture of ammonia and air.

Hydrofuel, Inc. demonstrated an ammonia fueled automobile in 2010.

Ammonia fuel cells can generate vehicle electricity directly.

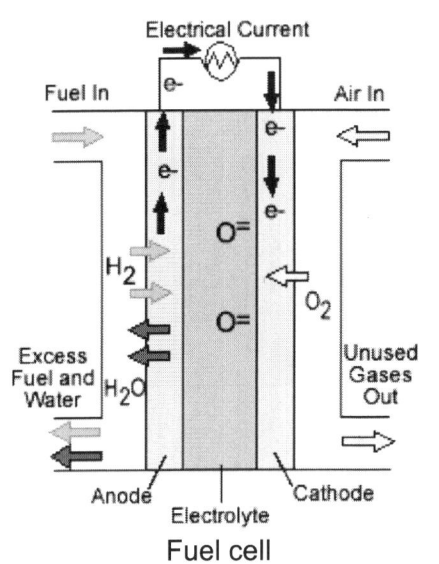

Hydrogen fuel cells first require dissociation of ammonia into hydrogen and nitrogen.

Direct ammonia fuel cells have no need to first crack ammonia into N_2 and H_2 fuel. Some use molten salt electrolytes.

High efficiency, high-temperature Solid Oxide Fuel Cells (SOFC) use proton-conducting ceramic electrolytes.

Fuel cell

Solid state ammonia synthesis cuts ammonia costs.

Today the Haber-Bosch ammonia production process annually manufactures 500 million tons of ammonia from natural gas, water, air, and electricity. This process alone accounts for 3-5% of world natural gas consumption. The carbon from the natural gas methane (CH_4) is emitted to the atmosphere as CO_2.

Sammes and Restuccia of the Colorado School of Mines have patented a solid state ammonia synthesis (SSAS) process fed by air, water, and electricity. Nitrogen is obtained from an air separation unit (ASU). Water supplies the hydrogen. There is never any separated explosive hydrogen gas. SSAS works like a solid oxide fuel cell, but in reverse, with a proton conducting ceramic membrane. The ceramic membranes are tubes, and the SSAS can be scaled up by using more tubes. In addition to electricity, LFTR can provide the 650°C steam heat for the SSAS cells.

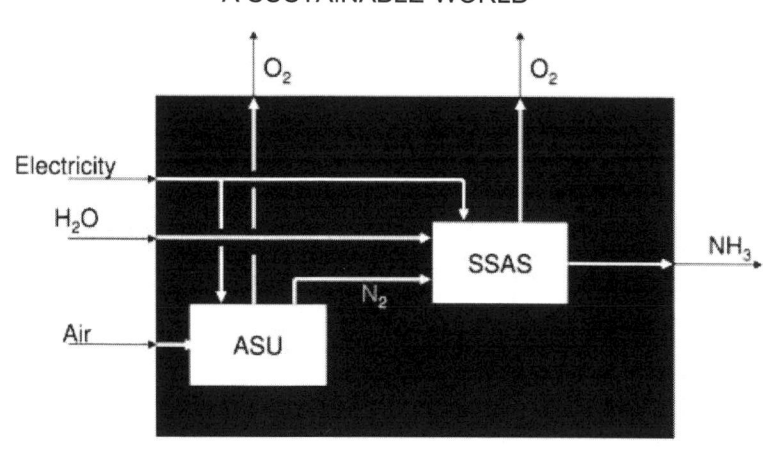

SSAS: 6 H2O + 2 N2 → 3 O2 + 4 NH3

With factory reactor production, LFTR electric power is projected to cost $0.03/kWh, leading to roughly-estimated ammonia costs of about $200 per tonne. This is half the cost of ammonia produced today from natural gas, and it avoids the release of carbon dioxide in the widespread Haber-Bosch manufacturing process. This new SSAS process has been demonstrated in the laboratory, but it requires considerable chemical engineering development before it can generate ammonia in commercial quantities.

The energy cost of nuclear ammonia is 1/3 that of gasoline.

The heat of combustion is the thermal energy that would be released in an internal combustion engine. Taking account of the different prices and heats of combustion of ammonia and gasoline illustrates that energy from ammonia is one-third the cost of energy from gasoline.

Energy content cost of ammonia and gasoline			
Fuel	Heat of combustion	Price	Energy cost
Nuclear ammonia	22 MJ/kg	$0.20/kg	$0.01/J
Gasoline	132 MJ/gal	$4/gal	$0.03/J

How might this lower energy cost translate into vehicle fuel costs?

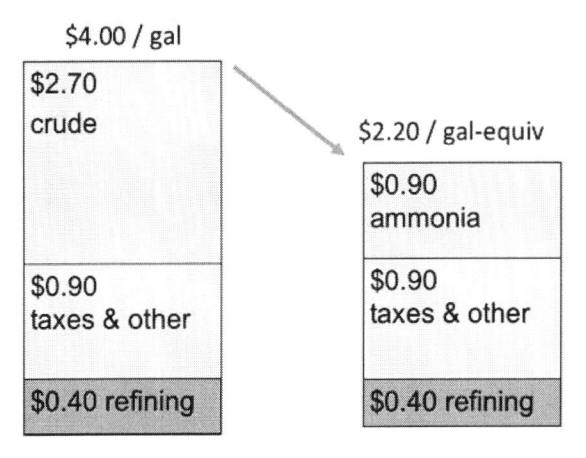

```
$4.00 / gal
┌──────────────┐
│ $2.70        │
│ crude        │      $2.20 / gal-equiv
│              │     ┌──────────────┐
│              │     │ $0.90        │
│              │     │ ammonia      │
├──────────────┤     ├──────────────┤
│ $0.90        │     │ $0.90        │
│ taxes & other│     │ taxes & other│
├──────────────┤     ├──────────────┤
│ $0.40 refining│    │ $0.40 refining│
└──────────────┘     └──────────────┘
```

Fuel cost components: energy source, taxes etc, refining

The left stack illustrates the typical cost components of gasoline in California. Most of the cost is for the crude petroleum that provides the energy content of the gasoline. The refining costs are only about 10%, even though refineries are complex, expensive investments. We don't really know the cost of SSAS chemical plants, but simply assume that the talented chemical engineers who built petroleum refineries can build similarly large ammonia production plants at about the same cost.

Ammonia can be handled safely.

Ammonia is the second most common industrial chemical, with US consumption of 20 million tons per year. In the US ammonia is distributed by a 3,000 mile network of pipelines, principally for agricultural use. Ammonia storage capacity is 5 million tons.

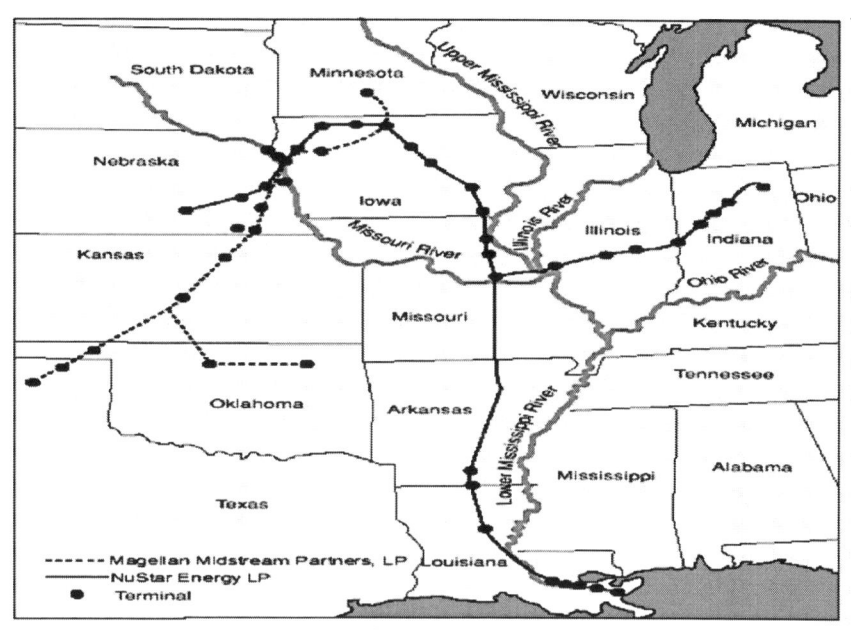

Midwest US ammonia pipelines

In a vehicle, ammonia would be liquid in pressure tanks at 200 psi, similar to propane (177 psi). Compare this to tanks needed for compressed natural gas (3000 psi) or hydrogen (5000 psi). In an accident, spill, or leak ammonia dissipates rapidly because it is lighter than air. Its pungent odor is alerting. Ammonia is difficult to ignite, with an ignition temperature of 650°C. Unlike gasoline, an ammonia fire can be extinguished with plain water.

Inhaling an ammonia concentration of one half percent for a half hour has a 50% fatality risk. Inhalation of 500 ppm is dangerous to health. Chronic exposures of 25 ppm are not cumulatively dangerous as humans and other mammals naturally excrete NH3 in the urea cycle. Ammonia is toxic to fish.

The hazards of ammonia are different but equivalent to those of gasoline. Ammonia is toxic and gasoline is explosive. A 2009 Iowa State University analysis concludes

> "In summary, the hazards and risks associated with the truck transport, storage, and dispensing of refrigerated anhydrous

ammonia are similar to those of gasoline and LPG. The design and siting of the automotive fueling stations should result in public risk levels that are acceptable by international risk standards. Previous experience with hazardous material transportation systems of this nature and projects of this scale would indicate that the public risk levels associated with the use of gasoline, anhydrous ammonia, and LPG as an automotive fuel will be acceptable."

In summary, nuclear ammonia is a suitable vehicle fuel. It emits no CO_2 when burned. Its production process can be CO_2 free. It would require fuel tanks twice as big as gasoline tanks on vehicles.

Nuclear Cement

Another process might be used to produce CO2 for carbonaceous synfuel manufacturing, yet be net carbon neutral.

Cement curing in concrete absorbs CO2.

The lime cycle has been used to make mortar for construction for millennia. Limestone is heated very hot to drive off CO2; it's not really "burned". Adding water to the resulting CaO makes calcium hydroxide, C(OH)2, used as the binding agent for mortar. As the mortar sets water is given off, and the setting mortar very slowly absorbs CO2 from the air to make strong calcium carbonate cement. This idealized cycle is carbon neutral.

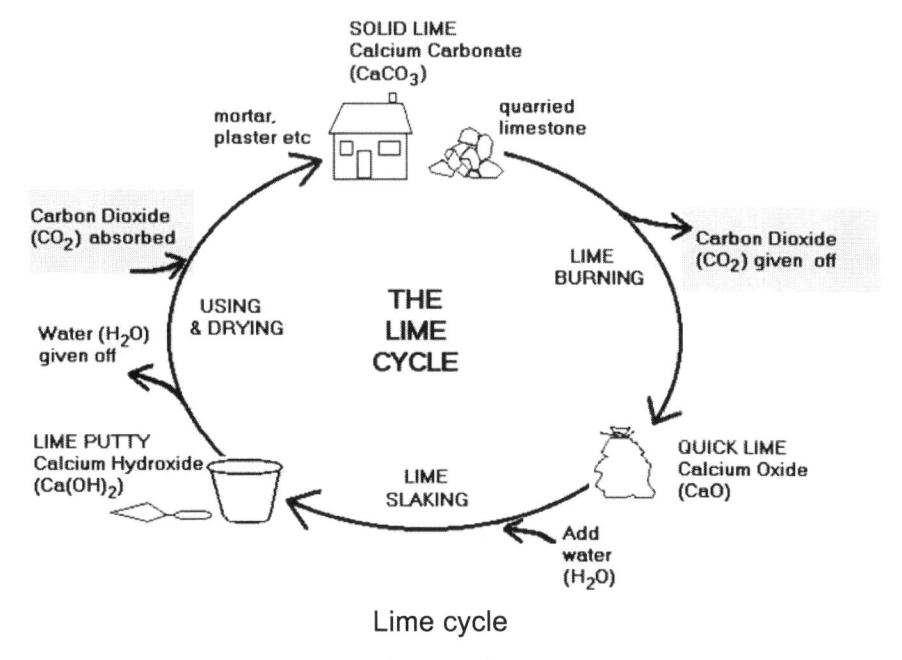

Lime cycle

In today's construction industry, lime mortar is replaced by Portland cement, produced by a similar cycle, but with sand added to the limestone, adding silicon to the chemistry, making a stronger cement of calcium silicates: 2CaO-SiO2 and 3CaO-SiO2. The CO2 cycle is the same.

In manufacturing Portland cement the heating of limestone and 1450°C sintering is accomplished by burning large quantities of fossil fuels. This process is the fourth largest contributor to atmospheric CO2 pollution after other uses of coal, oil, and natural gas.

Rather than simply letting the CO2 driven out of the CaCO3 limestone to escape into the atmosphere, it can be captured and used as a carbon feedstock for manufacturing synfuels.

Cement can be created with LFTR, rather than fossil fuel.

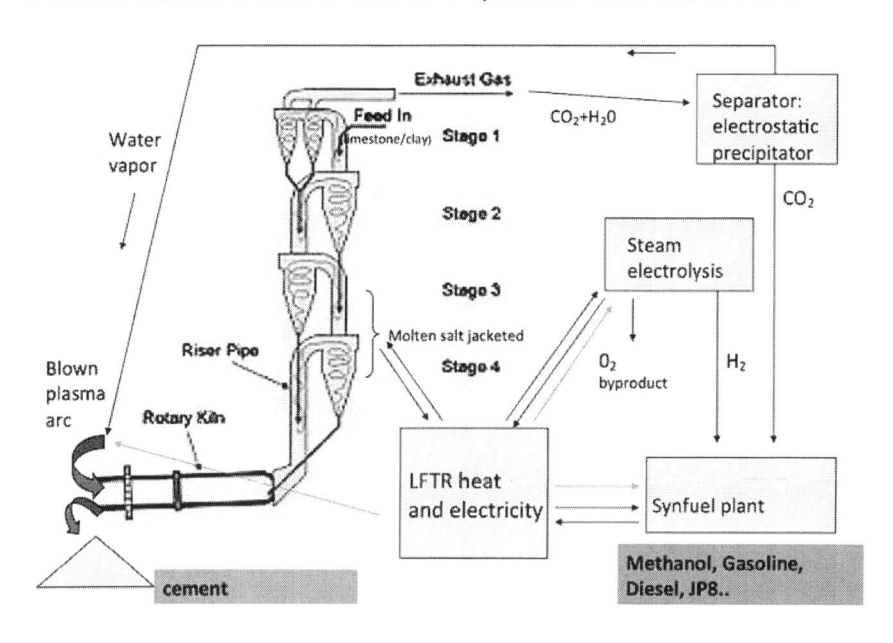

Cement kiln with heat from LFTR and plasma arc

This process is the conception of Darryl Siemer, a retired nuclear chemist from Idaho National Labs. Heat from a liquid fluoride thorium reactor (LFTR) would be transferred by molten salt to the cyclone preheaters to heat the sand and limestone to 700°C. The Portland cement process requires 1500°C, so that additional heat is supplied by a plasma arc powered by electricity from a LFTR. The exhaust gas from the kiln contains CO_2 and H_2O, also preheating the sand and limestone. The CO_2 is separated and fed to a synfuel plant where it is combined with H_2 from LFTR-powered high-temperature electrolysis to manufacture carbonaceous fuels.

When the synfuel is burned the carbon does escape as CO_2, but the cement absorbs the same amount of CO_2 as it is used in construction and cures over years. The CO_2 from the lime cycle is borrowed for a synfuel manufacturing and burning cycle. The energy for running today's cement manufacturing plants comes from fossil fuels. LFTR would replace that energy source and eliminate those CO_2 emissions.

Making nuclear cement could capture enough CO_2 to enable the production of synfuels that would replace 8% of today's US fuel consumption and would produce 390 Mt (megatonnes) of cement per year. US consumption is 106 Mt/a. Over half of world consumption of 3300 Mt is used in China.

This CO_2 source is an alternative or supplement to the other CO_2 capture processes presented: direct air capture and biomass farming.

HYDROGEN

The previous sections have discussed possible vehicle fuels that have no net increase in atmospheric CO2 emissions and could be produced from inexpensive LFTR heat and electricity.

Ammonia (NH3) was one example of a fuel that could be burned in an internal combustion engine, a rocket engine, or a fuel cell to produce electricity for motors. Ammonia could also be used simply as a hydrogen carrier, dissociating it with heat before burning the resulting H2 with oxygen from air.

Another example was synthetic carbonaceous fuels such as methanol (CH3OH) or dimethyl ether (H3COCH3). For net carbon neutrality these require atmospheric extractive carbon sources. The examples were Green Freedom CO2 extraction from air, biomass farming (for carbon not energy), and cement manufacturing.

Implementing these processes for fuel production and consumption on a global scale is truly challenging. There is another, potentially simple fuel – hydrogen. There is no natural free hydrogen on earth, because it reacts with oxygen to form H2O. Today hydrogen is produced by steam reforming natural gas, CH3, but that emits CO2. Hydrogen can be produced with LFTR supplied energy, by standard electrolysis, by high-temperature electrolysis, or by high-temperature catalytic dissociation.

A hydrogen-economy infrastructure does not yet exist.

The challenge for hydrogen is containing and transporting it. Hydrogen gas is a small molecule that can penetrate metal and embrittle it. Special coatings enable the safe transfer of compressed hydrogen gas in pipelines and in tanks for vehicle demonstration projects. For large tank storage hydrogen can be liquefied to be stored at lower pressure, but this requires refrigeration to -253°C (20°K) temperature and cryogenic insulation. Both pressurization and liquefaction of hydrogen are energy intensive, with an energy loss of roughly 30%.

Hydrogen safety is difficult to manage because hydrogen gas is colorless and odorless and explosive if mixed 20-60% with air. It can be ignited with very low energy, such as the static electricity spark from a rush of the gas. Because it is lighter than air, hydrogen disperses rapidly, but could collect under overhangs or in parking garages. Car fires involving escaping hydrogen gas are actually not as severe as car fires involving escaping gasoline, in tests.

Carbon-free hydrogen production could be accomplished in distributed or centralized modes. Distributed generation could be accomplished with ordinary electrolysis, storing the produced hydrogen locally for transfer to vehicles at fueling stations. High-temperature electrolysis has chemical/electrical conversion efficiency in the 50-80% range. So with electricity generated from the 40-50% efficient electrical/thermal conversion of LFTR heat, the overall chemical/thermal efficiency is near 30%. This is less efficient than with centralized production, but avoids hydrogen transportation.

Centralized production could be accomplished at purpose-built high-temperature electrolysis plants co-located with LFTRs that deliver 950-1000°C heat and electricity using the sulfur-iodine process at a chemical/thermal conversion efficiency near 50%. Using current materials technology an efficiency of 43% can be achieved at 530°C with a copper-chloride cycle.

A hydrogen-fueled car is available in California.

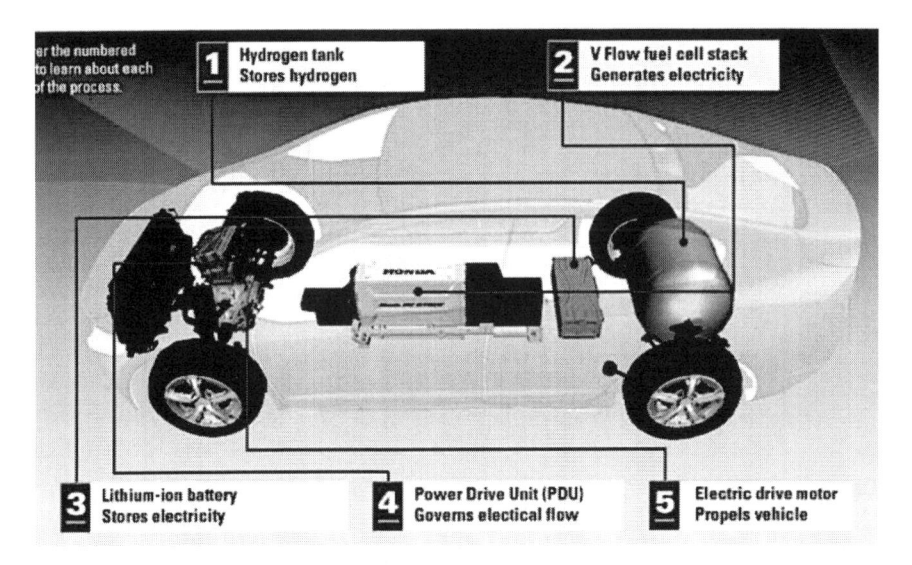

Honda FCX Clarity

Honda already sells a hydrogen fueled car, the FCX Clarity. Its fuel tank holds 4 kg of hydrogen compressed at 350 atmospheres of pressure, enough for 240 miles of driving. A 100 kW fuel cell converts hydrogen and oxygen to electricity that drives the electric motor, using an intermediate lithium-ion battery. For $600/month this vehicle can be leased in California, where there are 14 refueling stations along the California Hydrogen Highway.

Industrial gas supplier Linde has experience in manufacturing and transporting hydrogen gas in liquid and compressed forms. Linde worked with BMW, Ford, Daimler Chrysler, GM Opel, and others concerned with hydrogen vehicles. Linde has installed 19 hydrogen filling stations.

Hydrogen can power airplanes.

With extensive development, hydrogen may become a possible commercial airplane fuel. For the same amount of energy, hydrogen fuel has only 1/3 the weight of petroleum jet fuel, very advantageous to aircraft performance. Containing compressed hydrogen at 350 atmospheres of pressure (5000 psi) is possible with lightweight carbon-fiber tanks, but higher densities would require heavy steel tanks. At this pressure hydrogen's energy density of 2.8 MJ/liter compares unfavorably to jet fuel at 33 MJ/liter, so the volume occupied by hydrogen tanks will be 12 times more than jet fuel tanks, reducing cargo or shortening flights.

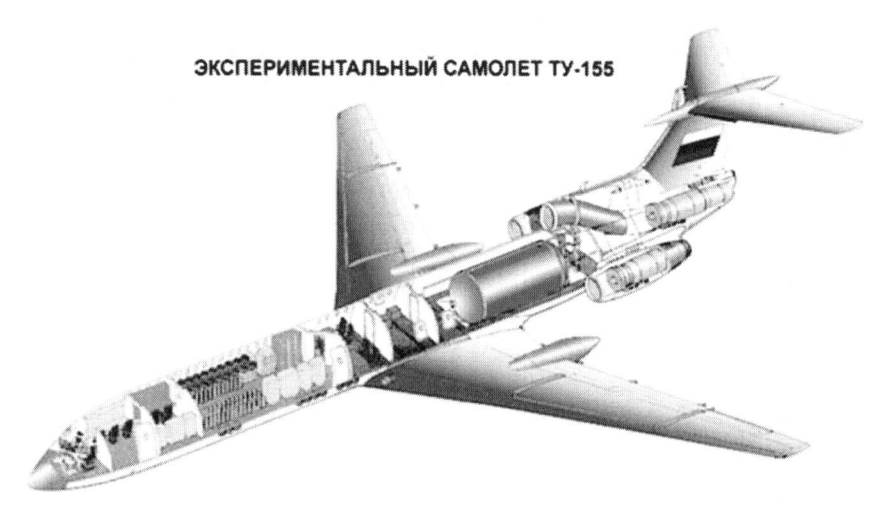

Experimental Tupolev TU-155

Russia demonstrated an airplane fueled by cryogenic, liquid hydrogen in 1989. Boeing used internal combustion engines on a hydrogen-fueled unmanned aircraft. Small aircraft powered by hydrogen fuel cells with electric motors have been demonstrated.

WATER AND DESALINIZATION

World water resources are stressed.

UNESCO reports that 8% of worldwide electric power is used for water pumping, purification, and wastewater treatment. The World Bank says 2.6 billion people have no access to sanitation, leading to illness that reduces GDP by 6%. Over a billion people have no access to electricity. Agriculture uses 70% of world water withdrawals, and food production must increase 70% in the next 40 years to sustain the population. The withdrawal of groundwater has revolutionized agriculture, but replenishment is insufficient for sustainability. Shrinking glaciers have temporarily added to water flows, but due to global warming these sources will diminish along with their buffering effects.

World energy production also competes for water resources. All thermal power plants require cooling, almost always accomplished with water by evaporative cooling or heating water in a river or ocean. Thermal power plants include nuclear, coal, natural gas, biomass, concentrated solar, and geothermal technologies. Even hydroelectric power consumes water by evaporation from reservoirs.

LFTR power can reduce global water stress.

High-temperature, air-cooled nuclear power plants such as LFTR will be especially valuable in water-stressed regions, because they do not compete for this scarce resource.

With electrical power, sanitation systems can economically treat wastewater for reuse in agriculture. Treated waste water represents a growing fraction of total water withdrawals in Mideast countries Saudi Arabia (1%), Oman (3%), Jordan (9%), and Qatar (10%).

Water desalination is becoming more efficient.

Today most of the daily 70 million cubic meters of potable water produced by desalination is made in plants that use petroleum fuels for energy, increasing CO_2 emissions. The desalination plants are mostly in the wealthy countries of the arid Mideast. The older, common multi-stage flash (MSF) steam distillation processes use about 25 kWh(t) per cubic meter of water produced. Cogeneration improves this; when the MSF facility is an integral part of the power plant cooling system the power requirements can be halved to roughly 10 kWh/m^3.

Reverse osmosis (RO) is most commonly used in new desalination plants. Reverse osmosis requires up to 6 kWh(e)/m^3, producing desalinated water at about \$0.50/m^3. The predominant cost for desalinated water is energy. Reducing the cost of energy with LFTR will reduce the cost of the water. Replacing petroleum fueled desalination plants with LFTRs will also reduce CO_2 emissions.

Multi-effect distillation (MED) is even more efficient, requiring only 1 kWh(t)/m^3 of power. Siemens has developed an electrolysis based desalination technology that uses 1.5 kWh(e)/m^3.

For LFTR with its high 700°C temperature, the Brayton power conversion cycle is highly efficient, minimizing waste rejected heat. In this case an advanced multi effect distillation (AMED) process can cogenerate an additional 1 m^3 of water for each 30 kWh of electric power produced.

Since fuel costs are very small for LFTR (and most nuclear power plants) they operate at full power, continuously. Electric power peak demand is typically about twice minimum demand. Cogenerating LFTR electric power plants can be designed to use excess power to desalinate water during off-peak periods.

POPULATION STABILITY

Ending energy poverty is key to achieving modest prosperity in the developing world. Microsoft founder and philanthropist Bill Gates remarked:

> "If you want to improve the situation of the poorest two billion on the planet, having the price of energy go down substantially is about the best thing you could do for them. ... Energy is the thing that allowed civilization over the last 220 years to dramatically change everything."

Ending energy poverty leads to a sustainable population.

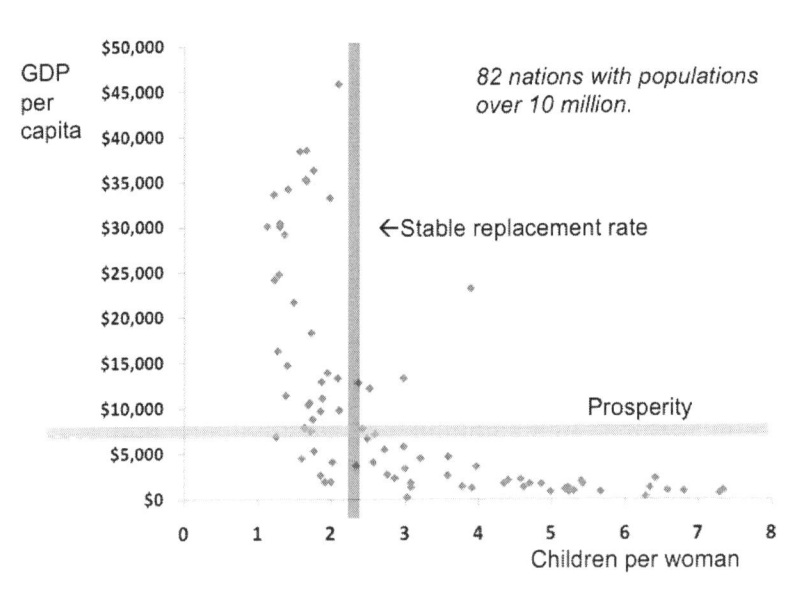

GDP, birthrates, and prosperity

The poor nations, below $7,500 GDP/person, are those that have the highest birthrates. Using inexpensive energy to improve the economic status of poor nations will lower birthrates, leading to a stable or shrinking world population.

8 Energy Policy

What are the goals of energy policy?

1. Stopping global warming?
2. Protecting the environment?
3. Protecting human health and safety?
4. Ensuring a sustainable world?
5. Ending energy poverty?
6. Furthering economic growth?
7. Assuring energy security?
8. Achieving energy hegemony?

Who should solve the world's energy and environmental crises?

1. A transnational organization such as the United Nations?
2. One nation such as the United States?
3. Multiple state or provincial governments?
4. Corporations?
5. Leadership individuals?

In the face of these dilemmas it is no wonder that US and global energy policies arc in unproductive pandemonium. Let's look at US energy policy.

US spends $21 billion on federal tax preferences for energy.

US energy policy is implemented by regulation, by direct grants, tax expenditures, and risk transfers -- at every level of government: federal, state, and sometime municipal.

Tax expenditures are also called tax preferences, which reduce the amount of income taxes paid. The federal government encourages business use of certain preferred energy sources by granting tax preferences, such as a 30% investment tax credit for the cost of building a solar power plant. In some cases credit is disbursed even if no taxes are payable. Another example is the power production tax credit of 2.2 cents per kWh for wind-generated power.

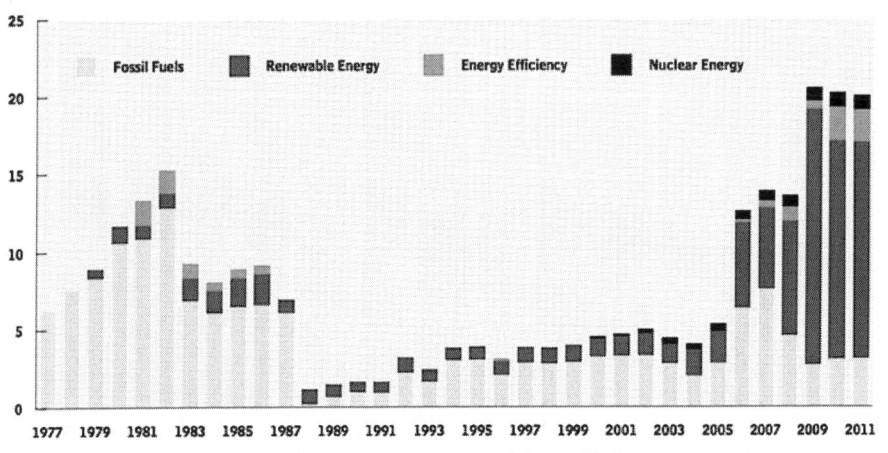

Tax preferences for fossil, renewable, efficiency, nuclear

The Congressional Budget Office prepared this chart of energy subsidies in the form of tax preferences, in billions of dollars. The increases in renewable energy investments starting in 2006 are for wind and solar power. In 2011 tax preferences totaled $20.5 billion, and DOE funding was $3.4 billion.

In response to a Congressional request, the Energy Information Agency developed a subsidy analysis for just the electric power sector for 2010, including all subsidy types.

2010 US subsidies and support for electric power by US DOE Energy Information Agency		
Energy source	Subsidy $ millions	Share %
Coal	1,189	10.0
Natural gas	654	5.5
Nuclear	2,499	21.0
Biomass	114	1.0
Geothermal	200	1.7
Hydro	215	1.8
Solar	968	8.2
Wind	4,986	42.0
Transmission	971	8.2
Total	11,873	100

$900 million of the nuclear power subsidy is reduced taxes on earnings of decommissioning funds.

DOE spends 3% of its budget on advanced nuclear power.

Excluding weapons work, DOE's 2012 budget is $17,700 million. DOE supports some research and development of advanced nuclear power, such as high-temperature gas-cooled reactors. This table includes just the line items for advanced nuclear power. LFTR and DMSR are not supported. DOE also budgeted $67 million to share license applications costs paid to NRC by businesses developing small modular reactors.

DOE Nuclear Engineering 2012 budget	
Advanced technology line items	**$ millions**
Nuclear energy enabling technologies	97
Reactor concepts R&D	125
Fuel cycle research and development	155
Idaho facilities management	150
Total for advanced nuclear technologies	527

The Advanced Research Projects Agency--Energy (ARPA-E) awards of $650 million included nothing for advanced nuclear power R&D.

The $33 billion energy allocation of the $800 billion American Recovery and Investment Act of 2009 provided nothing for advanced nuclear power R&D.

An analysis of DOE historical nuclear energy expenditures, by Management Information Services, shows a historical continuing decline in nuclear energy research, which was dominated by the cumulative $16 billion (2010$), spent for liquid metal cooled fast breeder reactor development ended in the Clinton administration.

2009 risk transfers were $31 billion in loans and guarantees.

Risk transfers occur when the federal government loans money to an enterprise or guarantees a loan. This enables risky projects that might not otherwise be undertaken. It also lowers the borrowing costs to the project. There is no cost if the loans are repaid, so the subsidy amount is not clear. The Pew Charitable Trust initiative, Subsidyscope, identifies $31 billion of such risk transfers for the 2009 fiscal year.

[Subsidyscope.org contains excellent analyses of federal subsidies to the energy sector and ten other US economic sectors.

Subsidyscope also contains a data base of grants and contracts, with interactive searching and reporting capabilities.]

The 50 US states have 50 additional energy policies.

In the US, states have been active in changing electric power generation rules and prices. The proffered reason for most changes is to check global warming by reducing CO_2 emission, even though a single state's small reductions, or even all of the states' reductions, can't make a dent in the global problem. The motivation seems to be to assuage pollution guilt or exhibit leadership in combating climate change, expecting others to follow suit. People feel good about taking any steps, however insignificant.

The national result is a mishmash of confusing and changing rules about electric power, which crosses state boundaries and should be managed with national scope.

The Regional Greenhouse Gas Initiative is a cap-and-trade market for limiting CO_2 emissions, started in 2008. Connecticut, Delaware, Maine, Maryland, Massachusetts, New Hampshire, New York, Rhode Island and Vermont cooperate by requiring utilities to bid for capped rights to emit CO_2 when generating power. The objective is to reduce CO_2 emissions 10% by 2018. The states require power generating utilities to pay for the CO_2 emitted; the market price in 2012 is roughly $2/ton. This will likely rise as the cap will decrease 2.5% per year beginning in 2014. The cap was set about 20% higher than actual emissions, so CO_2 reductions from this are nil. Quarterly auctions net about $40 million dollars; total to date is about $1 billion. The proceeds are divided among the participating states. The money is intended to be used for CO_2-reducing projects such as improving energy efficiency, but states are free to spend the money on other purposes. New Jersey has left RGGI and New Hampshire is debating leaving. The small cost of $2/ton of CO_2 has little effect on behavior; it is paid for by increased charges for electric power.

Investment tax credits for renewable energy projects exist at the state level as well as the federal level. In Vermont this was 30%, but this particular tax credit has been eliminated.

Feed-in tariffs are requirements forcing electric utilities to buy specified renewable-sourced power at above-market rates. In the US most states have a deregulated electric power market, where electric utilities buy power from independent companies – merchant generators. The utilities have responsibility for power transmission, distribution, and customer service. They buy power in a competitive marketplace from merchant generators who offer the lowest prices. Feed-in tariffs supersede this process in a market where price competition settles out at roughly 5 cents/kWh for hydro, nuclear, and natural-gas generated electricity. For example, in Vermont, the feed-in tariff for PV solar power was 30 cents/kWh when the first plants were built. The 2012 law now sets prices, not on CO_2 abatement, but on the cost of generating each type of renewable energy, for example (in cents/kWh): solar (27), hydro (12), farm methane (14), wind (11), small wind (25), biomass (12). Guaranteeing profitable prices reduces producer cost-reduction incentives. Feed-in tariffs also apply in states of the US where utilities generate power. Feed-in tariffs are common in Europe. Germany has reduced solar rates in 2012 to 23 to 30 cents/kWh. Greece pays up to 63 cents/kWh. Sunny Spain pays 27 cents/kWh. UK plans to reduce its home-scale solar feed-in tariff to 25 cents/kWh.

Production tax credits are paid to power producers for actual generation of power. In addition to the federal 2.2 cent/kWh program, Iowa pays at least 1 cent/kWh to wind power producers. Arizona, New Mexico, Oklahoma, and Maryland offer production tax credits.

Renewable energy certificates (RECs) represent a property right created by generating 1 MWh of CO_2-free electricity (except from nuclear power). Generating companies can sell the energy and certificates separately. Utilities can meet requirements for renewable energy by generating it or buying RECs in an open market. RECs are classified by energy source: wind, solar, biomass, etc. Massachusetts specifies a minimum price of 5.5 cents/kWh; elsewhere the auction market prices range from 0.1 to 3.0 cents/kWh. Companies seeking to reduce their advertised net carbon footprint can buy RECs; Intel bought 2.5 billion kWh of RECs in 2011 to offset over 85% of their electricity use.

Renewable portfolio standards (RPSs) are mandates that require electric utilities to obtain certain fractions of their power from specified renewable energy sources. Every state has different rules, requiring from 10% to 40% of electricity be obtained from various renewable sources by deadlines ranging from 2015 to 2030. Some states allow meeting RPS requirements by purchasing RECs. The US Congress is considering a federal RPS law.

Carbon taxes are taxes on CO_2 emitted to produce power. Small carbon taxes are enacted in Colorado (0.5 cents/kWh), California (4.4 cents/ton CO_2), and Maryland ($5/ton CO_2).

Administration of the mishmash of policies is expensive. The rules, exceptions, allowances, auctions, audits, and labor are very complex and volatile. Only clever business people can make use of the rats' nest of regulations. One solar power project in Vermont was able to be profitable because of a 30% federal investment tax credit, a 30% state investment tax credit, accelerated depreciation, a feed-in tariff guaranteeing sales at 30 cents/kWh, and opportunities to sell RECs.

Existing energy policies are failing.

Carbon dioxide emissions are still rising. In 2011 global CO_2 emissions rose 3.2% to 31.6 Gt, led by China and India. US emissions dropped 1.7% due to a mild winter and power generators switching from coal to natural gas. EU emissions dropped 1.9% due to a warm winter and industrial recession. Japan emissions rose 2.8% from shutting down nuclear power plants.

Reducing US CO_2 emissions can do little to check global warming, because the US represents just 17% of the problem. The DOE EIA projects 0.3% annual growth in US CO2 emissions. 1.3% for the world, and 2.6% for China and India.

Germany is shutting down nuclear power plants, burning more coal, building 17 new coal plants, and burning natural gas from Russia. The rising price of electricity has already bankrupted an aluminum company there.

ENERGY POLICY RECOMMENDATIONS

I recommend that the goals of US energy policy should be:

1. Stopping global warming.
2. Protecting the environment.
3. Protecting human health and safety.
4. Ensuring a sustainable world.
5. Ending energy poverty.
6. Furthering economic growth.
7. Assuring energy security.

I recommend that the agents of this pursuit be the federal government of the United States, enabling corporations to develop innovative energy sources, with leadership from politicians, philanthropists, and entrepreneurs.

Lead energy policy at the federal level, not the state level.

Energy flows across state lines, as do EPA-regulated emissions and DOT-regulated trucks. NRC continues strong effective control over all nuclear plants. Energy policy seems largely ceded to the states, which conceive and implement feed-in tariffs, RECs, RPSs, tax credits, etc in 50 different ways. There is a Federal Energy Regulatory Commission, but it is silent on these matters.

Audit energy policy with neutral experts.

The Congressional Budget Office assists the Congress by analyzing the financial impact of legislation. The CBO is well regarded as professional and neutral. Congress could benefit from similar reviews of existing energy policy by operational experts.

The integrated systems operators, ISOs, have regional utilities and generators as members. ISOs such as ERCOT (Texas) and ISO-NE (northeast US) manage day-to-day reliable operation of power generation and distribution and also oversee the administration of regional wholesale electricity markets. Their employees are neutral experts who understand the pricing and service impacts of intermittent power generation and the interaction of wind and solar power with hydro, nuclear, coal, and natural gas power plants. They would be ideal partners for a CBO study.

End subsidy-based energy policy.

Most energy subsidies are tax preferences. Federal and/or state governments do not make explicit payments but forego tax revenues. Tax preferences do not help innovative start-ups, which have no profits.

Electricity consumers pay another large subsidy for government-favored power sources through mandated feed-in tariffs at rates 300% over market. Wind and solar power could not compete without feed-in tariffs. The rationale for such consumer-paid subsidies is that with experience the cost of technologies for solar and wind will diminish, but that is not in evidence. Subsidies ruin economic competition and raise prices.

Reduce energy costs.

Energy costs are important to developed economies such as the US, but affordable electric power is crucial to the developing economies, where over a billion people have no access to electricity. Ending their poverty can reduce overpopulation and resource conflicts. Decreasing energy costs can improve economic productivity in OECD nations.

Reduce CO2 emissions.

This seems obvious, but this goal is often lost sight of. For example, states force consumers to buy wind power at three times the market price, even when the starting and stopping of natural gas backup generators largely cancel out the CO_2 savings of wind turbines when they operate. Policies should consider total CO_2 reductions. Japan and Germany dropped this goal and are now burning ever more fossil fuels.

End renewable-energy source favoritism.

Feed in tariffs, renewable energy credits, and renewable portfolio standards all call for a mix of government-selected generation technologies. The purpose is to stop global warming by ending CO2 emissions, but how can 50 state legislatures pick the right technology solutions? Even simple carbon taxes would be better than this unmanageable, lobbied favoritism.

Invest in energy cheaper than coal.

Innovative, disruptive energy sources that undersell coal will dissuade all nations from burning coal and other fossil fuels, in their own economic self-interest, thus reducing CO2 emissions that cause global warming. Ending coal mining and eliminating coal plant particulate emissions will preserve our landscapes and save millions of lives. Providing affordable power to the developing world can improve prosperity leading to a sustainable population. All economies will benefit from lowering, not raising, the cost of energy. Providing all nations with domestic energy sources will reduce wars.

Invest in innovative nuclear power R&D.

Current DOE R&D investment in advanced nuclear power is only about $500 million, with nothing for liquid fuel reactors. China is already spending $350 million on this area. The US could gainfully spend $1 billion per year to advance R&D for LFTR, DMSR, and PB-AHTR reactors. TerraPower and GE are now pursuing the path plowed by US R&D in LMFBRs.

I recommend that government-funded R&D be public domain, with resulting intellectual property available to all developers. Corporations can have an important role in such government R&D; for example, Union Carbide operated ORNL during MSR development.

Conducting R&D at several centers increases competition. Universities, corporations, and DOE national laboratories all can contribute. One problem of Weinberg's MSR is that all the expertise was at ORNL, with little knowledge of MSRs among government decision makers and advisors.

Invest in thorium energy cheaper than coal R&D.

Energy from a liquid fluoride thorium reactor can be cheaper than coal because of simple fuel handling, high thermal capacity of heat exchange fluids, atmospheric pressure core, inherent passive safety, the small components, efficient high-temperature turbine power conversion, factory production, and new technology already in development.

Invest in power conversion technology R&D.

There are two new power conversion technologies that efficiently take advantage of the high-temperature heat of advanced nuclear power such as LFTR. They have been demonstrated at laboratory scale, but not utility scale. The closed-cycle triple-reheat Brayton gas turbine uses the same technology as aircraft jet engines, except thermal power comes from heat transfer from molten salt, and the gas may be recirculated from exhaust to input. The supercritical CO_2 turbine uses recirculating CO_2 at such high temperature and pressure that it behaves like a compressible liquid. The small size augers low costs for this important power plant component.

Invest in high-temperature irradiated materials R&D.

The high-temperature heat of advanced nuclear power is important for two reasons: (1) increased efficiency and therefore lower costs for electric power generation, and (2) industrial process heat for extraction of oil from oil shale and tar sands, for cement manufacturing, and chemical and metals production. New metal alloys and silicon-carbon ceramic composites require extensive testing under neutron irradiation. The DOE has two high neutron flux reactors that can expose test materials to the radiation of years of power plant use, in months.

Invest in high-temperature hydrogen production.

Thermochemical cycles such as sulfur-iodine and copper-chlorine can generate H2 by splitting H2O molecules at chemical/thermal conversion efficiencies nearing 50%. Now demonstrated at laboratory scale, these technologies should be confirmed at pilot plant scale. Industrial scale hydrogen production will be required for fuel production in a post-carbon world.

Invest in hydrogen-energized synfuel pilot plants.

High-temperature reactors such as LFTR or NGNP can efficiently produce hydrogen by water dissociation. Existing chemical industry technologies such as Fischer-Tropsch can synthesize hydrocarbon fuels by combining hydrogen with carbon from sources such as coal plant flue gas, biomass. This will reduce both CO2 emissions and petroleum dependency. After successful demonstration of electrolysis and high-temperature hydrogen-dissociation pilot-plants, the petrochemical industry will become interested in building commercial synfuels production plants.

Facilitate corporate development of advanced nuclear power.

We will rely on the business and management skills of corporations to be able to mass produce affordable nuclear reactors the way Boeing produces airplanes. Both NASA and corporations were responsible for the Apollo mission successes. The federal government can facilitate LFTR development by lending its facilities, such as at the Savannah River Site for construction of prototype small modular reactors in parallel with obtaining NRC licenses. The Idaho National Laboratory might be similarly enabled. We must streamline and rationalize rules and regulations to let companies pursue our clean energy goals without deadly delays. We must inoculate development and construction processes against unreasonable injunctions sought by lawyerly opponents only seeking financial ruin by delay.

Fund the NRC to learn about advanced nuclear power.

In 2012 Congress provides the NRC with only $129 million in federal funds, in addition to fees collected from the existing LWR commercial power industry. The NRC regulations and staff are well able to oversee todays LWR nuclear power plants, but not new technologies such as LFTR. Today license applicants must pay over $250 per hour for all the hours for NRC staff to learn about new technologies and pass judgment on safety. It could take hundreds of millions of dollars to license new technologies, which even risk-taking ventures can simply not afford. NRC had initiated a concept of technology-neutral licensing, but that is not now funded. The NRC needs to hire and train nuclear-engineering-skilled staff able to understand and critique license applications for new technologies such as LFTR.

Invest in low-level radiation safety research.

The benefits of nuclear medicine and nuclear power are quantifiable and established. The risks are not so accepted. The US limits general public exposure to less than 1 mSv per year, but there is not evidence of harm from low-level radiation less than 100 mSv per year, and there is new evidence that it triggers protective responses. The US defunded low-level radiation research, but more is warranted to establish the true health risks of low-level ionizing radiation.

Educate the public about nuclear power.

Many people fear nuclear power. Politicians and the media use this fear to gain attention. Opponents make outrageous statements. The press still dwells on Fukushima but hardly reported Obama's March 26, 2012 endorsement

> "... let's never forget the astonishing benefits that nuclear technology has brought to our lives. Nuclear technology helps make our food safe. It prevents disease in the developing world. It's the high-tech medicine that treats cancer and finds new cures. And, of course, it's the energy—the clean energy—

that helps cut the carbon pollution that contributes to climate change."

The government and its leaders should learn more about the benefits and unexaggerated risks of nuclear power. We need a deliberate, accurate, strong education program to persuade more of the public that nuclear power plants are safer than any other energy source.

Prepare to compete with other nations.

The largest energy industry competitors are nations, not international corporations such as Exxon-Mobil. Exxon-Mobil has the largest revenue and market capitalization of any corporation, but produces just 3% of world oil and has fewer oil reserves than the national oil companies of Saudi Arabia, Iran, Iraq, Venezuela, Abu Dhabi, and Kuwait. Increasingly competition is among nations, not corporations.

Export restrictions now limit US companies' abilities to compete in the international marketplace for nuclear power. Restrictions should be revised, because LFTR represents an opportunity to achieve international superiority in the market for clean, safe, energy cheaper than coal.

Export LFTR nuclear power plants.

Simply generating inexpensive, nonpolluting LFTR power within the US is not enough to solve the global energy and environmental crises. The US should encourage exporting these small nuclear power plants because they can help the developing world end energy poverty, cut CO_2 emissions globally, and become a $70 billion export industry to help the US economy. Russia, China, South Korea, and India all plan nuclear power plant exports.

Lead!

Who will lead?

1. A transnational organization such as the United Nations?
2. One nation such as the United States?
3. Multiple state or provincial governments?
4. Corporations?
5. Leadership individuals?

The United Nations can not solve our energy/climate crises. Dozens of IPCC-sponsored meetings only end in promises to agree and contention between rich and poor nations. Few nations will sacrifice national energy sovereignty for global good.

The United States can lead in developing LFTR and thorium energy cheaper than coal. The US has the DOE national labs, the best university nuclear engineering programs, and the government/university/business tradition of entrepreneurism and commercialization.

Political leadership is lacking. At the executive, congressional, and state levels elected officials fail to grasp the realities of economics, energy, environmental pollution, and global resource contention. Instead these politicians capitalize on the crowd-sourced fears of all things nuclear, and they attract *feel good* voters by promoting *natural* wind and solar energy sources, hiding the true social costs in grants, subsidies, and tax preferences that only benefit select, savvy businessmen.

Yet there is an immense political opportunity for a leader to

1. satisfy liberals and environmentalists by checking global warming and ending energy poverty, and also
2. satisfy conservatives and businesses by avoiding carbon taxes, decreasing energy costs, and creating a new Boeing-size export industry.

Governments have an opportunity to incentivize corporations to undertake LFTR research and development. Once power-plant-scale LFTRs are successfully demonstrated, and once the legal system permits, corporations can then lead in mass production of LFTRs. We can then rely on economic self-interest of corporations to produce and install LFTRs as fast as Boeing sells airplanes. The corporations will succeed because they can rely on the economic self-interest of 7 billion people in 250 nations to choose the cheapest source of clean, safe energy. This will end CO_2-emitting energy from coal and reduce demand for energy from other fossil fuels.

Ultimately, individual leaders are the key. Rickover led nuclear power development. Eisenhower led atoms for peace. Weinberg led MSR development for "humankind's whole future". President Kennedy led the Apollo mission. Bill Gates is leading philanthropic efforts to end energy poverty. Jiang Mianheng is leading MSR development for China. Venturesome Kirk Sorensen is leading financial support for LFTR at Flibe Energy. An international businessman is quietly seeking to use LFTR to end energy poverty in Africa and beyond.

Kennedy's 1962 speech made LFTR development sound easy:

> *"But if I were to say, my fellow citizens, that we shall send to the moon, 240,000 miles away from the control station in Houston, a giant rocket more than 300 feet tall, the length of this football field, made of new metal alloys, some of which have not yet been invented, capable of standing heat and stresses several times more than have ever been experienced, fitted together with a precision better than the finest watch, carrying all the equipment needed for propulsion, guidance, control, communications, food and survival, on an untried mission, to an unknown celestial body, and then return it safely to earth, re-entering the atmosphere at speeds of over 25,000 miles per hour, causing heat about half that of the temperature of the sun--almost as hot as it is here today--and do all this, and do it right, and do it first before this decade is out--then we must be bold."*

Who will lead?

THORIUM: ENERGY CHEAPER THAN COAL

We can solve our global energy and environmental crises through technology innovation and free-market economics, using a disruptive technology – energy cheaper than coal.

If we offer to sell to all the world the capability to produce energy that cheaply, all the world will stop burning coal. We can rely on the economic self-interest of 7 billion people in 250 nations to choose cheaper, nonpolluting energy.

The US should fund rapid development of this innovative nuclear power technology that can deliver energy cheaper than coal. Thereafter corporations can achieve mass production of liquid fluoride thorium reactors. The US should enable corporations to develop, produce, and operate LFTRs quickly and safely.

This book, *THORIUM: energy cheaper than coal*, advocates lowering costs for energy – the market-based environmental solution.

$ 1 B	$ 5 B	$ 70 B per year industry
Develop	Scale up	Produce — Export
2012	2017	2022

- Cut 10 billion tons/year CO2 emissions to zero by 2060.
- Avoid carbon taxes.
- Stop deadly air pollution.
- Improve developing world prosperity.
- Foster a sustainable world population.
- Use inexhaustible thorium fuel, available in all nations.

Who will lead?

References

These references are also available on the book website, http://www.thoriumenergycheaperthancoal.com

FRONT MATTER, FOREWORD, INTRODUCTION

p 4 Aim High! http://rethinkingnuclearpower.googlepages.com/aimhigh
p 4 Ralph Moir: http://ralphmoir.com
p 23 Rethinking nuclear power course: http://rethinkingnuclearpower.googlepages.com
p 24 Energy from thorium blog and forum: http://energyfromthorium.com
p 24 American Scientist article: http://home.comcast.net/~robert.hargraves/public_html/2010Hargraves2.pdf
p 25 Kirk Sorensen, fire story; http://www.youtube.com/watch?v=L-T-WSWgBCc
p 26 Barack Obama March 26, 2012: http://www.whitehouse.gov/the-press-office/2012/03/26/remarks-president-obama-hankuk-university

ENERGY AND CIVILIZATION

p 30 Energy share of GDP: http://www.instituteforenergyresearch.org/2010/02/16/a-primer-on-energy-and-the-economy-energys-large-share-of-the-economy-requires-caution-in-determining-policies-that-affect-it/
p 31 Car: http://www.vanseodesign.com/web-design/visual-grammar-lines
p 32 Jaguar, Volvo flywheel hybrids: http://www.economist.com/node/21540386
p 32 Hurricane energy: http://www.aoml.noaa.gov/hrd/tcfaq/D7.html
p 32 Roller coaster: http://davidmanlysblog.blogspot.com/2011/05/ups-and-downs-of-physics.html
p 33 Clock weight escapement: http://commons.wikimedia.org/wiki/File:PSM_V29_D198_Gravity_clock_escapement_mechanism_aided_by_weight.jpg
p 34 Water wheel: http://chestofbooks.com/crafts/mechanics/Engineer-Mechanic-Encyclopedia-Vol2/Water-Wheel-Part-2.html

p 34 Chemical bonds:
http://www2.chemistry.msu.edu/faculty/reusch/VirtTxtJml/intro2.htm
p 37 Alhambra fountain: http://en.wikipedia.org/wiki/Alhambra
p 40 Gas molecules: http://en.wikipedia.org/wiki/File:Translational_motion.gif
p 47 Carnot heat engine:
http://en.wikipedia.org/wiki/File:Carnot_heat_engine_2.svg
p 48 Pulverized coal: http://web.mit.edu/mitei/docs/reports/beer-emissions.pdf
p 48 Pulverized coal temperature:
http://old.enea.it/attivita_ricerca/energia/sistema_elettrico/Centrali_carbone_rendimenti/RSE110.pdf
p 51 Taming of the Chloroplast:
http://evolutionaryroutes.wordpress.com/2011/08/31/the-taming-of-the-chloroplast/
p 52 Eukaryotic cell:
http://www.m2c3.com/chemistry/VLI/M2_Topic2/M2_Topic2_print.html
p 53 Richard Wrangham:
http://news.harvard.edu/gazette/story/2009/06/invention-of-cooking-drove-evolution-of-the-human-species-new-book-argues/
p 54 Horses: Vaclav Smil, Energies: An Illustrated Guide to the Biosphere and Civilization, MIT Press.
p 53 Quaker Oats; http://www.amazon.com/Quaker-Nutrition-Calories-Energy-Flakes/dp/B005DGWOK0
p 53 Rickover speech on energy and slavery:
http://desc.hinchey.house.gov/DODRickover1957Speech.doc
p 54 Brazil green steel: http://www.forestry-invest.com/2010/eucalyptus-charcoal-brazils-choice-for-the-steel-industry/268
p 55 Grinding flour: http://www.touregypt.net/featurestories/bread.htm
p 56 Waterwheel: http://www.top-alternative-energy-sources.com/water-wheel-design.html
p 57 Windmill: http://en.wikipedia.org/wiki/Windmill
p 58 England patent law and industrial revolution: http://www.amazon.com/The-Most-Powerful-Idea-World/dp/0226726347
p 58 Newcomer steam engine: http://en.wikipedia.org/wiki/Industrial_revolution
p 60 World GDP per capita:
http://econ161.berkeley.edu/TCEH/1998_Draft/World_GDP/Estimating_World_GDP.html
p 61 World energy consumption: http://www.eia.gov/forecasts/ieo/index.cfm
http://en.wikipedia.org/wiki/World_energy_consumption
p 63 World CO2 emissions: http://www.eia.gov/forecasts/ieo/index.cfm

AN UNSUSTAINABLE WORLD

p 66 Meadows' Limits to Growth at ASPO:
http://www.aspoitalia.it/images/stories/aspo5presentations/Meadows_ASPO5.pdf
p 66 Revisiting Limits to Growth, American Scientist:
http://www.americanscientist.org/issues/pub/2009/3/revisiting-the-limits-to-growth-after-peak-oil

NOTES AND REFERENCES

p 66 Meadows reviewed in Smithsonian:
http://www.smithsonianmag.com/science-nature/Looking-Back-on-the-Limits-of-Growth.html
p 68 Murphy interview: http://oilprice.com/Interviews/Tom-Murphy-Interview-Resource-Depletion-is-a-Bigger-Threat-than-Climate-Change.html
p 68 Murphy, Do the Math: http://physics.ucsd.edu/do-the-math/ p 70 CIA World Fact Book data: https://www.cia.gov/library/publications/the-world-factbook/docs/rankorderguide.html
p 74 EIA world energy; http://www.eia.doe.gov/oiaf/ieo/world.html
India energy use: http://www.world-nuclear-news.org/NP_Nuclear_the_fuel_for_energetic_Indian_growth_2202121.html
p 74 OECD Environmental Outlook to 2050:
%http://www.oecd.org/document/34/0,3746,en_21571361_44315115_49897570_1_1_1_1,00.html
p 76 NOAA climate charts: http://www.ncdc.noaa.gov/cmb-faq/anomalies.php
p 78 Hansen climate forcings:
http://www.columbia.edu/~jeh1/2010/201010_BluePlanet.ppt
p 79 IPCC projections:
http://www.ipcc.ch/publications_and_data/ar4/syr/en/main.html
p 80 Rongbuk glacier:
http://www.columbia.edu/~jeh1/2010/201010_BluePlanet.ppt
p 81 Coral bleaching:
http://news.nationalgeographic.com/news/bigphotos/10063392.html
p 81 NY Times ocean acidity:
http://www.nytimes.com/2009/01/31/science/earth/31ocean.htmlRising
p 81 Ocean acidity tutorial:
http://www.skepticalscience.com/Mackie_OA_not_OK_post_1.html
p 81 Ocean acidification:
http://www.sciencedaily.com/releases/2012/03/120301143735.htm
p 82 EPA sulfur dioxide emission:
http://www.epa.gov/airtrends/2007/report/sulfurdioxide.pdf
p 82 EPA 2011 air pollution rule:
http://yosemite.epa.gov/opa/admpress.nsf/d0cf6618525a9efb85257359003fb69d/cedd944b946fdc5f852578c60055e818!OpenDocument
p 83 Guardian ship emissions:
http://www.guardian.co.uk/environment/2009/apr/09/shipping-pollution
p 83 Gizmag ship emissions: http://www.gizmag.com/shipping-pollution/11526/
p 84 EPA ship emissions rule:
http://www.epa.gov/aging/press/epanews/2009/2009_1222_1.htm
p 86 Conflict over resources: Prof. Michael Klare, Hampshire College, author of "Resource Wars" and "Blood and Oil: The Dangers and Consequences of America's Growing Petroleum dependency"
p 89 China Daily News Oct 7, 2010:
http://pub1.chinadaily.com.cn/cdpdf/us/download.shtml?c=32073
p 89 Greenland ice sheet melt:
http://www.nasa.gov/topics/earth/features/greenland-melt.html
p 90 OECD Environmental Outlook to 2050:
http://www.oecd.org/dataoecd/32/53/49082173.pdf

ENERGY SOURCES

p 101 Energy Safari: http://pages.google.com/pages/energysafari.
p 102 EIA 2010 sources/uses:
http://www.eia.gov/totalenergy/data/annual/pdf/sec2_3.pdf
p 105 EIA 2010 Electric Power Annual: http://www.eia.gov/electricity/annual/
p 103 Data center power use:
http://online.wsj.com/article/SB10001424052702303610504577420251668850864.html
p 102 EIA 2012 Energy Outlook:
http://www.eia.gov/electricity/annual/http://www.eia.gov/pressroom/presentations/howard_01232012.pdf
p 105 EIA 2010 Electric Power Annual: http://www.eia.gov/electricity/annual/
p 106 EIA 2010 capital costs:
http://www.eia.gov/oiaf/beck_plantcosts/pdf/updatedplantcosts.pdf
p 107 EIA 2010 Power Generation Costs:
http://www.eia.gov/oiaf/beck_plantcosts/excel/table2.xls
p 108 EIA CO2 emissions: http://205.254.135.7/forecasts/ieo/emissions.cfm

Coal

p 109 IGCC efficiency: http://web.mit.edu/mitei/docs/reports/beer-emissions.pdf
p 110 New coal fired plants: http://www.netl.doe.gov/coal/refshelf/ncp.pdf
p 111 MIT CCS data base: http://sequestration.mit.edu/index.html
p 112 CBO CCS report: http://cbo.gov/publication/43357
p 113 China GreenGen: http://sequestration.mit.edu/tools/projects/greengen.html
p 113 Zobach Gorelick Earthquake triggering:
http://www.pnas.org/content/early/2012/06/13/1202473109.abstract?sid=f6da10e3-978d-4e86-9101-9079d428ba35
p 114 EIA coal costs: http://205.254.135.7/electricity/annual/pdf/table3.5.pdf
p 114 MIT Revised Cost of Nuclear Power:
http://web.mit.edu/nuclearpower/pdf/nuclearpower-update2009.pdf
p 114 MIT Future Nuclear Power Fuel Cycle Ch 1-3:
http://web.mit.edu/mitei/research/studies/documents/nuclear/nuclearpower-ch1-3.pdf
p 114 MIT Future Nuclear Power Fuel Cycle Ch 4-9:
http://web.mit.edu/mitei/research/studies/documents/nuclear/nuclearpower-ch4-9.pdf
p 114 MIT Future of Coal:
http://web.mit.edu/mitei/research/studies/documents/coal/The_Future_of_Coal.pdf
p 114 EIA World Energy Outlook:
http://www.eia.gov/forecasts/ieo/pdf/0484(2011).pdf
p 115 NY Times fossil fuel costs:
http://www.nytimes.com/2009/10/20/science/earth/20fossil.html
p 115 NAS fossil fuel hidden costs;
http://www.nytimes.com/2009/10/20/science/earth/20fossil.html

NOTES AND REFERENCES

p 115 Harvard Med School coal costs:
http://www.loe.org/images/content/110218/CoalPamphlet_Final_SingPg(2).pdf

Gas

p 117 Natural gas combustion turbine:
http://commons.wikimedia.org/wiki/File:Brayton_cycle.svg
p 121 Hydraulic fracturing: http://www.fraw.org.uk/ideas/fracking/index.html
p 121 Fugitive methane emissions:
http://www.sustainablefuture.cornell.edu/news/attachments/Howarth-EtAl-2011.pdf
p 122 Matt Ridley Shale Gas Shock:
http://marcellus.psu.edu/resources/PDFs/shalegas_GWPF.pdf
p 123 EIA Shale Gas:
http://www.eia.gov/analysis/studies/usshalegas/pdf/usshaleplays.pdf
p 122 EPA fracking emissions rule:
http://www.nytimes.com/2012/04/19/science/earth/epa-caps-emissions-at-gas-and-oil-wells.html
p 124 US nat gas pipelines:
http://www.eia.gov/pub/oil_gas/natural_gas/analysis_publications/ngpipeline/ngpipelines_map.html
p 126 Natural gas prices: http://www.ferc.gov/market-oversight/mkt-gas/overview/ngas-ovr-lng-wld-pr-est.pdf
p 127 Japan nuclear-free GDP drop 7%:
http://ajw.asahi.com/article/0311disaster/fukushima/AJ201206300053
p 128 Natural gas prices, EIA: http://www.eia.gov/dnav/ng/hist/rngc1d.htm
p 128 EIA 2012 Energy Outlook:
http://www.eia.gov/pressroom/presentations/howard_01232012.pdf
EIA Natural gas prices, Howard:
http://www.eia.gov/pressroom/presentations/howard_01232012.pdf
p 128 EIA Int'l natural gas outlook: http://www.eia.gov/forecasts/ieo/nat_gas.cfm
p 129 Pittinger shale gas prices: http://www.theoildrum.com/node/8212
p 130 Natural gas matches coal:
http://www.reuters.com/article/2012/06/27/utilities-coal-gas-eia-idUSL2E8HRG6820120627

Wind

p 132 Brazos wind farm: http://en.wikipedia.org/wiki/Brazos_Wind_Farm
p 133 US DOE wind map:
http://www.eere.energy.gov/windandhydro/windpoweringamerica/wind_maps.asp
p 135 Cape Wind prices: http://www.bostonglobe.com/metro/2011/12/28/after-court-ruling-cape-wind-poised-move-forward/cjtMPcMX47lYPDbtbH5fTK/story.html
p 135 NStar merger:
http://www.boston.com/Boston/businessupdates/2012/02/nstar-agrees-buy-cape-wind-power-win-state-okay-merger/38Tlb9N1uq7B8P3WHxfOOK/index.html
p 135 Deepwater Wind: http://www.reuters.com/article/2011/10/13/us-deepwater-wind-idUSTRE79C0YC20111013p 140 GE FlexEff/Wind combo:

http://theenergycollective.com/willem-post/59747/ge-flexefficiency-50-ccgt-facilities-and-wind-turbine-facilities
p 136 William Palmer, Ontario Coal/Wind: http://www.masterresource.org/2012/02/ontario-windpower-case-study-i/
p 139 EPA proposed CO2/kWh limit: http://epa.gov/carbonpollutionstandard/pdfs/20120327factsheet.pdf
p 140 Willem Post Wind/CO2: http://theenergycollective.com/willem-post/64492/wind-energy-reduces-co2-emissions-few-percent
p 141 Australia wind farm performance: http://windfarmperformance.info/
p 142 Irish grid wind CO2: http://www.clepair.net/IerlandUdo.html
p 142 Bentek study CO and TX: http://docs.wind-watch.org/BENTEK-How-Less-Became-More.pdf
p 142 Lang CO2 avoided by wind: http://bravenewclimate.files.wordpress.com/2009/08/peter-lang-wind-power.pdf

Solar

p 144 Passive solar: http://www.energysavers.gov/your_home/designing_remodeling/index.cfm/mytopic=10270
145 China solar hot water: http://www.easybizchina.com/freemember/products/3303/snxing_solar_energy_technology_co__ltd-1.html
p 146 IEA world solar: http://www.iea-shc.org/publications/downloads/Solar_Heat_Worldwide-2011.pdf p 146 AllEarth solar production: http://www.allearthrenewables.com/energy-production-report/detail/316#view=yearly&date=2011-01-01
p 146 AllEarth solar VT diocese: http://www.vermontbiz.com/news/january/largest-solar-installation-burlington-now-operating-rock-point
p 149 Albiasa Caceres: http://www.albiasasolar.com/pdfs/projects.pdf
p 149 Albiasa abandons Arizona: http://www.azinews.com/2011/09/01/albiasa-abandons-solar-project/
p 149 Abengoa solar cost: http://www.abengoasolar.com/corp/web/en/acerca_de_nosotros/sala_de_prensa/noticias/2011/solar_20110913.html
p 149 Brightsource CA solar cost: http://www.latimes.com/news/local/la-me-solar-tortoise-20120304,0,6145488.story
p 147 Andasol parabolic troughs: http://www.renewbl.com/2009/07/02/solar-millenium-officially-inaugurated-andasol-1-parabolic-trough-power-plant.html
p 148 Andasol solar molten salt: http://www.nrel.gov/csp/troughnet/pdfs/2007/martin_andasol_pictures_storage.pdf
p 148 Solar grand plan: http://www.scientificamerican.com/article.cfm?id=a-solar-grand-plan&page=1
p 150 MIT Intermittent Renewables: http://web.mit.edu/mitei/research/reports/intermittent-renewables-full.pdf

Biofuels

p 151 Wood composition:
http://marioloureiro.net/ciencia/ignicao_vegt/ragla91a.pdf
p 151 Wood moisture:
http://www.epa.gov/burnwise/workshop2011/WoodCombustion-Curkeet.pdf
p 152 EPA clean energy stats: http://www.epa.gov/cleanenergy/energy-resources/refs.html
p 151 USDA BTUs green wood: http://www.fpl.fs.fed.us/documnts/techline/fuel-value-calculator.pdf
p 151 EPA forest carbon sequestration:
http://www.epa.gov/sequestration/faq.html
p 152 NH biomass plant cost: http://supportnhbiomass.wordpress.com/press-releases/
p 153 Burlington McNeil wood chip cost:
https://www.burlingtonelectric.com/page.php?pid=75&name=mcneil
p 156 EROI ethanol: http://netenergy.theoildrum.com/node/6760
p 157 Biomass per gallon:
http://www1.eere.energy.gov/biomass/ethanol_yield_calculator.html
p 157 US Renewable Energy Labs Biomass: http://www.nrel.gov/biomass/
p 157 US cellulosic ethanol plant:
http://www.nytimes.com/2011/07/07/business/energy-environment/us-backs-plant-to-make-fuel-from-corn-waste.html
p 157 NREL biofuel brochure: http://www.nrel.gov/biomass/pdfs/40742.pdf
p 158 Corn prices: http://topics.nytimes.com/top/news/business/energy-environment/biofuels/index.html?scp=5&sq=corn%20prices&st=cse
p 158 Food fuel competition:
http://www.nytimes.com/2011/04/07/science/earth/07cassava.html
p 157 WSJ cellulosic ethanol mandate:
http://online.wsj.com/article/SB10001424052970204012004577072470158115782.html

Energy storage

p 159 Economist energy storage:
http://www.economist.com/node/21548495?frsc=dg%7Ca
p 160 Sadoway Mg-Sb liquid battery: http://sadoway.mit.edu/wordpress/wp-content/uploads/2011/10/Sadoway_Resume/141.pdf
p 160 MIT liquid flow battery: http://web.mit.edu/newsoffice/2011/flow-batteries-0606.html
p 160 Battery switching car: http://www.betterplace.com/
p 161 Utility scale batteries:
http://www.electrochem.org/dl/interface/fal/fal10/fal10_p049-053.pdf
p 161 Beacon Power flywheels:
http://www.beaconpower.com/files/EESAT_2011_Final.pdf
p 163 EPRI CAES:
http://my.epri.com/portal/server.pt?space=CommunityPage&cached=true&parentname=ObjMgr&parentid=2&control=SetCommunity&CommunityID=405
p 164 EPRI utility battery costs: http://gigaom.com/cleantech/5-things-you-need-to-know-about-energy-storage/

p 164 EPRI energy storage exec summary:
http://disgen.epri.com/downloads/EPRI%20CAES%20Demo%20Proj.Exec%20Overview.Deep%20Dive%20Slides.by%20R.%20Schainker.Auguat%202010.pdf
p 166 Siemens hydrogen storage:
http://www.technologyreview.com/energy/40001/?nlid=nldly&nld=2012-03-29

Conservation

p 170 EIA 2012 annual energy outlook: http://www.eia.gov/forecasts/aeo/er/
p 171 en.lighten energy saving: http://www.enlighten-initiative.org/portal/CountrySupport/CLAs/Energysavingbenefits/tabid/79099/Default.aspx
p 170 Energy intensity: http://www.eia.doe.gov/pub/international/iealf/tablee1p.xls
p 173 Hansen on meat: http://bravenewclimate.com/2012/03/24/dietary-gc-ignores-cc/

Other

p 168 Hydro: http://en.wikipedia.org/wiki/Hydroelectricity
p 169 Grand Inga Dam: http://www.internationalrivers.org/campaigns/grand-inga-dam-dr-congo
p 174 Desalination: http://en.wikipedia.org/wiki/Desalination#Cogeneration
p 174 Desalination Grand Cayman: http://www.desalination.com/
p 174 Nuclear power: http://www.world-nuclear.org/

LIQUID FLUORIDE THORIUM REACTOR

p 178 Periodic table: http://www.ptable.com/
p 182 NRC PWR: http://www.nrc.gov/reading-rm/basic-ref/students/animated-pwr.html
p 182 NRC BWR: http://www.nrc.gov/reading-rm/basic-ref/students/animated-bwr.html
p 183 Fuel rod cross section: http://jolisfukyu.tokai-sc.jaea.go.jp/fukyu/mirai-en/2008/5_3.html
p 187 Molten plutonium reactor:
http://fas.org/sgp/othergov/doe/lanl/pubs/00416628.pdf
p 189 Johnson thorium chemistry:
http://www.thoriumenergyalliance.com/downloads/TEAC3%20presentations/TEAC3_Johnson_KimLawrence.pdf
p 190 WNA on thorium: http://www.world-nuclear.org/info/inf62.html
p 200 Haubenreich, Engel, MSRE experience
http://energyfromthorium.com/pdf/NAT_MSREexperience.pdf
p 200 Wikipedia MSRE:
http://en.wikipedia.org/wiki/Molten_Salt_Reactor_Experiment
p 192 ORNL molten salt document repository:
http://www.energyfromthorium.com/pdf/

p 192 Hoglund's ORNL molten salt doc repository:
http://moltensalt.org/references/static/downloads/pdf/
p 192 MacPherson 1985 molten salt reactor adventure:
http://www.moltensalt.org/references/static/home.earthlink.net/bhoglund/mSR_Adventure.html
p 188 Aircraft reactor experiment:
http://moltensalt.org/references/static/downloads/pdf/NSE_ARE_Operation.pdf
\p 193 Wikipedia LFTR:
http://en.wikipedia.org/wiki/Liquid_fluoride_thorium_reactor
p 193 Forsberg et al advanced MSR high temp reactor:
www.ornl.gov/~webworks/cppr/y2001/pres/119930.pdf
p 198 MIT Steam/Brayton/SCO2 power conversion:
http://stuff.mit.edu/afs/athena/course/22/22.33/www/dostal.pdf
p 199 Forsberg open cycle Brayton:
https://www.ornl.gov/fhr/presentations/Forsberg.pdf
p 200 Haubenreich interview: http://energyfromthorium.com/msrp/paul-haubenreich/
p 200 Weinberg, Alvin; The First Nuclear Era: The life and times of a technological fixer
p 200 Martin, Richard; SuperFuel: Thorium, the green energy source for the future: http://www.amazon.com/SuperFuel-Thorium-Energy-Source-Future/dp/0230116477/
p 201 World spent fuel stocks:
https://iaea.org/NewsCenter/Features/UndergroundLabs/Grimsel/storageoverview.pdf
p 202 Liquid chloride fast reactor:
http://moltensalt.org/references/static/downloads/pdf/ANL-6792.pdf
p 203 Moir fission-fusion hybrid: http://ralphmoir.com/aFusFisHyb.htm
p 203 Moir Fusion thorium breeder:
http://ralphmoir.com/media/thBreedNProlifICENESdr7.pdf
p 204 US thorium reserves:
http://minerals.usgs.gov/minerals/pubs/commodity/thorium/myb1-2007-thori.pdf
p 204 Lemhi Pass thorium reserves:
http://www.thoriumenergy.com/index.php?option=com_content&task=view&id=43&Itemid=68
p 204 Thorium reserves: http://www.world-nuclear.org/info/inf62.html
p 206 David MSR waste:
http://www.europhysicsnews.org/index.php?option=article&access=standard&Itemid=129&url=/articles/epn/pdf/2007/02/epn07204.pdf
p 206 LeBrun et al MSBR radiotoxicity: http://hal.archives-ouvertes.fr/docs/00/04/14/97/PDF/document_IAEA.pdf

Denatured Molten Salt Reactor (DMSR)

p 208 ORNL DMSR 1971: http://www.energyfromthorium.com/pdf/ORNL-4541.pdf
p 208 ORNL 7207 scanned:
http://moltensalt.org/references/static/ralphmoir/ORNL-TM-7207.pdf
p 208 ORNL 7207 OCR Word, DMSR Engel et al 1980:
http://www.energyfromthorium.com/pdf/ORNL-TM-7207.pdf

p 208 ORNL MSRE design study:
http://moltensalt.org/references/static/downloads/pdf/ORNL-2796.pdf
p 208 ORNL DMSR 1978: http://www.energyfromthorium.com/pdf/ORNL-5388.pdf
p 208 ORNL 1979 development program:
http://moltensalt.org/references/static/downloads/pdf/ORNL-TM-6415.pdf
p 208 ORNL 1972 MSR noble metals:
http://moltensalt.org/references/static/downloads/pdf/ORNL-TM-3884.pdf
p 208 ORNL 1980 DMSR: http://www.energyfromthorium.com/pdf/ORNL-TM-7207.pdf
p 209 LeBlanc new beginning old idea:
http://www.energyfromthorium.com/forum/download/file.php?id=480&sid=d82b958034ccdcfbe4d859c75840036b
p 209 denatured molten salt reactors:
http://www.coal2nuclear.com/MSR%20-%20Denatured%20-%20CNSLeBlanc2010revised.pdf
p 209 LeBlanc MSRs:
http://www.torium.se/res/Documents/dleblancnewvisiongenivpdf.pdf
p 209 LeBlanc DMSR video: http://www.youtube.com/watch?v=_-BXg18fAlk&feature=player_embedded
p 210 Forsberg proliferation resistant fuel cycles:
http://www.ornl.gov/~webworks/cpr/misc/106598.pdf
p 210 Forsberg: MSR Options:
http://www.ornl.gov/~webworks/cppr/y2001/misc/120977.pdf
p 210 Uranium seawater collection:
http://www.physics.harvard.edu/~wilson/energypmp/2009_Tamada.pdf

Pebble bed advanced high-temperature reactor (PB-AHTR)

p 212 UC Berkeley PB-AHTR project home: http://pb-ahtr.nuc.berkeley.edu/
p 212 Peterson, Scarlat 2010 PB-AHTR presentation:
http://www.thoriumenergyalliance.com/downloads/TEAC3%20presentations/TEAC3_Scarlat_Raluca.pdf
p 215 Forsberg MIT/UCB/UW work:
http://web.mit.edu/nse/pdf/researchstaff/forsberg/FHR%20Project%20Presentation%20Nov%202011.pdf
p 215 TRISO fuel mfg B&W: https://www.ornl.gov/fhr/presentations/Nagley.pdf

LFTR energy cheaper than coal

p 217 Kasten MOSEL MSR cost:
http://www.moltensalt.org/references/static/brucehoglund/msrMOSELConcept_OCR.pdf
p 217 SL-1954 capital cost estimate:
http://moltensalt.org/references/static/downloads/pdf/SL-1954.pdf
p 217 Sargent and Lundy, Capital Investment for 1000 MW(e) Molten Salt Converter Reference Design Power Reactor, report SL 1994 (27 December 1962).
p 217 Oak Ridge TM1060 1965 cost estimate:
http://moltensalt.org/references/static/downloads/pdf/ORNL-TM-1060.pdf

NOTES AND REFERENCES

p 217 Moir MSR cost estimate: http://ralphmoir.com/media/coe_10_2_2001.pdf
p 217 ORNL LFTR fuel cycle costs:
http://moltensalt.org/references/static/downloads/pdf/CF-61-8-86.pdf
p 217 Moir 2008 MSR est costs: http://ralphmoir.com/media/moir_icenes_07.pdf
p 220 University of Chicago economic future nuclear power:
http://www.ne.doe.gov/np2010/reports/NuclIndustryStudy-Summary.pdf
p 221 Boeing 737 assembly line, photo k62904 copyright Boeing Aircraft

Development tasks

p 226 ORNL MSR development uncertainties:
http://www.energyfromthorium.com/doc/ORNL4541_sec16.html
p 226 Forsberg MSR technology gaps:
http://www.torium.se/res/Documents/124670.pdf
p 227 http://moltensalt.org/references/static/downloads/pdf/ORNL-TM-6415.pdf
p 227 ORNL MSR dev plan 1974: http://www.energyfromthorium.com/pdf/ORNL-5018.pdf
p 231 ORNL noble metal migration:
http://moltensalt.org/references/static/downloads/pdf/ORNL-TM-3884.pdf
p 231 Madden theoretical chemistry presentation:
http://www.itheo.org/sites/default/files/pdf/Paul_Madden.pdf
p 231 Madden flibe conductivity viscosity:
http://www.mendeley.com/research/conductivityviscositystructure-unpicking-the-relationship-in-an-ionic-liquid/
p 233 Messinger MIT off gass:
http://icapp.ans.org/icapp12/program/abstracts/12097.pdf
p 235 Heat exchanger diagram:
http://en.wikipedia.org/wiki/File:Spiral_heat_exchanger.png
p 235 Heat exchanger EfT forum:
http://energyfromthorium.com/forum/viewtopic.php?f=3&t=1017&sid=69b28d995589bc6238d49a4fc483bc65
p 235 ORNL materials testing plan: http://nuclear.inl.gov/deliverables/docs/intg-matls-plan.pdf
p 235 ORNL materials experience:
http://nuclear.inl.gov/deliverables/docs/gfr_matls_rd_plan_r1.pdf
p 235 Newsome, Snead, SiC neutron irradiation:
http://www.osti.gov/bridge/servlets/purl/903202-raGNdX/903202.pdf
p 236 ORNL tritium: http://www.energyfromthorium.com/pdf/ORNL-TM-5759.pdf
p 236 Sorensen Li-6 separation:
http://energyfromthorium.com/category/materials/lithium/
p 236 EfT forum Lithium-7:
http://energyfromthorium.com/forum/viewtopic.php?f=64&t=363
p 236 Ragheb isotopic separation:
https://netfiles.uiuc.edu/mragheb/www/NPRE%20402%20ME%20405%20Nuclear%20Power%20Engineering/Isotopic%20Separation%20and%20Enrichment.pdf
p 238 Brayton cycle reheat:
http://nuclear.inl.gov/deliverables/docs/genivihc_2006_milestone_report_7_1_2006_final.pdf
p 238 U Waterloo Brayton cycle tutorial:
http://www.mhtlab.uwaterloo.ca/courses/me354/lectures/pdffiles/web7.pdf

p 239 MIT supercritical CO2 power conversion:
http://stuff.mit.edu/afs/athena/course/22/22.33/www/dostal.pdf
p 239 Wright Sandia SCO2: http://www.barber-nichols.com/sites/default/files/wysiwyg/images/supercritical_co2_turbines.pdf
p 239 Wright, SCO2 interview: http://djysrv.blogspot.com/2012/05/supercritical-co2-turbine-being.html
p 239 Siemens 51% efficient steam turbine:
http://www.pennenergy.com/index/articles/pe-article-tools-template.articles.power-engineering-international.volume-13.issue-10.features.power-plant-control.finely-tuned.html
p 245 Bonometti program advice:
http://www.thoriumenergyalliance.com/downloads/TEAC3%20presentations/TEAC3_Bonometti_Joe.pdf
p 249 DOE plan destroy U-233:
http://www.em.doe.gov/PDFs/ProjectFiles/OakRidge.pdf
p 244 Magreb decay heat:
http://www.ewp.rpi.edu/hartford/~ernesto/F2011/EP/MaterialsforStudents/Petty/Ragheb-Ch8-2011.PDF
p 244 Sorensen spent fuel explorer:
http://www.energyfromthorium.com/javaws/SpentFuelExplorer.jnlp
p 245 Siemer, Nuclear Technology, June 2012, Improving the integral fast reactor's proposed salt waste management system

Builders

p 248 Moir, Restart MSR program:
http://ralphmoir.com/media/moir_icenes_07.pdf
p 248 ORNL docs, Energy from Thorium: http://www.energyfromthorium.com/pdf/
p 254 Transatomic Power: http://transatomicpower.com/
p 254 Transatomic Power money:
http://www.masshightech.com/stories/2012/05/28/daily28-Transatomic-secures-763K.html
p 255 Thorenco presentation:
http://www.thoriumenergyalliance.com/downloads/TEAC3%20presentations/TEAC3_Holden_Charles.pdf
p 257 ORLY Energy Group: http://www.orlygroup.com/lftr.html
p 259 China pebble bed reactor:
http://pebblebedreactor.blogspot.com/2007/03/china-has-built-pebble-bed-reactor.html
p 258 China AP1000 cost: http://www.world-nuclear.org/info/inf63.html
p 260 Chinese Academy of Sciences: http://energyfromthorium.com/2011/01/
p 263 International Thorium Energy Organization: http://itheo.org/
p 264 Merle-Lucotte Fast MSR start w plutonium: http://hal.archives-ouvertes.fr/in2p3-00135141_v1/
p 264 Merle-Lucotte iTheo 2010 TMSR overview:
http://www.itheo.org/sites/default/files/pdf/Elsa_Merle-Lucotte.pdf
p 264 Merle-Lucotte min fissile in fast MSR:
http://hal.in2p3.fr/docs/00/38/53/78/PDF/ANFM09-MSFR.pdf

NOTES AND REFERENCES

p 264 Merle-Lucotte transition 2nd 3rd gen to TMSR:
http://hal.in2p3.fr/docs/00/13/51/49/PDF/ICAPP07_final.pdf
p 265 Sustainable Nuclear Energy Technology MSR article:
http://www.snetp.eu/www/snetp/images/stories/Docs-SRA2012/sra_annex-MSRS.pdf
p 265 Mouney Pu management in LWR fuel cycle:
http://nuclear.tamu.edu/~ragusa/documents/courses/489_09A/lectures/projects/multi/Plutonium_and_minor_actinides_management_in_the_nuclear_fuel_cycle--_assessing_and_controlling_the_inventory.pdf
p 265 Czech LFTR joint venture: http://www.praguepost.com/business/10382-czechs-aussies-partner-on-energy.html
p 265 Uhlir Rez Czech R&D:
http://www.torium.se/res/Documents/uhlirfluorination1.pdf
p 266 Thorium Power Canada: http://www.thoriumpowercanada.com/
p 266 DBI Century Fuels: http://www.dauvergne.com/technology/technology-overview/
p 266 Thorium One Canada: http://www.thorium1.com/
p 267 Japan FUJI MSR: http://nextbigfuture.com/2007/12/fuji-molten-salt-reactor.html
p 267 Japan FUJI IAEA: http://www-pub.iaea.org/MTCD/publications/PDF/te_1536_web.pdf
p 268 Furukawa et al sustainable secure nuclear industry:
http://cdn.intechopen.com/pdfs/19683/InTech-New_sustainable_secure_nuclear_industry_based_on_thorium_molten_salt_nuclear_energy_synergetics_thorims_nes_.pdf

Contenders

p 271 US DOE NGNP:
https://inlportal.inl.gov/portal/server.pt/gateway/PTARGS_0_2_3310_277_2604_43/http%3B/inlpublisher%3B7087/publishedcontent/publish/communities/inl_gov/about_inl/gen_iv___technical_documents/a1_ngnp_fy07_external.pdf
p 272 NGNP Alliance docs: http://www.ngnpalliance.org/index.php/resources
p 272 NGNP 2010 status:
http://www.ngnpalliance.org/index.php/resources/download/czo4NDoiL2ltYWdIcy9nZW5lcmFsX2ZpbGVzL1N1bW1hcnlfZm9yX3RoZV9OZXh0X0dlbmVyYXRpb25fTnVjbGVhcl9QbGFudF9Qcm9qZWN0XzIwMTAucGRmIjs
p 272 INL NGNP: www.inl.gov/technicalpublications/Documents/4680340.pdf
p 272 NGNP schedule:
https://www.google.com/url?sa=t&rct=j&q=&esrc=s&source=web&cd=6&ved=0CHIQFjAF&url=https%3A%2F%2Finlportal.inl.gov%2Fportal%2Fserver.pt%2Fdocument%2F98008%2Fngnp_integrated_schedule_development_plan_pdf&ei=3QaoT9GyNer86QG8482fBA&usg=AFQjCNF5WT2T7IzxYHKUByby2m29Uj_LsA
p 272 INL NGNP fact sheet: http://www.inl.gov/research/next-generation-nuclear-plant/
p 276 Westinghouse AP1000:
http://www.westinghousenuclear.com/docs/AP1000_brochure.pdf
p 278 AP1000 in China: http://www.world-nuclear.org/info/inf63.html

Small modular reactors

p 280 B&W mPower: http://www.generationmpower.com/
p 281 NuScale: http://www.nuscale.com/index.php
p 282 Holtec presentation:
http://pbadupws.nrc.gov/docs/ML1120/ML112070201.pdf
p 283 Westinghouse SMR: http://www.westinghousenuclear.com/SMR/index.htm
p 283 NRC Westinghouse presentation:
http://pbadupws.nrc.gov/docs/ML1119/ML111920208.pdf
p 284 Gen4 Energy: http://www.gen4energy.com/

Liquid metal cooled fast breeder reactors

p 285 Fast neutron reactors: http://www.world-nuclear.org/info/default.aspx?id=540
p 287 EBR-II: http://en.wikipedia.org/wiki/EBR-II
p 289 Plutonium from UK magnox: http://atomicinsights.com/2010/07/proving-a-negative-why-modern-used-nuclear-fuel-cannot-be-used-to-make-a-weapon.html
p 289 GE Hitachi advanced recycling center:
http://www.usnuclearenergy.org/PDF_Library/_GE_Hitachi%20_advanced_Recycling_Center_GNEP.pdf
p 289 GEH Prism tech brief:
http://cfcc.edu/lrc/documents/PRISMTechnicalbriefR0.pdf
p 289 NRC GEH Prism pre application safety report NUREG-1368:
http://www.osti.gov/bridge/servlets/purl/10133164-2ZfTJr/native/10133164.pdf
p 290 Russian Alfa submarine: http://en.wikipedia.org/wiki/Alfa_class_submarine
p 290 Russian SVBR-100: http://www.world-nuclear-news.org/NN_Heavy_metal_power_reactor_slated_for_2017_2303122.html
p 291 TerraPower, Tyler Ellis et al: http://lumma.org/temp/Ellis_et_al-TWRs_A_Truly_Sustainable_Resource.pdf
p 292 TerraPower 500 MW, Charles Ahlfeld et al:
http://www.terrapower.com/Libraries/Article_Reprints/ICAPP_2011_Paper_11199.sflb.ashx
p 292 MIT Tech Rev of TWR; http://www.technologyreview.com/energy/38148/

Accelerator-driven subcritical reactors

p 296 McIntyre ADS: http://energy2050.se/uploads/files/rubbia2.pdf
p 296 WNA, ADS: http://www.world-nuclear.org/info/inf35.html
p 296 Subcritical reactors: http://en.wikipedia.org/wiki/Subcritical_reactor
p 298 ORNL spallation neutron source: http://neutrons.ornl.gov/facilities/SNS/
p 299 Thorium Energy Association: http://thorea.hud.ac.uk/
p 299 ThorEA 2010 report:
http://www.thorea.org/publications/ThoreaReportFinal.pdf
p 300 Rubbia, Aker Solutions, ADSR:
http://energy2050.se/uploads/files/rubbia2.pdf
p 300 ADNA ADSR 2010: http://www.phys.vt.edu/~kimballton/gem-star/workshop/presentations/bowman.pdf
p 300 Intl ADSR conferences [possible malware]:
http://www.ivsnet.org/ADS/ADS2011/

NOTES AND REFERENCES

p 300 iTheo 2010 ADSR and MSR presentations: http://www.itheo.org/thorium-energy-conference-2010
p 300 UK Daily Mail: Emma and thorium:
http://www.dailymail.co.uk/home/moslive/article-2001548/Electron-Model-Many-Applications-Technology-save-world.html#ixzz1P2lkjkiG

SAFETY

p 305 Madrigal 2010 accidents:
http://www.theatlantic.com/technology/archive/2011/03/25-other-energy-disasters-from-the-last-year/72814/
p 307 Paul Scherrer Insitut accidents:
http://gabe.web.psi.ch/pdfs/PSI_Report/ENSAD98.pdf
p 308 NRC SORCA: http://www.nrc.gov/about-nrc/regulatory/research/soar.html
p 310 Alpha particles etc diagram:
http://en.wikipedia.org/wiki/Ionizing_radiation#Ionizing_radiation_level_examples
p 311 Reactive oxygen species:
http://en.wikipedia.org/wiki/Reactive_oxygen_species
p 311 Idaho State U radioactivity in nature:
http://www.physics.isu.edu/radinf/natural.htm
p 311 Idaho State U radiation information network:
http://www.physics.isu.edu/radinf/
p 312 Post, radiation exposure: http://theenergycollective.com/willem-post/53939/radiation-exposure
p 312 Health Physics Society: http://www.radiationanswers.org/
p 312 IEM radiation tool box: http://www.iem-inc.com/toolset.html
p 312 Health physics society: http://www.hps.org/
p 315 NAS BEIR VII: http://www.nap.edu/catalog.php?record_id=11340
p 318 Levitt, Freakonomics: http://www.freakonomics.com/
p 318 Slovic, Bull Atomic Scientists: http://intl-bos.sagepub.com/content/68/3/67.full
p 318 Bulletin Atomic Scientists on LNT: http://intl-bos.sagepub.com/content/current
p 319 Furedi, Culture of fear: http://www.amazon.com/Culture-Fear-Revisited-Frank-Furedi/dp/0826493955/ref=sr_1_4?ie=UTF8&qid=1336081132&sr=8-4
p 321 Cohen, LNT validity: http://www.world-nuclear.org/sym/1998/cohen.htm
p 321 Cohen, LNT: http://www.phyast.pitt.edu/~blc/
p 321 Cohen, Nuclear energy option:
http://www.phyast.pitt.edu/~blc/book/BOOK.html
p 321 Craig, LNT validity URL collection: http://a-place-to-stand.blogspot.com/2010/03/low-level-radiation-evidence-that-it-is.html
p 327 Taiwan apartment Co-60 radiation: http://www.jpands.org/vol9no1/chen.pdf
p 324 Fukushima radiation: http://safetyfirst.nei.org/public-health/experts-say-health-effects-of-fukushima-accident-should-be-very-minor/
p 324 ANS Fukushima report:
http://fukushima.ans.org/report/Fukushima_report.pdf
p 325 DOE low dose radiation: http://lowdose.energy.gov/

p 325 US DOE low dose radiation:
http://lowdose.energy.gov/radiobio_slideshow.aspx
p 325 Cuttler Fukushima evacuation: http://www.ourenergypolicy.org/wp-content/uploads/2012/03/35766131k01w4103.pdf
p 325 Cuttler Fukushima presentation: http://atomicinsights.com/wp-content/uploads/Cuttler-2012_ANS-President-Session_Jun23-copy.pdf
p 325 New Mexico low background radiation experiment:
http://www.wipp.energy.gov/pr/2011/Low%20Background%20Radiation%20Experiment%20News%20Release.pdf
p 326 US DOE low dose research highlights:
http://lowdose.energy.gov/science_highlights.aspx
p 326 Lawrence Berkeley Lab DNA repair: http://newscenter.lbl.gov/news-releases/2011/12/20/low-dose-radiation/
p 326 Lawrence Berkeley Lab DNA repair:
http://www.pnas.org/content/early/2011/12/16/1117849108.full.pdf+html
p 327 MIT Engelward, Yanch, prolonged rad exposure: http://web.mit.edu/newsoffice/2012/prolonged-radiation-exposure-0515.html
p 328 Int'l Dose Response Society: http://www.dose-response.org/
p 328 Healthy worker effect:
http://www.ncbi.nlm.nih.gov/pmc/articles/PMC2889508/
p 328 Fukushima evacuation deaths:
http://www.yomiuri.co.jp/dy/national/T120204003191.htm
p 328 Zbigniew Jaworowski, APS newsletter, radiation ethics:
http://www.riskworld.com/Nreports/1999/jaworowski/NR99aa01.htm
p 329 Allison Radiation and Reason: http://www.radiationandreason.com/
p 329 Allison 100mSv/month:
http://www.youtube.com/watch?feature=player_embedded&v=Uj8Pl1AiOuA
p 330 ANS special session on LNT (big download):
http://www.new.ans.org/about/officers/docs/special-session-low-level-radiation-version1.4.pdf
p 331 Ragheb Gabon natural reactors:
https://netfiles.uiuc.edu/mragheb/www/NPRE%20402%20ME%20405%20Nuclear%20Power%20Engineering/Natural%20%20Nuclear%20Reactors,%20The%20Oklo%20Phenomenon.pdf
p 332 Sandia deep borehole disposal:
http://www.mkg.se/uploads/Bil_2_Deep_Borehole_Disposal_High-Level_Radioactive_Waste_-_Sandia_Report_2009-4401_August_2009.pdf
p 332 Economist, waste disposal: http://www.economist.com/node/21556100
p 334 WIPP: http://en.wikipedia.org/wiki/Waste_Isolation_Pilot_Plant
p 335 David MSR waste:
http://www.europhysicsnews.org/index.php?option=article&access=standard&Itemid=129&url=/articles/epn/pdf/2007/02/epn07204.pdf
p 336 Reed, Stillman, Nuclear Express:
http://www.amazon.com/gp/product/076033904X
p 336 NY Times, the bomb:
http://www.nytimes.com/2008/12/09/science/09bomb.html
p 339 LeBrun et al, MSBR radiotoxicity, proliferation resist: http://hal.archives-ouvertes.fr/docs/00/04/14/97/PDF/document_IAEA.pdf
p 339 U-232 decay: http://en.wikipedia.org/wiki/Uranium-233#U-232_impurity
p 340 Kang, von Hippel, proliferation resistance U-233:

http://scienceandglobalsecurity.org/archive/sgs09kang.pdf
p 340 Gamma ray detecting satellites:
http://imagine.gsfc.nasa.gov/docs/sats_n_data/gamma_missions.html
p 341 Hoglund molten salt references:
http://moltensalt.org/references/static/home.earthlink.net/bhoglund/index.html
p 341 Hoglund proliferation resistance:
http://www.moltensalt.org/references/static/home.earthlink.net/bhoglund/multiMissionMSR.html
p 341 Moir molten salt papers: http://ralphmoir.com/aMlt_slt.htmz
p 341 Moir U-232 proliferation resistance:
http://ralphmoir.com/media/lLNLReport2_2010_06_25.pdf
p 341 Hoglund Multi mission MSR:
http://www.moltensalt.org/references/static/home.earthlink.net/bhoglund/multiMissionMSR.html
p 341 Energy from Thorium proliferation discussion:
http://energyfromthorium.com/2010/10/02/lftr-discourages-weapons-proliferation
p 342 Pakistan nuclear weapons:
http://en.wikipedia.org/wiki/Pakistan_and_weapons_of_mass_destruction

A SUSTAINABLE WORLD

Coal

p 345 Holm thorium applications:
http://www.thoriumapplications.com/chapter_10_page_8.htm
p 345 1200 world's largest coal plants: http://carma.org/

Oil

p 347 Worldwatch sustainable world: http://ww.worldwatch.org/climate-energy
p 348 Shell Pearl gas to liquids:
http://www.shell.com/home/content/aboutshell/our_strategy/major_projects_2/pearl/ships_first_products/
p 348 Shell, van de Veer:
http://www.shell.com/home/content/media/speeches_and_webcasts/archive/2008/jvdv_two_energy_futures_25012008.html
p 349 Walter, alt transportation fuels: http://www.same-satx.org/briefs/090317-walters.pdf
p 350 Holm, coal2thorium: http://coal2thorium.com
p 353 Forsberg shale oil:
http://web.mit.edu/nse/pdf/faculty/forsberg/ANS%202011%20Transport%20Panel%20Nov%20Ext.pdf
p 350 Colorado Geo Survey, retort:
http://geosurvey.state.co.us/energy/Oil%20Shale/Pages/OilShale.aspx
p 350 RAND report on shale oil:
http://www.rand.org/pubs/monographs/2005/RAND_MG414.pdf

p 350 Shale oil extraction: http://en.wikipedia.org/wiki/Shale_oil_extraction
p 351 Shell electric heating: http://ostseis.anl.gov/guide/oilshale/
p 352 Exxon Mobil ElectroFrac: http://208.88.130.69/August-2008-Shale-oil-pilot-projects-proliferate.html
p 352 ElectroFrac test results: http://ceri-mines.org/documents/29thsymposium/papers09/Paper_03-4_Symington-Bill.pdf
p 353 Forsberg shale oil: http://web.mit.edu/nse/pdf/faculty/forsberg/ANS%202011%20Transport%20Panel%20Nov%20Ext.pdf
p 353 Oil Drum EROI shale oil tar sands: http://www.theoildrum.com/node/3839
p 354 Alberta Oil Sands: http://www.world-nuclear.org/info/inf49a_Alberta_Tar_Sands.html
p 355 Gasoline: http://en.wikipedia.org/wiki/Gasoline
p 360 Uhrig et al hydrogen economy synfuels: www.tbp.org/pages/publications/Bent/Features/Su07Uhrig.pdf
p 360 Bogart et al production liquid synfuels: ICAPP ' http://www.osti.gov/energycitations/product.biblio.jsp?osti_id=21016358
p 359 SRI coal plus natural gas synfuels: http://www.sri.com/news/releases/122011.html
p 360 Green Freedom: http://www.lanl.gov/news/newsbulletin/pdf/Green_Freedom_Overview.pdf
p 360 Green Freedom presentation: http://www.coal2nuclear.com/Green%20Freedom%20-%20Martin_AEC_2008_revised.pdf
p 360 David Keith air capture: http://keith.seas.harvard.edu/AirCapture.html
p 362 Olah et al Recycling CO2: https://wiki.ornl.gov/sites/carboncapture/Shared%20Documents/Background%20Materials/Alternative%20Methods/G.%20Olah.pdf
p 357 Copper chlorine cycle: http://en.wikipedia.org/wiki/Copper-chlorine_cycle
p 362 Biomass to diesel, Seiler, Hohwiller: http://www.wcce8.org/doc/090803_CH_Technico_economy_of_ScBtL.pdf
p 362 Biomass to diesel: http://www-ist.cea.fr/publicea/exl-doc/200500001687.pdf
p 362 DOE biomass study: http://www.eere.energy.gov/biomass/pdfs/final_billionton_vision_report2.pdf
p 362 Entrained flow gasifier: http://www.biofuelstp.eu/btl.html

Ammonia

p 365 Hargraves, Siemer, Nuclear Ammonia: http://www.itheo.org/sites/default/files/pdf/Nuclear%20Ammonia;%20Thorium's%20Killer%20App%20-%20Robert%20Hargraves%20-%20Dartmouth%20College%20-%20ThEC11.pdf
p 366 NH3 Fuel Association: http://www.nh3fuelassociation.org/
p 367 Free piston engine: http://pubs.acs.org/doi/pdfplus/10.1021/ef800217k
p 367 Sturman hydraulic engine: http://www.stevesturgess.com/2011/08/no-cam-no-crank-no-carbon-engine.html
p 367 Hydrofuel Inc NH3 vehicles: http://www.nh3fuel.com/
p 368 Apollo Fuel Cells: http://www.electricauto.com/prod_00.html
p 368 Solid state ammonia synthesis:

http://www.energy.iastate.edu/Renewable/ammonia/ammonia/2008/Sammes_20
08.pdf
p 370 Calif. gasoline costs:
http://energyalmanac.ca.gov/gasoline/margins/index.php
p 371 Ammonia hazard analysis:
http://www.energy.iastate.edu/Renewable/ammonia/downloads/NH3_RiskAnalysi
s_final.pdf
p 372 Hargraves Nuclear Ammonia:
http://energyfromthorium.com/2011/10/29/nuclear-ammonia/

Nuclear cement

p 373 Hargraves Nuclear Cement:
http://energyfromthorium.com/2011/11/07/nuclear-cement/

Hydrogen

p 376 Forsberg Nuclear Hydrogen:
http://www.ornl.gov/~webworks/cppr/y2001/pres/124155.pdf
p 376 Forsberg Hydrogen markets:
www.ornl.gov/~webworks/cppr/y2001/pres/122902.pdf
p 377 Copper chloride cycle:
http://en.wikipedia.org/wiki/Copper%E2%80%93chlorine_cycle
p 378 Honda Clarity: http://automobiles.honda.com/fcx-clarity/
p 378 How Honda Clarity works: http://automobiles.honda.com/fcx-clarity/how-
fcx-works.aspx
p 378 Linde US hydrogen: http://www.linde-
gas.com/en/innovations/hydrogen_energy/index.html
p 378 Linde hydrogen:
http://www.lindegaz.com.tr/international/web/lg/com/likelgcom30.nsf/docbyalias/n
av_hydrogen
p 378 Hydrogen economy: http://en.wikipedia.org/wiki/Hydrogen_economy
p 378 Hydrogen car fire: http://evworld.com/library/Swainh2vgasVideo.pdf
p 378 Hydrogen powered aircraft: http://en.wikipedia.org/wiki/Hydrogen_aircraft
p 379 Tupolev aircraft: http://www.tupolev.ru/English/Show.asp?SectionID=82

Water

p 380 UNESCO water report:
http://www.unesco.org/new/fileadmin/MULTIMEDIA/HQ/SC/pdf/WWDR4%20Vol
ume%201-Managing%20Water%20under%20Uncertainty%20and%20Risk.pdf
p 380 UN Water Under Pressure:
http://unesdoc.unesco.org/images/0021/002156/215644e.pdf
p 381 Wikipedia desalination: http://en.wikipedia.org/wiki/Desalination
p 381 Siemens desalination:
http://www.siemens.com/innovation/en/news/2011/desalinating-seawater-with-
minimal-energy-use.htm
p 381 Peterson, Zhao advanced multi effect distillation: http://pb-
ahtr.nuc.berkeley.edu/papers/05-003_HTR_MED_Desalt_E.pdf

Energy Policy

p 391 CBO 2012 energy subsidies and support:
http://www.cbo.gov/sites/default/files/cbofiles/attachments/03-06-FuelsandEnergy_Brief.pdf
p 385 EIA 2010 subsidies: http://www.eia.gov/analysis/requests/subsidy/
p 385 DOE 2012 budget:
http://www.cfo.doe.gov/budget/12budget/Content/FY2012Highlights.pdf
p 386 MIS subsidies analysis for NEI:
http://www.nei.org/filefolder/60_Years_of_Energy_Incentives_-_Analysis_of_Federal_Expenditures_for_Energy_Development_-_1950-2010.pdf
p 386 Pew Char Trust energy subsidies:
http://subsidyscope.org/energy/summary/
p 387 RGGI website: http://www.rggi.org/
p 387 Sourcewatch RGGI:
http://www.sourcewatch.org/index.php?title=Regional_Greenhouse_Gas_Initiative
p 388 Vermont feed-in tariffs: http://vermontspeed.squarespace.com/
p 388 Feed-in tariffs survey: http://en.wikipedia.org/wiki/Feed_in_tariff
p 388 Iowa production tax credits:
http://www.state.ia.us/iub/energy/renewable_tax_credits.html
p 388 State prod tax credits: http://eetd.lbl.gov/ea/EMS/reports/51465.pdf
p 388 State energy incentives database: http://www.dsireusa.org/
p 388 EPA renew energy cert: http://www.epa.gov/greenpower/gpmarket/rec.htm
p 388 REC market: http://www.srectrade.com/
p 389 Carbon taxes: http://en.wikipedia.org/wiki/Carbon_tax
p 389 ORNL CO2 information analysis center: http://cdiac.ornl.gov/
p 389 IEA 2011 CO2:
http://www.iea.org/newsroomandevents/news/2012/may/name,27216,en.html
p 390 EIA projections: http://www.eia.gov/forecasts/ieo/
p 390 Tindale, thorium MSR policy for EU:
http://www.cer.org.uk/sites/default/files/publications/attachments/pdf/2011/pb_thorium_june11-153.pdf
p 390 Germany energy policy:
http://www.nytimes.com/2012/05/29/world/europe/29iht-letter29.html?_r=2
p 390 Europe energy prices: http://www.energy.eu

Appendices

p 421 Hargraves/Moir article: http://ralphmoir.com/media/hargraves2010_2.pdf
p 449 Moir/Teller article: http://ralphmoir.com/media/moir_teller.pdf
p 467 Video about Teller: http://motherboard.vice.com/2012/3/7/motherboard-tv-doctor-teller-s-strange-loves-from-the-hydrogen-bomb-to-thorium-energy--2

APPENDIX A

AMERICAN SCIENTIST JUNE/JULY 2010
LIQUID FLUORIDE THORIUM REACTOR

An old idea in nuclear power gets re-examined.

Robert Hargraves and Ralph Moir

What if we could turn back the clock to 1965 and have an energy do-over? In June of that year, the Molten Salt Reactor Experiment (MSRE) achieved criticality for the first time at Oak Ridge National Laboratory (ORNL) in Tennessee. In place of the familiar fuel rods of modern nuclear plants, the MSRE used liquid fuel— hot fluoride salt containing dissolved fissile material in a solution roughly the viscosity of water at operating temperature. The MSRE ran successfully for five years, opening a new window on nuclear technology. Then the window banged closed when the molten-salt research program was terminated.

Knowing what we now know about climate change, peak oil, Three Mile Island, Chernobyl, and the Deepwater Horizon oil well gushing in the Gulf of Mexico in the summer of 2010, what if we could have taken a different energy path? Many feel that there is good reason to wish that the liquid-fuel MSRE had been allowed to mature. An increasingly popular vision of the future sees liquid-fuel reactors playing a central role in the energy economy, utilizing relatively abundant thorium instead of uranium, mass producible, free of carbon emissions, inherently safe and generating a trifling amount of waste.

Figure 1. Thorium is a relatively abundant, slightly radioactive element that at one time looked like the future of nuclear power. It was supplanted when the age of uranium began with the launching of the nuclear-powered *USS Nautilus,* whose reactor core was the technological ancestor of today's nuclear fleet. Thorium is nonfissile but can be converted to fissile uranium-233, the overlooked sibling of fissile uranium isotopes. The chemistry, economics, safety features and nonproliferation aspects of the thorium/uranium fuel cycle are earning it a hard new look as a potential solution to today's problems of climate change, climbing requirements for energy in the developing world, and the threat of diversion of nuclear materials to illicit purposes. Shown are monazite crystals, a plentiful ore containing thorium. India, which has the task of developing a long-range program to convert India to thorium-based power over the next fifty years, making the most of India's modest uranium reserves and vast thorium reserves.

Of course we can't turn back the clock. Maddeningly to advocates of liquid-fuel thorium power, it is proving just as hard to simply restart the clock. Historical, technological and regulatory reasons conspire to make it hugely difficult to diverge from our current path of solid-fuel, uranium-based plants. And yet an alternative future that includes liquid-fuel thorium-based power beckons enticingly. We'll review the history, technology, chemistry and economics of thorium power and weigh the pros and cons of thorium versus uranium. We'll conclude by asking the question we started with: What if?

The Choice

The idea of a liquid-fuel nuclear reactor is not new. Enrico Fermi, creator in 1942 of the first nuclear reactor in a pile of graphite and uranium blocks at the University of Chicago, started up the world's first liquid-fuel reactor two years later in 1944, using uranium sulfate fuel dissolved in water. In all nuclear chain reactions, fissile material absorbs a neutron, then fission of the atom releases tremendous energy and additional neutrons. The emitted neutrons, traveling at close to 10 percent of the speed of light, would be very unlikely to cause further fission in a reactor like Fermi's Chicago Pile-1 unless they were drastically slowed--moderated--to speeds of a few kilometers per second. In Fermi's device, the blocks of graphite between pellets of uranium fuel slowed the neutrons down. The control system for Fermi's reactor consisted of cadmium-coated rods that upon insertion would capture neutrons, quenching the chain reaction by reducing neutron generation. The same principles of neutron moderation and control of the chain reaction by regulation of the neutron economy continue to be central concepts of nuclear reactor design.

In the era immediately following Fermi's breakthrough, a large variety of options needed to be explored. Alvin Weinberg, director of ORNL from 1955 to 1973, where he presided over one of the major research hubs during the development of nuclear power, describes the situation in his memoir, *The First Nuclear Era:*

> In the early days we explored all sorts of power reactors, comparing the advantages and disadvantages of each type. The number of possibilities was enormous, since there are many possibilities for each component of a reactor—fuel, coolant, moderator. The fissile material may be U-233, U-235, or Pu-

239; the coolant may be: water, heavy water, gas, or liquid metal; the moderator may be: water, heavy water, beryllium, graphite—or, in a fast-neutron reactor, no moderator....if one calculated all the combinations of fuel, coolant, and moderator, one could identify about a thousand distinct reactors. Thus, at the very beginning of nuclear power, we had to choose which possibilities to pursue, which to ignore.

Among the many choices made, perhaps the most important choice for the future trajectory of nuclear power was decided by Admiral Hyman Rickover, the strong-willed Director of Naval Reactors. He decided that the first nuclear submarine, the *USS Nautilus*, would be powered by solid uranium oxide enriched in uranium-235, using water as coolant and moderator. The *Nautilus* took to sea successfully in 1955. Building on the momentum of research and spending for the *Nautilus* reactor, a reactor of similar design was installed at the Shippingport Atomic Power Station in Pennsylvania to become the first commercial nuclear power plant when it went online in 1957.

Rickover could cite many reasons for choosing to power the *Nautilus* with the S1W reactor (S1W stands for submarine, 1st generation, Westinghouse). At the time it was the most suitable design for a submarine. It was the likeliest to be ready soonest. And the uranium fuel cycle offered as a byproduct plutonium-239, which was used for the development of thermonuclear ordnance. These reasons have marginal relevance today, but they were critical in defining the nuclear track we have been on ever since the 1950s. The down sides of Rickover's choice remain with us as well. Solid uranium fuel has inherent challenges. The heat and radiation of the reactor core damage the fuel assemblies, one reason fuel rods are taken out of service after just a few years and after consuming only three to five percent of the energy in the uranium they contain. Buildup of fission products within the fuel rod also undermines the efficiency of the fuel, especially the accumulation of xenon-135, which has a spectacular appetite for neutrons, thus acting as a fission poison by disrupting the neutron economy of the chain reaction. Xenon-135 is short-lived (half-life of 9.2 hours) but it figures importantly in the management of the reactor. For example, as it burns off, the elimination of xenon-135 causes the chain reaction to accelerate, which requires control rods to be reinserted in a carefully managed cycle until the reactor is stabilized. Mismanagement of this procedure contributed to the instability in the Chernobyl core that led to a runaway reactor and the explosion that followed.

Other byproducts of uranium fission include long-lived transuranic materials (elements above uranium in the periodic table), such as plutonium, americium, neptunium and curium. Disposal of these wastes of the uranium era is a problem that is yet to be resolved.

Thorium

When Fermi built Chicago Pile-1, uranium was the obvious fuel choice: Uranium-235 was the only fissile material on Earth. Early on, however, it was understood that burning small amounts of uranium-235 in the presence of much larger amounts of uranium-238 in a nuclear reactor would generate transmuted products, including fissile isotopes such as plutonium-239. The pioneers of nuclear power (Weinberg in his memoir calls his cohorts "the old nukes") were transfixed by the vision of using uranium reactors to breed additional fuel in a cycle that would transform the world by delivering limitless, in-expensive energy.

Figure 2. In a reactor core, fission events produce a controlled storm of neutrons that can be absorbed by other elements present. Fertile isotopes are those that can become fissile (capable of fission) after successive neutron captures. Fertile Th-232 captures a neutron to become Th-233, then undergoes beta decay—emission of an electron with the transformation of a neutron into a proton. With the increase in proton number,

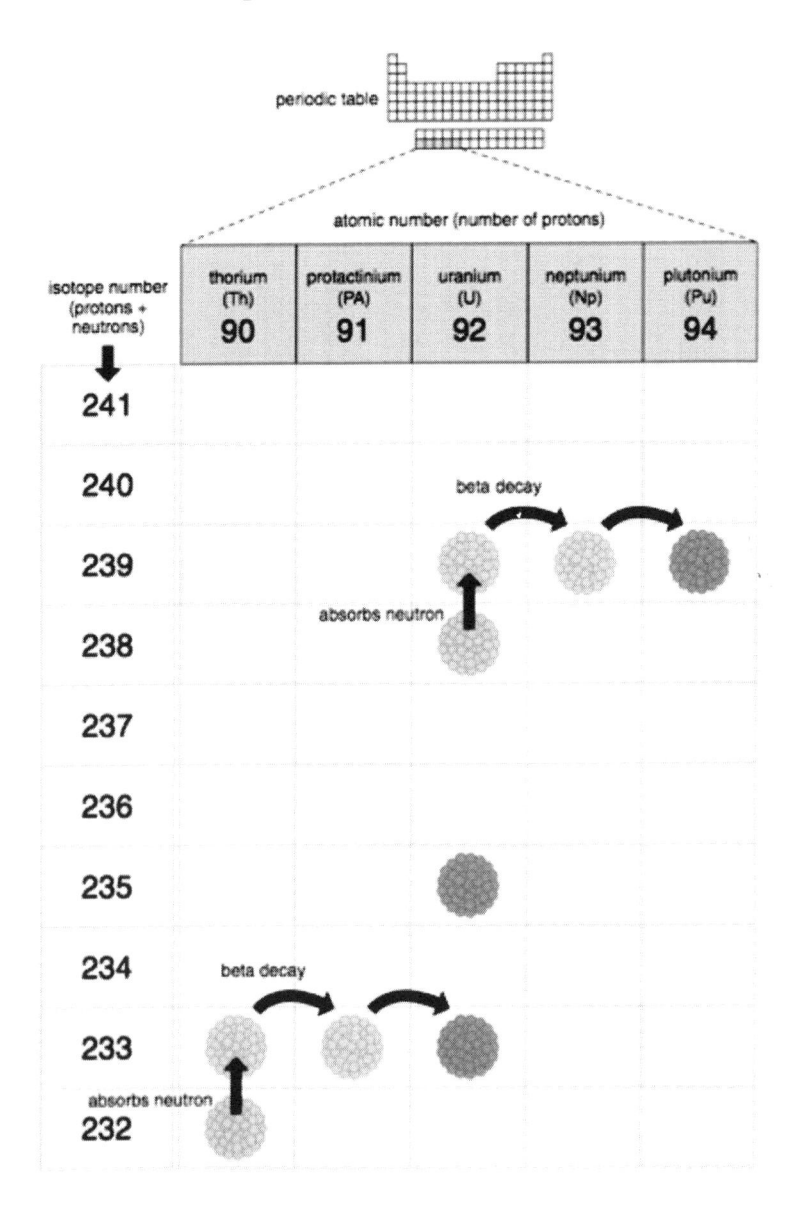

Th-233 transmutes into Pa-233, then beta decay of Pa-233 forms fissile U-233. Most U-233 in a reactor will absorb a neutron and undergo fission; some will absorb an additional neutron before fission occurs, forming U-234 and so on up the ladder. Comparing the transmutation routes to plutonium in thorium- and uranium-based reactors, many more absorption and decay events are required to reach Pu-239 when starting from Th-232, thus leaving far less plutonium to be managed, and possibly diverted, in the thorium fuel and waste cycles.

By the same alchemistry of transmutation, the nonfissile isotope thorium-232 (the only naturally occurring isotope of thorium) can be converted to fissile uranium-233. A thorium-based fuel cycle brings with it different chemistry, different technology and different problems. It also potentially solves many of the most intractable problems of the uranium fuel cycle that today produces 17 percent of the electric power generated world- wide and 20 percent of the power generated in the U.S.

Thorium is present in the Earth's crust at about four times the amount of uranium and it is more easily extracted. When thorium-232 (atomic number 90) absorbs a neutron, the product, thorium-233, undergoes a series of two beta decays—in beta decay an electron is emitted and a neutron becomes a proton-forming uranium-233 (atomic number 91). Uranium-233 is fissile and is very well suited to serve as a reactor fuel. In fact, the advantages of the thorium/uranium fuel cycle compared to the uranium/plutonium cycle have mobilized a community of scientists and engineers who have resurrected the research of the Alvin Weinberg era and are attempting to get thorium-based power into the mainstream of research, policy and ultimately, production. Thorium power is sidelined at the moment in the national research laboratories of the U.S., but it is being pursued intensively in India, which has no uranium but massive thorium reserves. Perhaps the best known research center for thorium is the Reactor Physics Group of the Laboratoire de Physique

Subatomique et de Cosmologie in Grenoble, France, which has ample resources to develop thorium power, although their commitment to a commercial thorium solution remains tentative. (French production of electricity from nuclear power, at 80 percent, is the highest in the world, based on a large infrastructure of traditional pressurized water plants and their own national fuel-reprocessing program for recycling uranium fuel.)

Figure 3a. At its most schematic, the uranium-fueled light

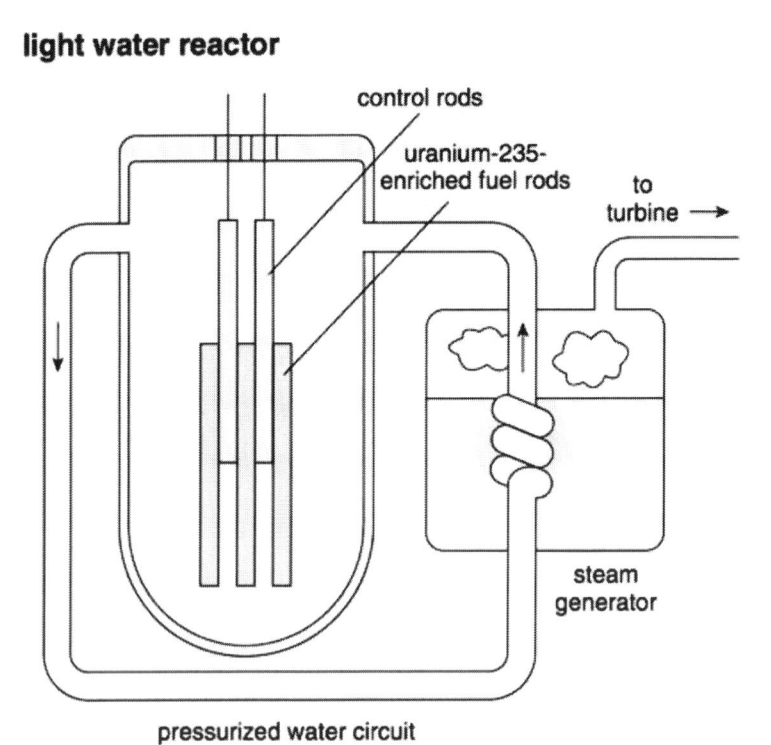

light water reactor

water reactor (all of the U.S. reactor fleet) consists of fuel rods, control rods, and water moderator and coolant.

Figure 3b. The liquid fluoride thorium reactor (LFTR) consists

liquid fluoride thorium reactor

thorium-232 + neutron ⟶ protactinium-233 ⟶ uranium-233
(short-lived)

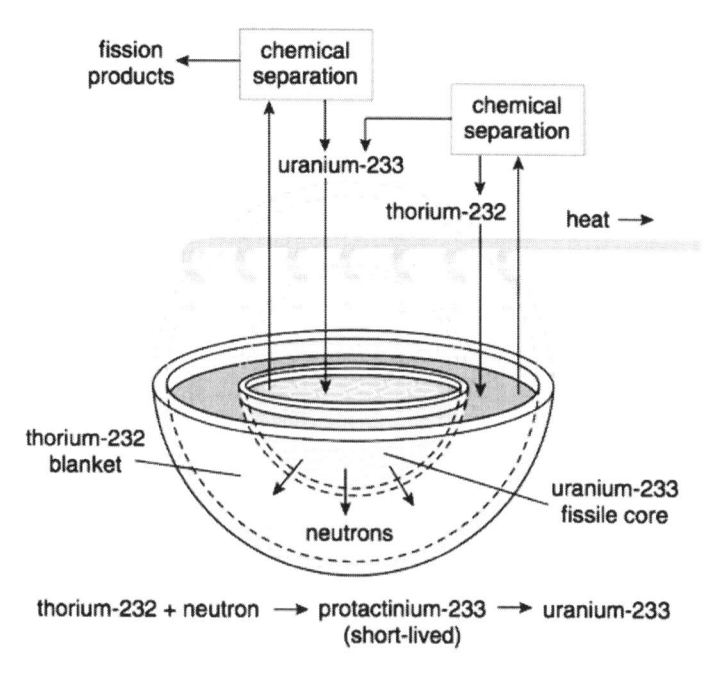

of a critical core containing fissile uranium-233 in a molten fluoride salt, surrounded by a blanket of molten fluoride salt containing thorium-232. Excess neutrons produced by fission in the core are absorbed by thorium-232 in the blanket, generating uranium-233 by transmutation. The uranium-233 and other fission products are recovered by chemical separation and the newly bred and recovered uranium-233 is directed to the core, where it sustains the chain reaction.

The key to thorium-based power is detaching from the well-established picture of what a reactor should be. In a nutshell, the liquid fluoride thorium reactor (LFTR, pronounced "lifter") consists of a core and a "blanket," a volume that surrounds the core. The blanket contains a mixture of thorium tetrafluoride in a fluoride salt containing lithium and beryllium, made molten by the heat of the core. The core consists of fissile uranium-233 tetrafluoride also in molten fluoride salts of lithium and beryllium within a graphite structure that serves as a moderator and neutron reflector. The uranium-233 is produced in the blanket when neutrons generated in the core are absorbed by thorium-232 in the surrounding blanket. The thorium-233 that results then beta decays to short-lived protactinium-233, which rapidly beta decays again to fissile uranium-233. This fissile material is chemically separated from the blanket salt and transferred to the core to be burned up as fuel, generating heat through fission and neutrons that produce more uranium-233 from thorium in the blanket.

Advantages of Liquid Fuel

Liquid fuel thorium reactors offer an array of advantages in design, operation, safety, waste management, cost and proliferation resistance over the traditional configuration of nuclear plants. Individually, the advantages are intriguing. Collectively they are compelling. Unlike solid nuclear fuel, liquid fluoride salts are impervious to radiation damage. We mentioned earlier that fuel rods acquire structural damage from the heat and radiation of the nuclear furnace. Replacing them requires expensive shutdown of the plant about every 18 months to swap out a third of the fuel rods while shuffling the remainder.

Figure 4. Uranium fuel rods are removed after just four percent or so of their potential energy is consumed. Noble

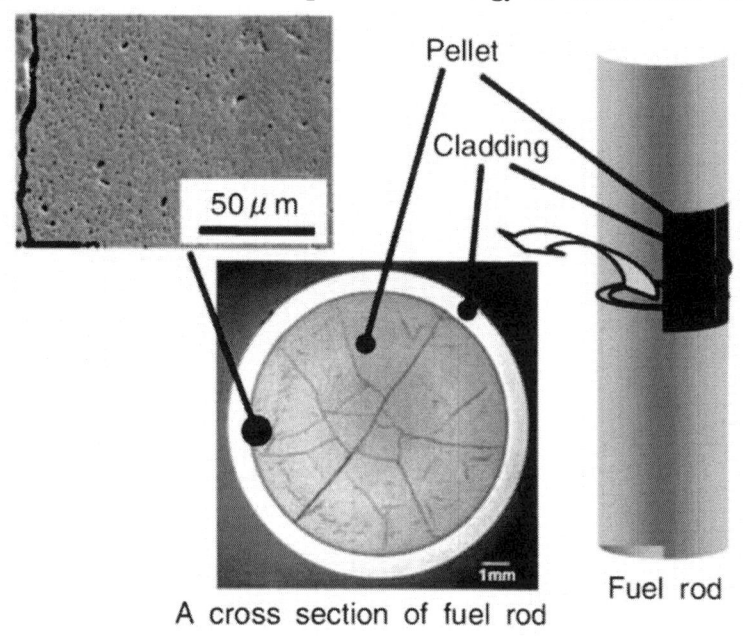

Pellet
Cladding
50 μm
1mm
A cross section of fuel rod
Fuel rod

gases such as krypton and xenon build up, along with other fission products such as samarium that accumulate and absorb neutrons, preventing them from sustaining the chain reaction. The solid is stressed by internal temperature differences, by radiation damage that breaks the covalent bonds of uranium dioxide, and by fission products that disturb the solid lattice structure. As the solid fuel swells and distorts, the irradiated zirconium cladding tubes must contain the fuel and all fission products within it, both in the reactor and for centuries thereafter in a waste storage repository.

Fresh fuel is not very hazardous, but spent fuel is intensely radioactive and must be handled by remotely operated equipment. After several years of storage underwater to allow highly radioactive fission products to decay to stability, fuel rods can be safely transferred to dry-cask storage. Liquid fluoride fuel is not subject to the structural stresses of solid fuel and its ionic bonds can tolerate unlimited levels of radiation dam- age, while eliminating the (rather high) cost of fabricating fuel elements and the (also high) cost of periodic shutdowns to replace them.

More important are the ways in which liquid fuel accommodates chemical engineering. Within uranium oxide fuel rods, numerous transuranic products are generated, such as plutonium-239, created by the absorption of a neutron by uranium-238, followed by beta decay. Some of this plutonium is fissioned, contributing as much as one-third of the energy production of uranium reactors. All such transuranic elements could eventually be destroyed in the neutron flux, either by direct fission or transmutation to a fissile element, except that the solid fuel must be removed long before complete burnup is achieved. In liquid fuel, transuranic fission products can remain in the fluid fuel of the core, transmuting by neutron absorption until eventually they nearly all undergo fission.

In solid-fuel rods, fission products are trapped in the structural lattice of the fuel material. In liquid fuel, reaction products can be relatively easily removed. For example, the gaseous fission poison xenon is easy to remove because it bubbles out of solution as the fuel salt is pumped. Separation of materials by this mechanism is central to the main feature of thorium power, which is formation of fissile uranium-233 in the blanket for export to the core. In the fluoride salt of the thorium blanket, newly formed uranium-233 forms soluble uranium tetrafluoride (UF_4). Bubbling fluorine gas through the blanket solution converts the uranium tetrafluoride into gaseous uranium hexafluoride (UF_6), while not chemically affecting the less-reactive thorium tetrafluoride. Uranium hexafluoride comes out of solution, is captured, then is reduced back to soluble UF_4 by hydrogen gas in a reduction column, and finally is directed to the core to serve as fissile fuel.

Figure 5. Among the many differences between the thorium/uranium fuel cycle and the enriched uranium/plutonium cycle is the volume of material handled from beginning to end to generate comparable amounts of electric power. Thorium is extracted in the same mines as rare earths, from which it is easily separated. In contrast, vast

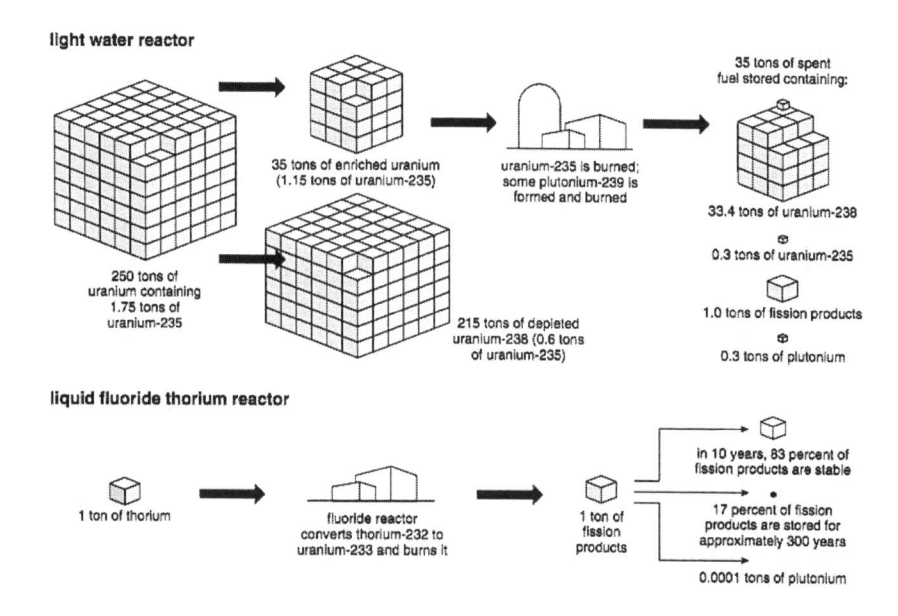

light water reactor

250 tons of uranium containing 1.75 tons of uranium-235

35 tons of enriched uranium (1.15 tons of uranium-235)

uranium-235 is burned; some plutonium-239 is formed and burned

215 tons of depleted uranium-238 (0.6 tons of uranium-235)

35 tons of spent fuel stored containing:

33.4 tons of uranium-238

0.3 tons of uranium-235

1.0 tons of fission products

0.3 tons of plutonium

liquid fluoride thorium reactor

1 ton of thorium

fluoride reactor converts thorium-232 to uranium-233 and burns it

1 ton of fission products

in 10 years, 83 percent of fission products are stable

17 percent of fission products are stored for approximately 300 years

0.0001 tons of plutonium

amounts of uranium ore must be laboriously and expensively processed to get usable amounts of uranium enriched in the fissile isotope uranium-235. On the other end of the fuel cycle, the uranium fuel cycle generates many times the amount of waste by mass, which must be stored in geological isolation for hundreds of centuries. The thorium fuel cycle generates much less waste, of far less long-term toxicity, which has to be stored for just three centuries or so.

Other fission products such as molybdenum, neodymium and technetium can be easily removed from liquid fuel by fluorination or plating techniques, greatly prolonging the viability and efficiency of the liquid fuel.

Liquid fluoride solutions are familiar chemistry. Millions of metric tons of liquid fluoride salts circulate through hundreds of aluminum chemical plants daily, and all uranium used in today's reactors has to pass in and out of a fluoride form in order to be enriched. The LFTR technology is in many ways a straightforward extension of contemporary nuclear chemical engineering.

Waste Not

Among the most attractive features of the LFTR design is its waste profile. It makes very little. Recently, the problem of nuclear waste generated during the uranium era has become both more and less urgent. It is more urgent because as of early 2009, the Obama administration has ruled that the Yucca Mountain Repository, the site designated for the permanent geological isolation of existing U.S. nuclear waste, is no longer to be considered an option. Without Yucca Mountain as a strategy for waste disposal, the U.S. has no strategy at all. In May 2009, Secretary of Energy Steven Chu, Nobel laureate in physics, said that Yucca Mountain is off the table. What we're going to be doing is saying, let's step back. We realize that we know a lot more today than we did 25 or 30 years ago. The [Nuclear Regulatory Commission] is saying that the dry-cask storage at current sites would be safe for many decades, so that gives us time to figure out what we should do for a long-term strategy.

The waste problem has become some-what less urgent because many stakeholders believe Secretary Chu is correct that the waste, secured in huge, hardened casks under adequate guard, is in fact not vulnerable to any foreseeable accident or mischief in the near future, buying time to develop a sound plan for its permanent disposal. A sound plan we must have. One component of a long-range plan that would keep the growing problem from getting worse while meeting growing power needs would be to mobilize nuclear technology that creates far less waste that is far less toxic. The liquid fluoride thorium reactor answers that need.

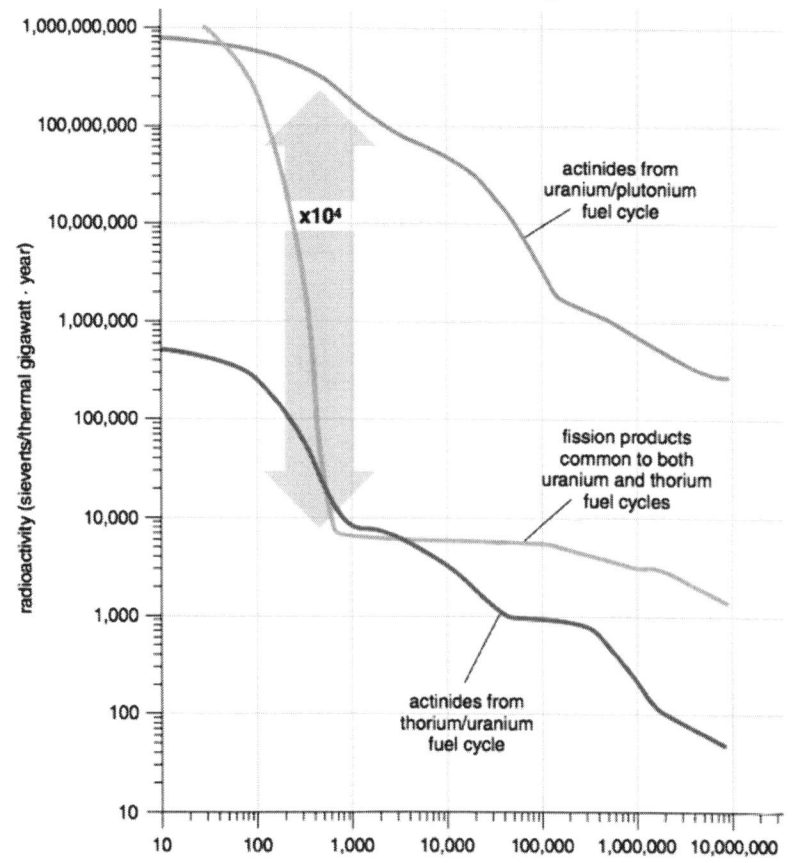

Figure 6. Switching to liquid fluoride thorium reactors would go a long way toward neutralizing the nuclear waste storage issue. The relatively small amount of waste produced in LFTRs requires a few hundred years of isolated storage versus the few hundred thousand years for the waste generated by the uranium/plutonium fuel cycle. Thorium- and uranium-fueled reactors produce essentially the same fission products, whose radiotoxicity is displayed in the dark bottom line on this diagram of radiation dose versus time. The top line is actinide waste from a light-water reactor, and the light bottom line is actinide waste from a LFTR. After 300 years the radiotoxicity of the thorium fuel cycle waste is 10,000 times less than that of

the uranium/plutonium fuel cycle waste. The LFTR scheme can also consume fissile material extracted from light-water reactor waste to start up thorium/uranium fuel generation.

Thorium and uranium reactors produce essentially the same fission (breakdown) products, but they produce a quite different spectrum of actinides (the elements above actinium in the periodic table, produced in reactors by neutron absorption and transmutation). The various isotopes of these elements are the main contributors to the very long-term radio- toxicity of nuclear waste.

The mass number of thorium-232 is six units less than that of uranium-238, thus many more neutron captures are required to transmute thorium to the first transuranic. Figure 6 shows that the radiotoxicity of wastes from a thorium/uranium fuel cycle is far lower than that of the currently employed uranium/plutonium cycle--after 300 years, it is about 10,000 times less toxic.

By statute, the U.S. government has sole responsibility for the nuclear waste that has so far been produced and has collected $25 billion in fees from nuclear-power producers over the past 30 years to deal with it. Inaction on the waste front, to borrow the words of the Obama administration, is not an option. Many feel that some of the $25 billion collected so far would be well spent kickstarting research on thorium power to contribute to future power with minimal waste.

Safety First

It has always been the dream of reactor designers to produce plants with inherent safety--reactor assembly, fuel and power-generation components engineered in such a way that the reactor will, without human intervention, remain stable or shut itself down in response to any accident, electrical outage, abnormal change in load or other mishap. The LFTR design appears, in its present state of research and design, to possess an extremely high degree of inherent safety. The single most volatile aspect of current nuclear reactors is the pressurized water. In boiling light-water, pressurized light-water, and heavy-water reactors (accounting for nearly all of the 441 reactors worldwide), water serves as the coolant and neutron moderator. The heat of fission causes water to boil, either directly in the core or in a steam generator, producing steam that drives a turbine. The water is maintained at high pressure to raise its boiling temperature. The explosive pressures involved are contained by a system of highly engineered, highly expensive piping and pressure vessels (called the "pressure boundary"), and the ultimate line of defense is the massive, expensive containment building surrounding the reactor, designed to withstand any explosive calamity and prevent the release of radioactive materials propelled by pressurized steam.

A signature safety feature of the LFTR design is that the coolant--liquid fluoride salt--is not under pressure. The fluoride salt does not boil below 1400 degrees Celsius. Neutral pressure reduces the cost and the scale of LFTR plant construction by reducing the scale of the containment requirements, because it obviates the need to contain a pressure explosion. Disruption in a transport line would result in a leak, not an explosion, which would be captured in a noncritical configuration in a catch basin, where it would passively cool and harden.

Another safety feature of LFTRs, shared with all of the new generation of LWRs, is its *negative temperature coefficient of reactivity*. Meltdown, the bogey of the early nuclear era, has been effectively designed out of modern nuclear fuels by engineering them so that power excursions--the industry term for runaway reactors--are self-limiting. For example, if the temperature in a reactor rises beyond the intended regime, signaling a power excursion, the fuel itself responds with thermal expansion, reducing the effective area for neutron absorption--the temperature coefficient of reactivity is negative--thus suppressing the rate of fission and causing the temperature to fall. With appropriate formulations and configurations of nuclear fuel, of which there are now a number from which to choose among solid fuels, runaway reactivity becomes implausible.

In the LFTR, thermal expansion of the liquid fuel and the moderator vessel containing it reduces the reactivity of the core. This response permits the desirable property of load following-- under conditions of changing electricity demand (load), the reactor requires no intervention to respond with automatic increases or decreases in power production.

As a second tier of defense, LFTR designs have a freeze plug at the bot- tom of the core--a plug of salt, cooled by a fan to keep it at a temperature below the freezing point of the salt. If temperature rises beyond a critical point, the plug melts, and the liquid fuel in the core is immediately evacuated, pouring into a subcritical geometry in a catch basin. This formidable safety tactic is only possible if the fuel is a liquid. One of the current requirements of the Nuclear Regulatory Commission (NRC) for certification of a new nuclear plant design is that in the event of a complete electricity outage, the reactor remain at least stable for several days if it is not automatically deactivated. This setup is the ultimate in safe power-outage response. Power isn't needed to shut down the reactor, for example by manipulating control elements. Instead power is needed to *prevent* the shutdown of the reactor.

Cost Wise

In terms of cost, the ideal would be to compete successfully against coal without subsidies or market-modifying legislation. It may well be possible. Capital costs are generally higher for conventional nuclear versus fossil-fuel plants, whereas fuel costs are lower. Capital costs are outsized for nuclear plants because the construction, including the containment building, must meet very high standards; the facilities include elaborate, redundant safety systems; and included in capital costs are levies for the cost of decommissioning and removing the plants when they are ultimately taken out of service. The much-consulted MIT study *The Future of Nuclear Power,* originally published in 2003 and updated in 2009, shows the capital costs of coal plants at $2.30 per watt versus $4 for light-water nuclear. A principal reason why the capital costs of LFTR plants could depart from this ratio is that the LFTR operates at atmospheric pressure and contains no pressurized water. With no water to flash to steam in the event of a pressure breach, a LFTR can use a much more close-fitting containment structure. Other expensive high-pressure coolant-injection systems can also be deleted. One concept for the smaller LFTR containment structure is a hardened concrete facility below ground level, with a robust concrete cap at ground level to resist aircraft impact and any other foreseeable assaults.

Other factors contribute to a favorable cost structure, such as simpler fuel handling, smaller components, markedly lower fuel costs and significantly higher energy efficiency. LFTRs are high-temperature reactors, operating at around 800 degrees Celsius, which is thermodynamically favorable for conversion of thermal to electrical energy—a conversion efficiency of 45 percent is likely, versus 33 percent typical of coal and older nuclear plants. The high heat also opens the door for other remunerative uses for the thermal energy, such as hydrogen production, which is greatly facilitated by high temperature, as well as driving other industrial chemical processes with excess process heat. Depending on the siting of a LFTR plant, it could even supply heat for homes and offices.

Figure 7. Nuclear power plants provide 20 percent of U.S. electricity and 70 percent of low-emissions energy supply.

Every 750 megawatts of installed nuclear reactor capacity could avoid the release of one million metric tons of CO_2 per year versus similar electricity output obtained from natural gas.

Thorium must also compete economically with energy-efficiency initiatives and renewables. A mature decision process requires that we consider whether renewables and efficiency can realistically answer the rapidly growing energy needs of China, India and the other tiers of the developing world as cheap fossil fuels beckon--at terrible environmental cost. Part of the cost calculation for transitioning to thorium must include its role in the expansion of prosperity in the world, which will be linked inexorably to greater energy demands. We have a pecuniary interest in avoiding the environmental blowback of a massive upsurge in fossil-fuel consumption in the developing world. The value of providing an alternative to that scenario is hard to monetize, but the consequences of not doing so are impossible to hide from.

Figure 8. Boeing produces one $200 million plane per day in massive production lines that could be a model for mass production of liquid fluoride thorium reactors. Centralized mass production offers the advantages of specialization among workers, product standardization, and optimization of quality control, as inspections can be conducted by highly trained workers using installed, specialized equipment.

Perhaps the most compelling idea on the drawing board for pushing thorium-based power into the main- stream is mass production to drive rapid deployment in the U.S. and export elsewhere. Business economists observe that commercialization of any technology leads to lower costs as the number of units increases and the experience curve delivers benefits in work specialization, refined production processes, product standardization and efficient product redesign. Given the diminished scale of LFTRs, it seems reasonable to project that reactors of 100 megawatts can be factory produced for a cost of around $200 million. Boeing, producing one $200 million airplane per day, could be a model for LFTR production.

Modular construction is an important trend in current manufacturing of traditional nuclear plants. The market-leading Westinghouse AP-1000 advanced pressurized-water reactor can be built in 36 months from the first pouring of concrete, in part because of its modular construction. The largest module of the AP1000 is a 700-metric- ton unit that arrives at the construction site with rooms completely wired, pipe-fitted and painted. Quality benefits from modular construction because inspection can consist of a set of protocols executed by specialists operating in a dedicated environment.

One potential role for mass-produced LFTR plants could be replacing the power generation components of existing fossil-fuel fired plants, while integrating with the existing electrical-distribution infrastructure already wired to those sites. The savings from adapting existing infrastructure could be very large indeed.

Nonproliferation

Cost competitiveness is a weighty consideration for nuclear power development, but it exists on a somewhat different level from the life-and-death considerations of waste management, safety and nonproliferation. Escalating the role of nuclear power in the world must be anchored to decisively eliminating the illicit diversion of nuclear materials.

When the idea of thorium power was first revived in recent years, the focus of discussion was its inherent proliferation resistance (see the September–October 2003 issue of *American Scientist;* Mujid S. Kazimi, "Thorium Fuel for Nuclear Energy"). The uranium-233 produced from thorium-232 is necessarily accompanied by uranium-232, a proliferation prophylactic. Uranium-232 has a relatively short half-life of 73.6 years, burning itself out by producing decay products that include strong emitters of high-energy gamma radiation. The gamma emissions are easily detectable and highly destructive to ordnance components, circuitry and especially personnel. Uranium-232 is chemically identical to and essentially inseparable from uranium-233.

The neutron economy of LFTR designs also contributes to securing its inventory of nuclear materials. In the LFTR core, neutron absorption by uranium-233 produces slightly more than two neutrons per fission--one to drive a subsequent fission and another to drive the conversion of thorium- 232 to uranium-233 in the blanket solution. Over a wide range of energies, uranium-233 emits an average of 2.4 neutrons for each one absorbed. However, taking into account the overall fission rate per capture, capture by other nuclei and so on, a well-designed LFTR reactor should be able to direct about 1.08 neutrons per fission to thorium transmutation. This delicate poise doesn't create excess, just enough to generate fuel indefinitely. If meaningful quantities of uranium-233 are misdirected for nonpeaceful purposes, the reactor will report the diversion by winding down because of insufficient fissile product produced in the blanket.

Only a determined, well-funded effort on the scale of a national program could overcome the obstacles to illicit use of uranium-232/233 produced in a LFTR reactor. Such an effort would certainly find that it was less problematic to pursue the enrichment of natural uranium or the generation of plutonium. In a world where widespread adoption of LFTR technology undermines the entire, hugely expensive enterprise of uranium enrichment—the necessary first step on the way to plutonium production—bad actors could find their choices narrowing down to unusable uranium and unobtainable plutonium.

Prospects

What kind of national effort will be required to launch a thorium era? We are watching a rehearsal in the latter half of 2010 with the unfolding of the Department of Energy's (DOE) flagship $5 billion Next Generation Nuclear Plant (NGNP) project. Established by the Energy Policy Act of 2005, NGNP was charged with demonstrating the generation of electricity and possibly hydrogen using a high-temperature nuclear energy source. The project is being executed in collaboration with industry, Department of Energy national laboratories and U.S. universities. Through fiscal year 2010, $528 million has been spent. Proposals were received in November 2009 and designs are to be completed by September 30, 2010. Following a review by the DOE's Nuclear Energy Advisory Committee, Secretary Chu will announce in January 2011 whether one of the projects will be funded to completion, with the goal of becoming operational in 2021.

There are two major designs under consideration, the pebble bed and prismatic core reactors, which are much advanced versions of solid-fuel designs from the 1970s and 1980s. In both designs, tiny, ceramic-coated particles of enriched uranium are batched in spheres or pellets, coupled with appropriate designs for managing these fuels in reactors. These fuel designs feature inherent safety features that eliminate meltdown, and in experiments they have set the record for fuel burnup in solid designs, reaching as high as 19 percent burnup before the fuel must be replaced. Thorium is not currently under consideration for the DOE's development attention.

> Figure 9a. Thorium is more common in the earth's crust than tin, mercury, or silver. A cubic meter of average crust yields the equivalent of about four sugar cubes of thorium, enough to supply the energy needs of one person for more than ten years if completely fissioned. Lemhi Pass on the Montana-Idaho border is estimated to contain 1,800,000 tons of high-grade thorium ore. Five hundred tons could supply all U.S. energy needs for one year.

Figure 9b. Due to lack of current demand, the U.S. government has returned about 3,200 metric tons of refined thorium nitrate to the crust, burying it in the Nevada desert. Image courtesy of the National Nuclear Security Administration Nevada Site Office.

If the DOE is not promoting thorium power, who will? Utilities are constrained by the most prosaic economics when choosing between nuclear and coal, and they are notoriously risk averse. The utilities do not have an inherent motive, beyond an unproven profit profile, to make the leap to thorium. Furthermore, the large manufacturers, such as Westinghouse, have already made deep financial commitments to a different technology, massive light-water reactors, a technology of proven soundness that has already been certified by the NRC for construction and licensing. Among experts in the policy and technology of nuclear power, one hears that large nuclear-plant technology has already arrived--the current so-called Generation III+ plants have solved the problems of safe, cost-effective nuclear power, and there is simply no will from that quarter to inaugurate an entirely new technology, with all that it would entail in research and regulatory certification--a hugely expensive multiyear process. And the same experts are not overly oppressed by the waste problem, because current storage is deemed to be stable. Also, on the horizon we can envision burning up most of the worst of the waste with an entirely different technology, fast-neutron reactors that will consume the materials that would otherwise require truly long-term storage.

But the giant preapproved plants will not be mass produced. They don't offer a vision for massive, rapid conversion from fossil fuels to nuclear, coupled with a nonproliferation portfolio that would make it reasonable to project the technology to developing parts of the world, where the problem of growing fossil-fuel consumption is most urgent.

The NGNP project is not the answer. There is little prospect that it can gear up on anything close to the timescale needed to replace coal and gas electricity generation within a generation or two. Yet its momentum may crowd out other research avenues, just as alternative nuclear technologies starved support of Alvin Weinberg's Molten Salt Reactor Project. We could be left asking, What if? Or we can take a close look at thorium as we rethink how we will produce the power consumed by the next generation. These issues and others are being explored at the online forum http://energyfromthorium.com, an energetic, international gathering of scientists and engineers probing the practical potential of this fuel.

References

David, S., E. Huffer and H. Nifenecker. 2007. Revisiting the thorium-uranium nuclear fuel cycle. *Europhysics News* 38(2):24–27.

International Atomic Energy Agency. 2005. Thorium fuel cycle: Potential benefits and challenges. IAEA-Tecdoc-1450.

Kazimi, M. S. 2003. Thorium fuel for nuclear energy. *American Scientist* 91:408–415.

MacPherson, H. G. 1985. The molten salt reactor adventure. *Nuclear Science and Engineering* 90:374–380.

Mathieu, L., et al. 2006. The thorium molten salt reactor: Moving on from the MSBR. *Progress in Nuclear Energy* 48:664–679.

Sorensen, K. 2010. Thinking nuclear? Think thorium. *Machine Design* 82 (May 18):22–26.

Weinberg, A. M. 1997. *The First Nuclear Era: The Life and Times of a Technological Fixer*. New York: Springer.

U.S. Department of Energy, Office of Nuclear Energy. 2010. *Next Generation Nuclear Plant: A Report to Congress*. www.ne.doe.gov/pdf-Files/NGNP_ReporttoCongress_2010.pdf

Appendix B

UNDERGROUND POWER PLANT BASED ON MOLTEN SALT TECHNOLOGY

RALPH W. MOIR* and EDWARD TELLER†

Lawrence Livermore National Laboratory, P.O. Box 808, L-637 Livermore, California 94551

Originally published in NUCLEAR TECHNOLOGY, VOL. 151, SEP. 2005

Received August 9, 2004

Accepted for Publication December 30, 2004

* email: ralph@ralphmoir.com

† We are sorry to inform our readers that Edward Teller is deceased September 9, 2003.

This paper addresses the problems posed by running out of oil and gas supplies and the environmental problems that are due to greenhouse gases by suggesting the use of the energy available in the resource thorium, which is much more plentiful than the conventional nuclear fuel uranium. We propose the burning of this thorium dissolved as a fluoride in molten salt in the *minimum viscosity mixture of LiF and BeF2 together with a small amount of ^{235}U or plutonium fluoride to initiate the process to be located at least 10 m underground. The fission products could be stored at the same underground location. With graphite replacement or new cores and with the liquid fuel transferred to the new cores periodically, the power plant could operate for up to 200 yr with no transport of fissile material to the reactor or of wastes from the reactor during this period. Advantages that include utilization of an abundant fuel, inaccessibility of that fuel to terrorists or for diversion to weapons use, together with good economics and safety features such as an underground location will diminish public concerns. We call for the construction of a small prototype thorium-burning reactor.*

1. POWER PLANT DESIGN

This paper brings together many known ideas for nuclear power plants. We propose a new combination including non-proliferation features, undergrounding, limited separations, and long-term, but temporary, storage of reactor products also underground. All these ideas are intended to make the plant economical, resistant to terrorist activities, and conserve resources in order to be available to greatly expand nuclear power if needed as envisioned by Generation IV reactor requirements.

We propose the adoption of the molten salt thorium reactor that uses flowing molten salt both as the fuel carrier and as a coolant. The inventors of the molten salt reactor were E. S. Bettis and R. C. Briant, and the development was carried out by many people under the direction of A. Weinberg at Oak Ridge National Laboratory.[1] The present version of this reactor is based on the Molten Salt Reactor Experiment[2-4] that operated between 1965 and 1969 at Oak Ridge National Laboratory at 7-MW(thermal) power level and is shown in Fig. 1. The solvent molten salt is lithium fluoride (LiF, ~70 mol%) mixed with beryllium fluoride (BeF2, 20%), in which thorium fluoride (ThF4, 8%) and uranium fluorides are dissolved (1% as ^{238}U and 0.2% as ^{235}U in the form of UF4 and UF3, UF3/UF4 < 0.025}. [a] This mixture is pumped into the reactor at a temperature of ~560°C and is heated up by fission reactions to 700°C by the time it leaves the reactor core, always near or at atmospheric pressure. The materials for the vessel, piping, pumps, and heat exchangers are made of a nickel alloy. [5,6,b] The vapor pressure of the molten salt at the temperatures of interest is very low (<10^{-4} atm), and the projected boiling point at atmospheric pressure is very high (~1400°C). This heat is transferred by a heat exchanger to a nonradioactive molten fluoride salt coolant [c] with an inlet temperature of 450°C and the outlet liquid temperature of 620°C that is pumped to the conventional electricity-producing part of the power plant located above-ground. This heat is converted to electricity in a modern steam power plant at an efficiency of ~43%.

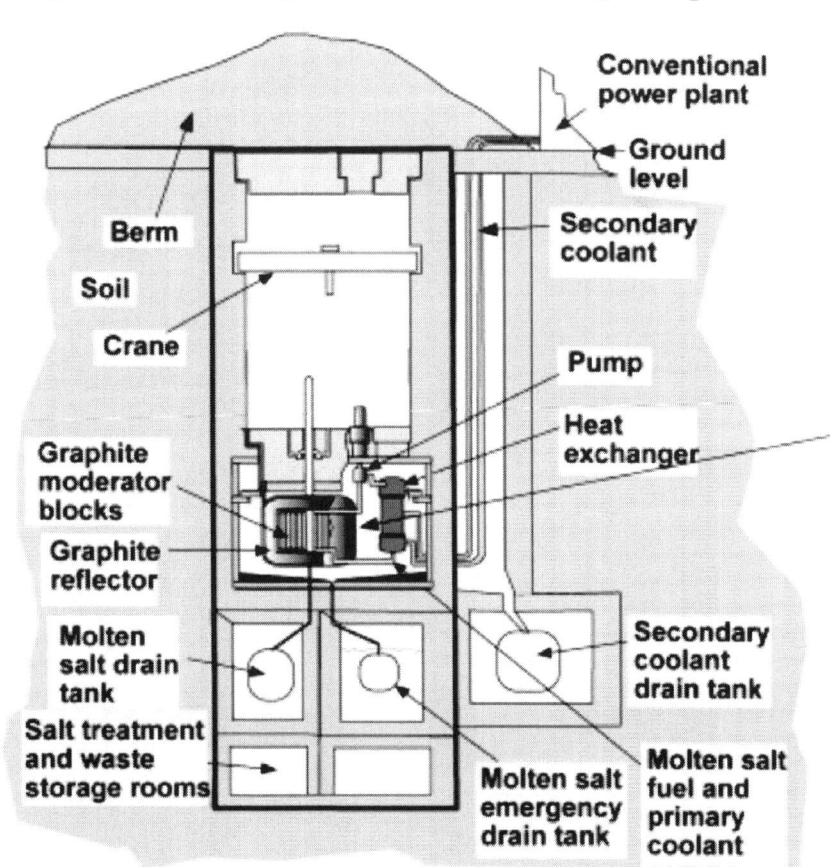

Fig. 1. The nuclear part of the molten salt power plant [7] is

illustrated belowground with the nonradioactive conventional part aboveground; many rooms and components are not shown.

New cores would be installed after each continuous operating period of possibly 30 yr or the graphite in the cores can be replaced.

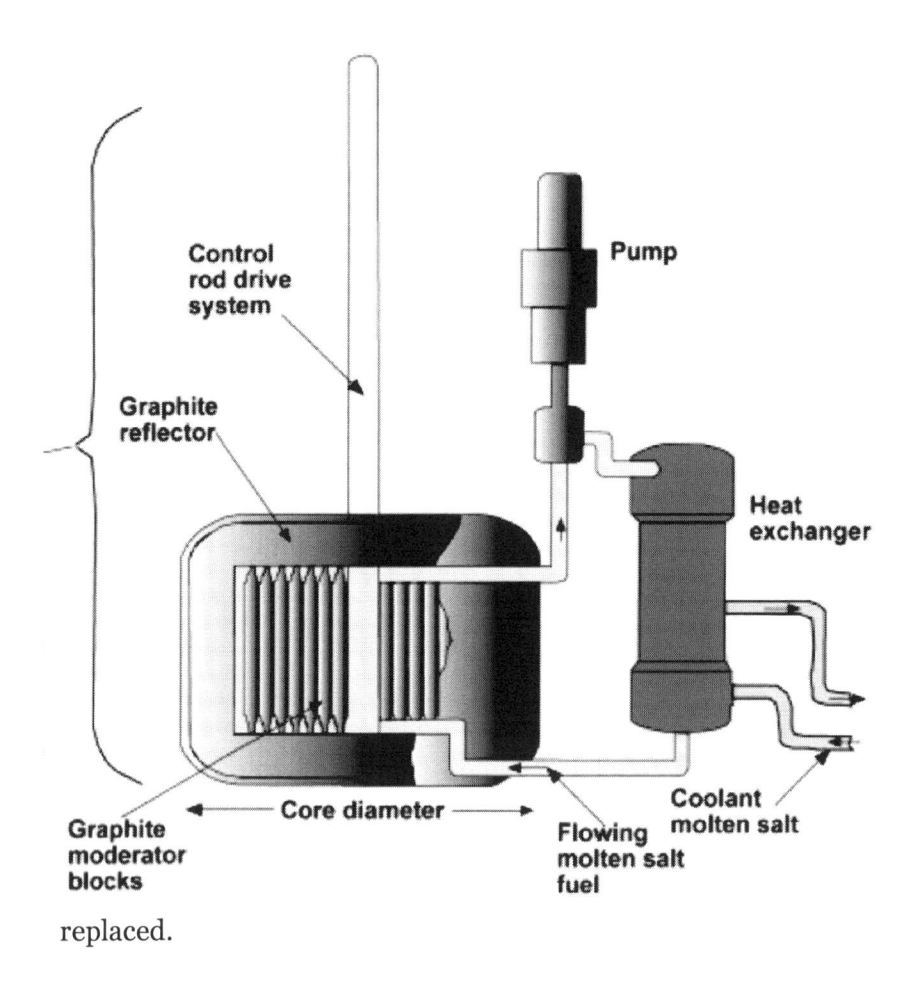

The fluid circulates at a moderate speed of 0.5 m/s in 5-cm-diam channels amounting to between 10 and 20% of the volume within graphite blocks of a total height of a few meters.

The graphite slows down the fast neutrons produced by the fission reaction. The slowed neutrons produce fission and another generation of neutrons to sustain the chain reaction.

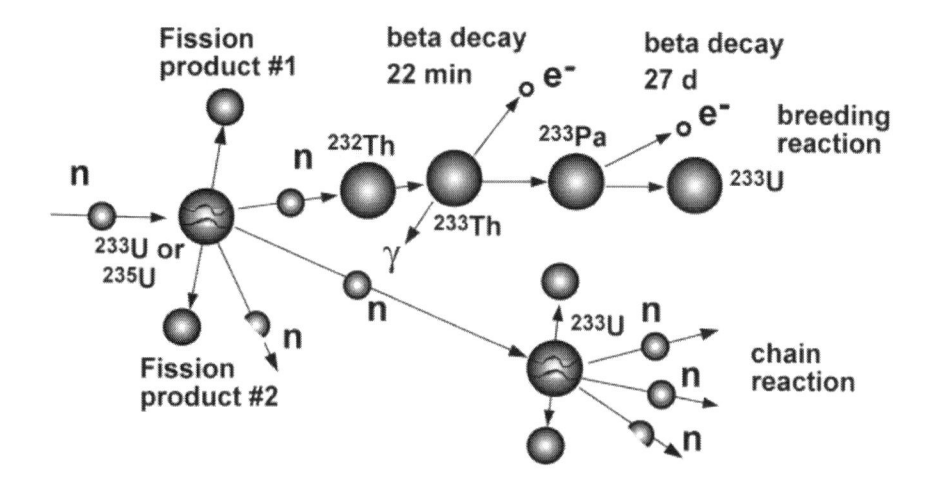

Fig. 2. Illustration of the process of breeding or producing new fuel, ^{233}U, from neutron capture in ^{232}Th as a part of the chain reaction. Each fission reaction produces two or three neutrons (about 2.5 on average as illustrated by the "half neutron" above).

One of the slowed neutrons is absorbed in ^{232}Th producing ^{233}Th, which undergoes a 22-min beta decay to ^{233}Pa. The ^{233}Pa undergoes a month-long beta decay into ^{233}U, which with a further neutron produces fission and repeats the cycle. The reactions are illustrated in Fig. 2. Note that the cycle does not include ^{235}U, which is used only to initiate the process. The result is a drastic reduction of the need for mined uranium.

The initial fuel to start up the reactor can be mined and enriched ^{235}U [~3500 kg for 1000 MW(electric)]. An alternative might be to start up on discharged light water reactor (LWR) spent fuel, particularly ^{239}Pu. Actually, ^{239}Pu is contained in a waste of transuranium elements that people might actually pay to give this fuel away. As the plant operates, the plutonium and higher actinides and ^{235}U would be fissioned and replaced with ^{233}U produced from thorium, which is even a better fuel than ^{235}U, because nonfission thermal neutron captures are about half as likely.

An important feature of our proposal is to locate everything that is radioactive at least 10 m underground—where all fissions occur—while the electric generators are located in the open, being fed by hot, nonradioactive liquids. The reactor's heat-producing core is constructed to operate with a minimum of human interaction and limited fuel additions for decades. Of the three underground options [8] excavation into mountains with tunnel or vertical access or surface excavation with a berm covering, we prefer the berm as illustrated in Fig. 1. Undergrounding will preclude the possibility of radioactive contamination in case of airplane disasters. A combination of 10 m of concrete and soil is enough mass to stop most objects. It would eliminate tornado hazards and, most particularly, contribute to defense against terrorist activities. In case of accidents, undergrounding, in addition to the usual containment structures, enhances containment of radioactive material. The 10-m figure is a compromise between safety and plant construction expense. We anticipate the cost to construct underground with only 10 m of overburden using the berm technique will add <10% to the cost.

The molten salt reactor that operated in the 1960s had a big advantage in the removal of many fission products without much effort. Gases (Kr and Xe) simply bubble off aided by helium gas bubbling, where these gases are separated from the helium and stored in sealed tanks to decay. Noble and seminoble metals [d] precipitated. In the planned reactor, the old method of removing the gases may be repeated. The precipitation process might conceivably be enhanced by using a centrifuge and filtering rather than the old uncontrolled method of precipitation. In this way, the need to remove the remaining fission products, e.g., the rare-earth elements (Sm, Pm, Nd, Pr, Eu, and Ce) and alkali-earth elements with valence two and three fluoride formers, is reduced and may be postponed to intervals, perhaps as long as once every 30 yr. The accumulation of these elements has a small effect on neutron economy and on chemistry such as corrosion. Experience is needed on these long-term effects.

Most fission products have half-lives of ~30 yr or less. These "short-lived" fission products can be stored and monitored at the plant site for hundreds of years, while their hazard decreases by three orders of magnitude or more by the natural process of radioactive decay. Three elements are notable because they need to be separated for special treatment because of their extra long lives: ^{99}Tc, ^{129}I, and ^{135}Cs [with half-lives of 210 000 yr, 1.6 million yr, and 2.3 million yr] capture cross sections of 20, 30, and 9 barns (10^{-24} cm^2); and production rates of 23, 3.8, and 34 kg/GW(electric)-yr, respectively]. New ways should be found for separating these long-lived products (>=30-yr half-life) from short-lived products (<=30-yr half-life).

After a period of operation, perhaps as long as 30 yr, the reactor is shut down, owing to the swelling of the graphite blocks as shown in Fig. 3. The criterion [4,9] used here is 30 yr for a 10-m-diam core at 1000 MW(electric), for a neutron dose of $<3 \times 10^{26}$ n/m^2 for $E >$ 50 keV and a swelling of 3 vol% at 750°C for a capacity factor of 85%. Robotic technology is developing so rapidly that graphite replacement might be a quick and a low-cost operation. Another process that might be life limiting is corrosion. At the time that a new or refurbished power-generating graphite core is put into operation and the corroded parts are replaced, the fuel dissolved in the molten salt is transferred to the new core in a liquid state. This fuel transfer and core refurbishment allows the power station to continue operating for several more decades. At this time, the remaining fission products in solution can be removed by the chemical process known as *reductive extraction* to limit the neutron loss to absorption. The bulk of materials (lithium, beryllium, and thorium fluorides) may last for several hundred years before they are transmuted to other elements by nuclear reactions.

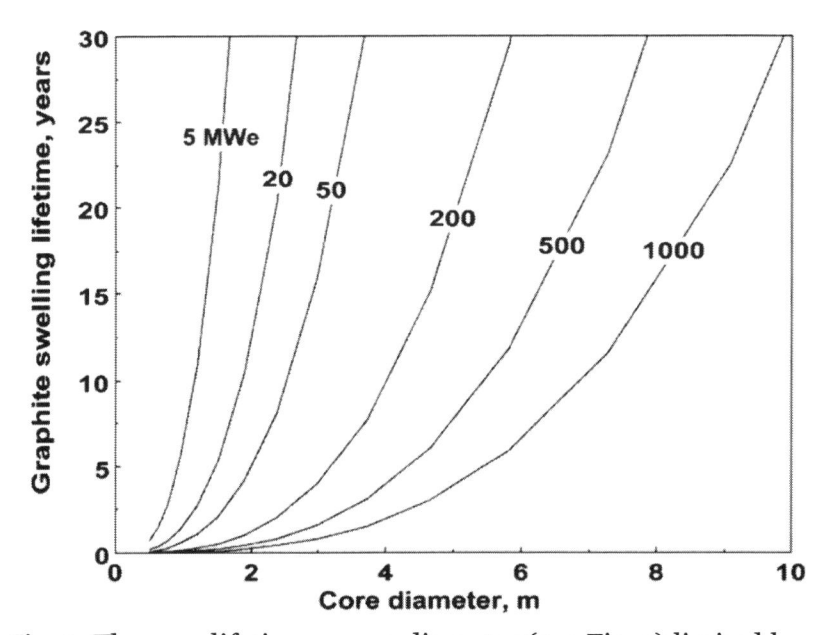

Fig. 3. The core lifetime versus diameter (see Fig. 1) limited by graphite swelling is shown for a wide range of output power.

This process might conceivably be continued as long as we operate the power station, perhaps even hundreds of years, making operations and ownership similar to a dam but with less impact. The fission products will be separated and stored at the power plant site in a suitable form under careful supervision or they will be transported to a permanent disposal site. We propose the twofold argument for the safe interim storage of radioactive material: first, that the location will be underground, and second, that the storage will be at the site of operating reactors, which require carefully planned defense anyway.

When the site with its collection of reactors is to be shut down, careful considerations will have to be used in the choice between whether the accumulated radioactivity should be transported to a permanent storage site or whether continuation of established supervision is safer and less expensive. The idea is to transport only mildly radioactive fuel to the power plant but have a minimum of transport of highly radioactive fission products and fuel away from the plant, thus minimizing the chance of accidents or terrorist activities. One conclusion is obvious: It will become important to find useful applications of radioactivity such as radioactive tracers, thereby converting a serious worry into a potential asset.

II. WASTE FORM: SUBSTITUTED FLUORAPATITE

A possible waste form for the molten salt reactor might be based on the naturally occurring mineral that has been found to contain ancient actinides in the natural reactor in Africa in mineral deposits called fluorapatite [10,11] $Ca_5(PO_4)_3F$. This low solubility mineral is much like fluoridated tooth enamel. If we substitute the fission product ions, for example, Sm, for the Ca ions, we call this substituted fluorapatite

$$SmF_3 + 4.5Ca_3(PO_4)_2 \rightarrow 3(Sm_{0.33}Ca_{4.5})(PO_4)_3F. \quad (1)$$

The result of this reaction is a ceramic powder that can be melted into bricks for long-term storage either at the power plant site or at a repository.

It might be preferable to transport it in a more compact fluoride form and produce the more stable but larger mass and volume form of material at the permanent repository site. The stored fluoride wastes could be melted and transferred in liquid form to a shipping container much like that used for sulfur shipping except more massive and shipped to a permanent storage site where again they are transferred in liquid form to be made into substituted fluorapatite bricks. If permanent storage is decided upon, we estimate the space needed in a Yucca Mountain–like repository for molten salt wastes to be ten and maybe closer to 100 times less than for once-through LWR spent fuel based on the heat generation rate of the wastes.

III. SAFETY

The molten salt reactor is designed to have a negative temperature coefficient of reactivity. This means the reactor's power quickly drops if its temperature rises above the operating point, which is an important and necessary safety feature. The molten salt reactor is especially good in this respect—it has little excess reactivity because it is refueled frequently online and has a high conversion rate that automatically replaces fuel consumed. Failure to provide makeup fuel is fail-safe as the reactivity is self-limiting by the burnup of available fuel. A small amount of excess reactivity would be compensated by a temporary interruption of adding makeup fuel on-line. Present reactors have ~20% excess reactivity. Control rods and burnable poisons are used not only in accident control but also to barely maintain criticality. In the molten salt reactor, control rods are used to control excess reactivity of perhaps only 2%, which is necessary to warm the salt from the cooler start-up temperature to the operating temperature (i.e., overcome the negative temperature coefficient). That is, only enough fissile fuel is in the core to maintain a chain reaction and little more.

Gaseous fission products are continually removed and stored separately from the reactor in pressurized storage tanks. By contrast, in conventional reactors the gaseous fission products build up in the Zr-clad fuel tubes to a high pressure that presents a hazard and can cause trouble. If an unforeseen accident were to occur, the constant fission-product removal means the molten salt reactor has much less radioactivity to potentially spread.

The usual requirement of containing fission products within three barriers is enhanced by adding a fourth barrier. The primary vessel and piping boundary, including drain tanks, constitute one barrier. These components are located in a room that is lined with a second barrier, including an emergency drain or storage tank for spills. The third barrier is achieved by surrounding the entire reactor building in a confinement vessel. A fourth safety measure is locating the reactor underground, which itself is one extra "gravity barrier" aiding confinement. A leakage of material would have to move against gravity for 10 m before reaching the atmosphere.

In case of accidents or spills of radioactive material, the rooms underground would remain isolated. However, the residual decay heat that continues to be generated at a low rate would be transferred through heat exchangers that passively carry the heat to the environment aboveground, while retaining the radioactive material belowground. This passive heat removal concept perhaps using heat pipes will be used to cool the stored fission products as well.

The initial fuel needed including the amount circulating outside the core is considerably less than half that of other breeding reactors such as the liquid metal–cooled fast reactor. This is a consequence of fast reactors having much larger critical mass than thermal reactors and for the molten salt case, avoiding the need for extra fuel at beginning of life to account for burnup of fuel.

IV. FUEL CYCLE WITHOUT FUEL PROCESSING AND WITHOUT WEAPONS-USABLE MATERIAL

When the 1000-MW(electric) reactor is started up, the initial fissile fuel is 20% enriched uranium (20% ^{235}U and 80% ^{238}U) along with thorium, actually 3.5 tons of ^{235}U, 14 tons of ^{238}U, and 110 tons of thorium. This low enrichment makes the uranium undesirable as weapons material without isotope separation, and therefore it does not have to be guarded so vigorously. An important side product is a small amount of ^{232}U produced by (n,2n) and (gamma,n) reactions on ^{233}U producing ^{232}U. Uranium-232 is highly radioactive and has unusually strong and penetrating gamma radiation (2.6 MeV), making diversion of this fuel for misuse extra difficult and easier to detect if stolen; the resulting weapons would be highly radioactive and therefore dangerous to those nearby as well as making detection easier.

The uranium in the core starts at 20% fissile and drops so it is never weapons usable. [e] The plutonium produced from neutron capture in ^{238}U rather quickly develops higher isotopes of plutonium, making it a poor material for weapons. [f] Safeguarding is still necessary but less important. The advantage of this fuel cycle is that 80% of its fuel is made in the reactor, and the fuel shipments to the plant during its operation are nonweapons usable.

$$\text{Conversion ratio} = \frac{^{233}\text{U and fissile Pu production rate}}{\text{all fissile consumption rate}} \quad (2)$$

The conversion ratio starts out at 0.8, and after 30 yr of operation drops to 0.77 [Ref. 4]. Today's LWRs [g] each require 5700 tons of mined uranium in 30 yr. Our molten salt reactor example also at 1000-MW(electric) size, in 30 yr of operation at 75% capacity factor would consume by fissioning, 17 tons of thorium, 3.8 tons of ^{238}U, and 6.7 tons of ^{235}U. This requires 1500 tons of mined uranium. [h] Our worries about the consumption of uranium are reduced by a factor of 4 relative to today's reactors while the depletion of thorium remains entirely negligible.

In our example, 14% of the heavy atoms that have been transported to the reactor are burned up or fissioned in 30 yr of operation. [i] If we include the 1500 tons of mined uranium that went into the depletion process and was not used in the reactor, then the percentage of burnup is 1.3%. This compares to our present-day reactor example with once-through fueling of 0.5% burnup of mined uranium with the assumptions in footnote g.

V. ALTERNATIVE FUEL CYCLE

If we decide in future versions of the molten salt reactor to move toward the pure thorium-^{233}U cycle with fuel processing then the conversion ratio approaches unity and the use of mined uranium will drop by over an order of magnitude or be eliminated once started up. This cycle would start up the reactor with only ^{235}U and thorium dissolved in the molten salt. [j] Neutrons absorbed in thorium would produce ^{233}U. Although this fuel is highly radioactive, after chemical separation it is directly usable in nuclear weapons and therefore poses a danger that would have to be guarded against with extra measures. We should avoid designs that permit separation of protactinium because it decays into ^{233}U without the highly radioactive ^{232}U "spike" previously mentioned.

The strong advantage of this fuel cycle is that it breeds essentially all of its own fuel, thus removing the need for transportation of weapons-usable material to the reactor site once it is started up. Also it makes no further demands for mined uranium for several hundred years although the graphite had to be changed a number of times. [4] For example, a present-day reactor would use 38 000 tons of mined uranium over 200 yr, while the molten salt reactor once started up on ^{235}U and thorium would need only 600 tons of mined uranium and could operate for 200 yr (see footnote g again). One hundred thirty-seven tons of thorium would be fissioned. [k] The burnup of the 600 tons of uranium and 137 tons of thorium would be ~18%.

Even a small amount of fissile material removed from the reactor would cause it to cease operation, and this mitigates the danger of diversion from the plant site. Diversion of the material for weapons use would be an interruption of normal procedures, which could be carried out only by insiders. It is clear that continuous operation would be needed. Thus, it should be easily noticed unless carried out by separating small amounts for a long period.

We advocate full compliance and even strengthened international safeguard agreements including inspection regimes and technical means for monitoring the reactor and all its operations. Monitoring devices including cameras and transceivers possibly in miniature or even subgram sizes might aid monitoring systems to find out whether all components in the system are in place and operating normally. It is difficult to exclude the possibility that considerable quantities of components of nuclear explosives might be produced in reactors, and therefore information on the production of these materials should be readily available. This requirement should be considered a crucial part of a policy of openness (to be introduced gradually), which, in a general sense, will be necessary to insure the stability of the world. Openness is not an easy condition to fulfill but perhaps better than any obvious alternative.

VI. ECONOMIC COMPETITIVENESS

Our economic goal is to achieve a cost of electrical energy averaged over the life of the power station to be no more than that from burning fossil fuels at the same location. Past studies have shown a potential for the molten salt reactor to be somewhat lower in cost of electricity than both coal and LWRs [Refs. 4 and 12]. There are several reasons for substantial cost savings: low pressure operation, low operations and maintenance costs, lack of fuel fabrication, easy fuel handling, low fissile inventory, use of multiple plants at one site allowing sharing of facilities, and building large plant sizes. The cost of undergrounding the nuclear part of the plant obviously needs to be determined and will likely not offset the cost advantages of a liquid-fueled low-pressure reactor.

VII. WHY HAS THE MOLTEN SALT REACTOR NOT ALREADY BEEN DEVELOPED?

If the molten salt reactor appears to meet our criteria so well, why has it not already been developed since the molten salt reactor experiment operated over 30 yr ago?

Several decades ago an intense development was undertaken to address the problem of rapid expansion of reactors to meet a high growth rate of electricity while the known uranium resources were low. The competition came down to a liquid-metal fast breeder reactor (LMFBR) on the uranium-plutonium cycle and a thermal reactor on the thorium-^{233}U cycle, the molten salt breeder reactor. The LMFBR had a larger breeding rate, a property of fast reactors having more neutrons per fission and less loss of neutrons by parasitic capture, and won the competition. This fact and the plan to reduce the number of candidate reactors being developed were used as arguments to stop the development of the molten salt reactor rather than keep an effort going as a backup option. In our opinion, this was an excusable mistake.

As a result there has been little work done on the molten salt reactor during the last 30 yr. As it turned out, a far larger amount of uranium was found than was thought to exist, and the electricity growth rate has turned out to be much smaller than predicted. High excess breeding rates have turned out not to be essential. A reactor is advantageous that once started up needs no other fuel except thorium because it makes most or all its own fuel.

Studies of possible next-generation reactors, called Generation IV, have included the molten salt reactor among six reactor types recommended for further development. In addition the program called Advanced Fuel Cycle Initiative has the goal of separating fission products and recycling for further fissioning.

VIII. DEVELOPMENT REQUIREMENTS AND CONCLUSIONS

In conclusion, we believe a small prototype plant should be built to provide experience in all aspects of a commercial plant. The liquid nature of the molten salt reactor permits an unusually small plant that could serve the role just so that the temperatures, power densities, and flow speeds are similar to that in larger plants. A test reactor, e.g., 10 MW(electric) or maybe even as small as 1 MW(electric) would suffice and still have full commercial plant power density and therefore the same graphite damage or corrosion limited lifetime. Supporting research and development would be needed on corrosion of materials, process development, and waste forms, all of which, however, are not needed for the first prototype.

We give some examples of development needs. We need to show adequate long corrosion lifetime for nickel alloy resistant to the tellurium cracking observed after the past reactor ran for only 4 yr. If carbon composites are successful, corrosion will likely become less important. We want to prove feasible extraction of valence two and three fluorides, especially rare-earth elements, which will then allow the fuel to burn far longer than 30 yr (200 yr). We need to study and demonstrate an interim waste form suggested to be solid and liquid fluorides and substitute fluorapatite for the permanent waste form of fission products with minimal carryover of actinides during the separation process. This solution holds the promise to diminish the need for repository space by up to two orders of magnitude based on waste heat generation rate. We need a study to show the feasibility of passive heat removal from the reactor after-heat and stored fission products to the atmosphere without material leakage and at reasonable cost. Another study needs to show that all aspects of the molten salt reactor can be done competitively with fossil fuel. The cost for such a program would likely be well under $1 billion with operation costs likely on the order of $100 million per year. In this way a very large-scale nuclear power plan could be established, including even the developing nations, in a decade.

ACKNOWLEDGMENT

This work was performed under the auspices of the U.S. Department of Energy by University of California Lawrence Livermore National Laboratory under contract W-7405-Eng-48.

FOOTNOTES

[a] Instead of the Be and Li combination, we might consider sodium and zirconium fluorides in some applications to reduce hazards of Be and tritium production from lithium.

[b] It seems likely all these components could be made of composite carbon-based materials instead of nickel alloy that would allow raising the operating temperature so that a direct cycle helium turbine could be used rather than a steam cycle (~900°C) and hydrogen could be made in a thermochemical cycle (~1050°C). A modest size research and development program should be able to establish the feasibility of these high-temperature applications.

[c] A secondary coolant option is the molten salt, sodium fluoroborate, which is a mixture of $NaBF_4$ and NaF. Other coolants are possible depending on design requirements such as low melting temperature to avoid freeze-up.

APPENDIX B: MOIR AND TELLER, MOLTEN SALT 465

[d] Noble and seminoble metals are Zn, Ga, Ge, As, Nb, Mo, Tc, Ru, Rh, Pd, Ag, Cd, In, Sn, and Sb. Seminoble here means they do not form fluorides but rather precipitate in elemental form.

[e] For example after 15 yr of operation, the isotopes of the uranium in the molten salt are ~0.02% ^{232}U; 8% ^{233}U; 2% ^{234}U; 4% ^{235}U; 3% ^{236}U; and 83% ^{238}U.

[f] After operation for 15 yr, the plutonium in the molten salt has the following isotopes: 7% ^{238}Pu; 36% ^{239}Pu; 21% ^{240}Pu; 15% ^{241}Pu; and 20% ^{242}Pu.

[g] The assumption on LWR fuel usage can be seen:

$$\frac{1000 \text{ MW(electric)} \cdot 0.75 \cdot 365 \text{ day/yr} \cdot 30 \text{ yr} \cdot 5\%}{0.32 \frac{\text{MW(electric)}}{\text{MW}} \cdot 50\,000 \text{ MWd}/T \cdot 0.45\%} = 5700 \text{ tons}$$

of mined uranium in 30 yr with tails of 0.25%. (5700 tons x 200 yr) / 30 yr) = 38 000 tons in 200 yr.

Burnup of heavy atoms =

$$\frac{1000 \text{ MW(electric)} \cdot 235 \text{ amu} \cdot 1.67 \cdot 10^{-27} \text{ kg/amu} \cdot 365 \text{ days/yr} \cdot 24 \text{ h/day} \cdot 3600 \text{ s/h}}{0.32 \text{ MW(electric)}/\text{MW} \cdot 195 \text{ MeV} \cdot 1.6 \cdot 10^{-19} \text{ J/eV}}$$

= ~1240 kg / full power year.

Burnup fraction = 1.24 tons x 30yr x 0.75 / 5700tons = 0.49%.

Mined uranium for the molten salt reactor to start up is 3.5 tons ^{235}U/0.0045 = 780 tons of mined uranium. For the alternative fuel cycle, the start-up is 2.8 tons ^{235}U/0.0045 = 620 tons of mined uranium.

[h] 6.7 tons of ^{235}U/0.0045 = tons of mined uranium where we assume the ^{235}U content of 0.7% of uranium can be used with tails of 0.25%.

[i] Burnup of heavy atoms =

$$\frac{1000 \text{ MW(electric)} \cdot 233 \text{ amu} \cdot 1.67 \cdot 10^{-27} \text{ kg/amu } 365 \text{ days/yr} \cdot 24 \text{ h/day} \cdot 3600 \text{ s/h}}{0.43 \text{ MW(electric)}/\text{MW} \cdot 195 \text{ MeV} \cdot 1.6 \cdot 10^{-19} \text{ J/eV}}$$

= 915 kg per full power year.

Burnup fraction = atoms burned (fissioned) in 30 yr / all heavy atoms =

$$\frac{915 \text{ kg} \cdot 0.75 \cdot 30 \text{ yr}}{110\,000 \text{ kg Th} + 32\,400 \text{ kg }^{238}\text{U} + 7900 \text{ kg }^{235}\text{U}} = \frac{20\,600 \text{ kg}}{150\,300 \text{ kg}} = 13.7\% \ .$$

We use 30-yr period and 75% capacity factor consistently for all cases, so that relative comparisons are unaffected by this assumption. The fissile consumption is then 0.75 x 915 = 690 kg/yr.

[j] Uranium-233 for start-up fuel could be produced externally from accelerator or thermonuclear fusion produced neutrons absorbed in thorium if these technologies become developed successfully. This fissile source or use of discharge fuel from current fission reactor designs would virtually eliminate the need for further uranium mining but would introduce proliferation issues that could and would have to be dealt with.

[k] Burnup of heavy atoms in 200 yr = 0.915 tons/yr per full power year x 0.75 capacity factor x 200 yr = 137 tons in 200 yr. Burnup in 200 yr = (137 tons Th) / (620 tons mined U + 137 tons Th) = 18%.

REFERENCES

1. H. G. MacPHERSON, "The Molten Salt Reactor Adventure," *Nucl. Sci. Eng.*, **90,** 374 (1985).

2. P. N. HAUBENREICH and J. R. ENGEL, "Experience with the Molten-Salt Reactor Experiment," *Nucl. Appl. Technol.*, **8,** 118 (1970).

3. M. PERRY, "Molten-Salt Converter Reactors," *Ann. Nucl. Energy,* **2,** 809 (1975).

4. J. R. ENGEL, H. F. BAUMAN, J. F. DEARING, W. R. GRIMES, E. H. McCOY, and W. A. RHOADES, "Conceptual Design Characteristics of a Denatured Molten-Salt Reactor with Once-

Through Fueling," ORNLoTM-7207, Oak Ridge National Laboratory (July 1980).

5. C. W. FORSBERG, P. F. PETERSON, and P. S. PICKARD, "Molten-Salt-Cooled Advanced High-Temperature Reactor for Production of Hydrogen and Electricity," *Nucl. Technol.*, **144**, 289 (2003).

6. P. F. PETERSON, "Multiple-Reheat Brayton Cycles for Nuclear Power Conversion with Molten Coolants," *Nucl. Technol.*, **144**, 279 (2003).

7. K. FURUKAWA, K. MITACHI, and Y. KATO, "Small MSR with a Rational Th Fuel Cycle," *Nucl. Eng. Des.*, **136**, 157 (1992).

8. F. C. FINLAYSON, W. A. KRAMMER, and J. BENVENISTE, "Evaluation of the Feasibility, Economic Impact, and Effectiveness of Underground Nuclear Power Plants," ATR-78 (7652-14)-1, The Aerospace Corporation, Energy and Transportation Division, Aerospace (May 1978).

9. R.C. ROBERTSON et al., "Conceptual Design Study of a Single-Fluid Molten-Salt Breeder Reactor," ORNL-454, Oak Ridge National Laboratory (1971).

10. S. M. McDEAVITT, Argonne National Laboratory, Private Communications (2002).

11. D. LEXA, "Preparation and Physical Characteristics of a Lithium-Beryllium Substituted Fluorapatite," *Metall. Mat. Trans. A*, **30A**, 147 (1999).

12. R. W. MOIR, "Cost of Electricity from Molten Salt Reactors," *Nucl. Technol.*, **138**, 93 (2002).

A 2012 12-minute video produced by Motherboard TV reviews the controversial work of Edward Teller and his association with Ralph Moir that led to the publication of this article.

http://motherboard.vice.com/2012/3/7/motherboard-tv-doctor-teller-s-strange-loves-from-the-hydrogen-bomb-to-thorium-energy--2

Bibliography

Allison, Wade; Radiation and Reason: The impact of science on a culture of fear

Allwood, Julian and Cullen, Jonathan; Sustainable Materials: with both eyes open

Blees, Tom; Prescription for the Planet: The painless remedy for our energy and environmental crises

Bryce, Robert: Power Hungry

Cohen, Bernard; The Nuclear Energy Option: An alternative for the 90s

Cravens, Gwyneth; Power to Save the World: The truth about nuclear energy

Domenici, Pete; A Brighter Tomorrow: Fulfilling the promise of nuclear energy

Herbst, Alan and Hopley, George; Nuclear Energy Now: Why the time has come for the world's most misunderstood energy source

Hogerton, John; The Atomic Energy Deskbook

Lomborg, Bjorn; Cool It: The skeptical environmentalist's guide to global warming

Lovelock, James; The Revenge of Gaia: Earth's climate crisis and the fate of humanity

MacKay, David; Sustainable Energy: without the hot air

Martin, Richard; SuperFuel: Thorium, the green energy source for the future

Morris, Robert; The Environmental Case for Nuclear Power: Economic, medical, and political considerations

Muller, Richard; Physics for Future Presidents: The science behind the headlines

Nuttall, W J; Nuclear Renaissance: Technologies and policies for the future of nuclear power

Olah, George et al: Beyond Oil and Gas: The methanol economy

Reed, Thomas and Stillman, Danny; The Nuclear Express: A political history of the bomb and its proliferation

Richter, Burton; Beyond Smoke and Mirrors: Climate change and energy in the 21st century

Romm, Joseph; The Hype about Hydrogen: Fact and fiction in the race to save the climate

Ropeik, David: How Risky Is It, Really?: Why our fears don't always match the facts

Sachs, Jeffrey; Common Wealth: Economics for a crowded planet

Smil, Vaclav: Energies; An illustrated guide to the biosphere and civilization

Till, Charles and Chang, Yoon Il; Plentiful Energy: The story of the integral fast reactor

Tucker, Todd; Atomic America: How a deadly explosion and a feared admiral changed the course of nuclear history

Tucker, William; Terrestrial Energy: How nuclear power will lead the green revolution and end America's energy odyssey

Weinberg, Alvin; The First Nuclear Era: The life and times of a technological fixer

Wilson, Richard and Spengler, John; Particles in Our Air: Concentrations and health effects

Wilson, Richard et al; Health Effects of Fossil Fuel Burning: Assessment and mitigation

Yergin, Daniel; The Quest: Energy, security, and the remaking of the modern world

Index

A Better Place, 160
Abengoa, 149, 406
Abu Dhabi, 397
accelerator driven reactor, 296
Accelerator driven subcritical reactor, 296
ADNA, 300, 414
ADSR, 297, 298, 299, 300, 301, 414, 415
ADSRs, 298, 300
Advanced Research Projects Agency, 386
AEC, 176, 199, 418
AECL, 266
Agriculture, 55, 56, 80, 380
Aim High, 4, 24
air pollution, 83, 92, 100, 161, 217, 329, 345, 400, 403
Aker Solutions, 300, 414
Alabama, 163, 253
ALARA, 329
Alaska, 81, 125, 161, 174
Alberta, 266, 350, 354
Albiasa, 149, 406
Algeria, 337, 338
AllEarth, 146, 148, 149, 406
Allison, 321, 329, 416, 468
Ameren, 284
Ammonia, 365, 366, 367, 368, 369, 370, 371, 372, 376, 418, 419
Andasol, 148, 406

AP1000, 258, 274, 275, 276, 277, 278, 279, 283, 413, 442
Areva, 258, 272, 278, 284
Argonne, 250, 289, 467
Argonne National Laboratories, 250
Arizona, 149, 388, 406
Atomic Energy Commission, 176
Atoms for Peace, 181, 338
ATP, 52, 53
Australia, 141, 174, 204, 265, 266, 406
Avatar, 320
Babcock & Wilcox, 187, 280
batteries, 32, 38, 155, 159, 160, 161, 162, 165, 166, 356, 407
Beacon Power, 161, 407
Becquerel, 311
BEIR, 315, 415
Belgium, 332, 367
Bergius, 347
Berkeley, 215, 262, 326, 328, 416
Berlin, 152
beryllium, 21, 188, 194, 236, 238, 262, 424, 430, 450, 455
Bettis Atomic Power Laboratory, 251
Big Bang, 45, 64
biodiesel, 158, 367

biofuel, 23, 29, 151, 155, 156, 158, 363, 407
biofuels, 29, 144, 151, 155, 157, 158, 364, 407
bitumen, 347, 354
Blue Ribbon Commission, 4, 333
BMW, 378
BN-600, 260, 286, 290
BN-800, 260, 290
Boeing, 169, 221, 338, 355, 379, 395, 398, 399, 411, 441
Boice, 324
Bq, 311, 312, 313
Brayton, 198, 215, 226, 237, 238, 267, 295, 381, 394, 409, 411, 467
Brazil, 156
Brightsource, 149, 406
Brzezinski, 319
Burlington, 148, 153, 407
burning cigar, 291
Bush, 319
CAES, 163, 164, 407
California, 149, 370, 378, 389, 449, 464
Callaway, 284
Cameron, 320
Canada, 85, 191, 237, 248, 249, 258, 266, 267, 312, 332, 337, 338, 347, 348, 354, 413
Canadian Nuclear Safety Commission, 267
cancer, 26, 308, 312, 314, 315, 316, 318, 321, 323, 324, 326, 396
CANDU, 191, 258, 266, 267
Capacity factors, 141, 169
cap-and-trade, 88, 387

Cape Wind, 134, 143
carbon taxes, 24, 29, 88, 91, 100, 110, 216, 389, 393, 398, 400, 420
Caribbean, 81, 174
Carlsbad, 331, 334
CAS, 260
Cattle, 54
Cavendish, 186
CCGT, 21, 106, 119, 120, 129, 130, 138, 139, 140, 163, 217
CCS, 110, 111, 112, 113, 114, 120, 404
Cellulosic ethanol, 156
centrifuge enrichment, 336, 338, 342
CERN, 190
Charles River Ventures, 292
Chemical energy, 59
Chemical potential energy, 34, 35
chemists' reactor, 200
Chernobyl, 175, 201, 232, 307, 312, 321, 328, 421, 425
Chevron's, 353
China, 28, 75, 76, 83, 84, 89, 93, 98, 109, 113, 126, 146, 148, 158, 171, 191, 215, 258, 259, 260, 261, 263, 273, 277, 278, 290, 305, 306, 332, 337, 338, 348, 375, 389, 390, 393, 397, 399, 403, 404, 406, 412, 413, 440
Chinese Academy of Sciences, 260, 262, 412
Chu, 252, 434, 444
Chubu Electric Power, 269
CIA, 70, 72, 403
Clean Air Task Force, 82

Clean Coal, 110
Clinch River, 280
Clinton, 289, 386
CNG, 127
coefficient of performance, 50
Cogeneration, 174, 381, 408
Cohen, 321, 415, 468
Colorado, 125, 142, 348, 389
Colorado School of Mines, 368
computer, 4, 35, 66, 79, 103, 220, 226, 274, 276, 293, 308
Congressional Budget Office, 384, 391
Connecticut, 152, 387
conservation, 46, 170, 171, 172, 173, 408
copper-chloride, 357, 377
Cross-State Air Pollution Rule, 82
Cumbria, 290
cyanobacteria, 51
Czech Republic, 248, 249, 262, 265, 266
Daimler Chrysler, 378
Dartmouth, 4, 23, 66, 101, 174
death, 29, 82, 83, 268, 315, 318, 320, 321, 322, 442
deaths, 65, 70, 73, 82, 83, 84, 87, 109, 175, 322, 323, 324
deep boreholes, 332, 334
Deepwater Wind, 135, 143
Delaware, 387
Desalination, 174, 381, 408
Deuterium, 186
developing nations, 27, 30, 61, 69, 72, 74, 75, 78, 87, 88, 91, 92, 93, 97, 98, 172, 173, 216, 336, 344, 464
Dewan, 254
Diesel, 48, 83, 98, 108, 111, 155, 158, 161, 222, 278, 348, 355, 356, 359, 363, 364, 366, 367, 418
dimethyl ether, 222, 356, 366, 376
DNA, 311, 318, 323, 326, 327, 416
Dostal, 239
Dow Chemical, 272
Driscoll, 239
EBR-II, 286, 287, 289, 292, 293, 295, 414
Economist, 103, 407, 416
Edison, 174
Einstein, 39, 320
Eisenhower, 181, 338, 399
Electric energy, 43, 93, 161, 164
electric/thermal efficiency, 117, 118, 138, 140, 152, 153, 198, 293
ElectroFrac, 353, 418
energy density, 55, 154, 155, 355, 356, 362, 365, 366, 379
Energy from Thorium, 261
energy poverty, 27, 92, 93, 95, 169, 382, 383, 391, 397, 398, 399
Energy Return on Investment, 156
energy security, 28, 97, 155, 211, 246, 383, 391
energy storage, 135, 147, 159, 162, 163, 164, 165, 407, 408
Engel, 217

Engelward, 327, 416
Entergy, 272
Environmental Protection Agency, 82, 92, 345
EPA, 82, 83, 84, 92, 109, 122, 139, 152, 312, 314, 316, 318, 391, 403, 405, 406, 407, 420
ERCOT, 391
EROI, 67, 156, 347, 349, 351, 353, 407, 418
ethanol, 29, 68, 151, 155, 156, 157, 158, 362, 363, 367, 407
eutectic, 236, 290
EVOL, 265
Fairbanks, 161
Federal Energy Regulatory Commission, 126, 391
Feed-in tariffs, 154, 388, 420
Feed-in-tariffs, 388
Fermi, 186, 250, 423, 425
Finland, 278, 332
FlexEfficiency, 140
FLiBe, 21, 194, 196, 210, 231, 236, 253
Flibe Energy, 253, 254
fluorescent, 171
flywheel, 32
Flywheels, 32, 161
Ford, 378
Fornalski's, 328
fossil fuel, 30, 50, 68, 74, 101, 127, 131, 139, 145, 161, 329, 351, 374, 404, 464
fracking, 121, 405
France, 28, 57, 244, 248, 249, 264, 265, 266, 286, 332, 337, 338, 428
FUJI, 248, 267, 268, 269, 413

Fukushima, 127, 240, 278, 298, 303, 307, 312, 324, 325, 327, 329, 396, 415
Furedi, 319, 415
Furukawa, 267, 268
Gabon, 331, 416
gas turbine, 21, 111, 119, 127, 129, 130, 136, 137, 140, 163, 166, 198, 394
gasoline, 29, 32, 35, 44, 47, 48, 98, 108, 127, 155, 222, 251, 347, 348, 355, 356, 358, 359, 362, 364, 366, 367, 369, 370, 371, 372, 377, 418, 419
Gates, 71, 161, 291, 292, 382, 399
GDP, 59, 70, 71, 72, 73, 74, 88, 93, 171, 380, 401, 402
GE, 119, 130, 140, 236, 239, 275, 278, 289, 290, 295, 393, 405, 414
Gen4 Energy, 284, 414
General Electric, 338
General Synfuels, 353
geothermal, 50, 189, 380
Germany, 166, 215, 258, 259, 265, 273, 332, 388, 390, 392, 420
Gilleland, 291
global warming, 5, 27, 31, 65, 68, 76, 80, 82, 88, 92, 94, 95, 99, 115, 122, 130, 131, 155, 319, 325, 329, 336, 338, 345, 359, 383, 387, 389, 391, 393, 398, 468
GM, 127, 171, 378
Google, 4, 249, 274
Gore, 336
Grand Cayman, 174, 408
Grand Central Station, 316

Grand Inga, 169, 408
gravitational potential energy, 33, 34, 44, 136
Gray, 313
Green Freedom, 251, 360, 362, 376, 418
Green Revolution, 366
Green River, 348, 354
Grenoble, 248, 264, 428
Gy, 313
Haber-Bosch, 368, 369
Hanford, 325
Hansen, 76, 78, 173, 403, 408
Hargraves, 3, 4, 254, 418, 419, 421
Harvard, 115, 405
Hastelloy, 188, 193, 198, 233, 234, 235, 256
Heat Death, 45
heat engine, 47, 48, 49, 146, 402
heat exchanger, 185, 188, 192, 194, 197, 214, 236, 237, 254, 255, 271, 279, 286, 293, 451
heat exchangers, 222, 235, 238, 252, 265, 293, 451, 459
heat of combustion, 116, 359, 369
heat pump, 50, 145
Heat pumps, 49
Hejzlar, 239
Helium, 233
HFIR, 250
Hiroshima, 321, 322, 329
Hitachi, 289, 295, 414
Holdren, 255, 256
Holtec, 282, 283, 414
Honda, 127, 378, 419

hormesis, 328
horsepower, 44, 54
hydraulic fracturing, 116, 121, 122, 128, 347, 353, 405
hydro, 23, 34, 50, 101, 107, 109, 116, 118, 129, 135, 136, 161, 162, 164, 165, 166, 168, 169, 258, 307, 385, 388, 392, 408
Hydroelectric, 106, 136
IAEA, 242, 294, 312, 339, 341, 413, 446
ICRP, 329
Idaho National Laboratory, 251, 281, 286, 395
Idaho National Labs, 272, 374
IFR, 288, 289, 290, 292, 293, 295
IGCC, 110, 111, 112, 114, 360, 404
incandescent, 171
India, 55, 72, 73, 75, 80, 83, 93, 126, 171, 191, 204, 286, 300, 336, 337, 338, 343, 389, 390, 397, 403, 422, 427, 440
Industrial Revolution, 5, 31, 58, 59, 61, 62, 64, 77, 82, 103, 108, 152
inexhaustible, 97, 100, 131, 175, 191, 204, 293, 400
INL, 272, 281, 413
integrated systems operators, 391
Intergovernmental Panel on Climate Change, 79
International Atomic Energy Agency, 242, 446

International Thorium Energy Organization, 263, 412
Investment tax credits, 388
ionizing radiation, 236, 309, 311, 312, 315, 326, 328, 341, 396
IPCC, 79, 80, 152, 398, 403
Iraq, 86, 319, 397
Ireland, 142
isobreeder, 228, 229
ISO-NE, 391
isotopes, 189, 236, 242, 243, 244, 289, 293, 310, 341, 342, 422, 425, 426, 436, 459, 465
Israel, 337, 338
iTheo, 263, 412, 415
Japan, 28, 59, 126, 127, 161, 248, 249, 268, 269, 286, 312, 322, 328, 332, 389, 392, 413
Jaworowski, 328, 329, 416
Jiang Mianheng, 260, 399
Jiang Zemin, 260
Jordan, 380
Kamei, 269
Kazakhstan, 275
Kennedy, 176, 177
kerogen, 230, 235, 348, 349, 350, 353
Keystone, 347, 354
Khosla Ventures, 292
kinetic energy, 22, 31, 32, 33, 34, 35, 36, 40, 44, 46, 47, 48, 50, 53, 57, 58, 146, 179, 297
Kingman, 149
Knolls Atomic Power Laboratory, 251
Kubic, 360, 361, 362

Kuwait, 86, 174, 397
Kyoto, 29, 88
Lavoisier, 151
Lawrence, 248, 251, 297, 326, 416, 449, 464
Lawrence Livermore National Laboratory, 251, 449, 464
LeBlanc, 267, 410
Lemhi Pass, 97, 205, 444
Leukemia, 322
Levitt, 318, 415
Lightbridge, 190
lime cycle, 373, 375
Limits to Growth, 65, 402
Linde, 378, 419
liquid fluoride thorium reactor, 4, 22, 24, 91, 92, 95, 97, 100, 177, 191, 193, 202, 217, 269, 338, 344, 374, 394, 429, 434
liquid fuel nuclear reactors, 185
Liquid fueled nuclear reactors, 185
Liquid metal fast breeder reactors, 285
lithium, 21, 32, 35, 194, 236, 237, 262, 378, 411, 430, 450, 455, 464
Lithium-6, 236
Lithium-7, 236, 411
LMFBR, 200, 285, 290, 295, 462
LNT, 315, 316, 321, 325, 415
Los Alamos, 187, 284, 336, 360
Los Alamos National Laboratory, 251, 360
Lovins, 170

Low Dose Radiation Research, 326
Madden, 231, 411
Madrigal, 305, 415
Maine, 387
Manhattan Project, 202, 248, 337
Marcellus, 123, 125
Martin, 306, 360, 361, 362, 409, 468
Maryland, 387, 388, 389
Massachusetts, 34, 134, 135, 328, 387, 389
Massie, 254
McIntosh, 163, 164
Meadows, 66, 67, 402, 403
metabolism, 53, 311
methane, 76, 78, 111, 116, 120, 121, 122, 123, 131, 179, 217, 347, 348, 356, 359, 368, 388, 405
methanol, 222, 356, 362, 376, 469
Mexico, 85, 158, 306, 334, 388, 421
Microsoft, 71, 161, 291, 382
MiniFUJI, 268
Ministry of Electricity and Water, 174
Missouri, 284
MIT, 113, 114, 150, 160, 215, 239, 249, 254, 327, 402, 404, 406, 407, 409, 412, 414, 416, 438
Mitsubishi, 238
Moir, 4, 24, 217, 233, 245, 248, 249, 251, 299, 411, 412, 417, 421, 467, 481
MOX, 202, 251, 266, 290
mPower, 280, 414
MSRE, 192, 193, 229, 231, 252, 262, 410, 421
Multi effect distillation, 381
Murphy, 68, 403
Myhrvold, 291
Nacogdoches, 153
Nagasaki, 321, 322, 329, 337
nameplate, 106, 132, 134, 135, 142
NASA, 246, 248, 253, 275, 395
National Academy of Sciences, 115, 315
National Council on Radiation Protection and Measurements, 324
National Grid, 135
National Nuclear Security Administration, 325, 445
National Oceanic and Atmospheric Administration, 76
National Renewable Energy Laboratory, 157
Nautilus, 247, 251, 275, 346, 422, 424
NCSS, 250
Netherlands, 186, 248, 249, 265
New Hampshire, 387
New Mexico, 331
New York, 161, 174, 387, 447
Newcomen, 58
NGK, 161
NGNP, 251, 271, 272, 273, 298, 413, 443, 446, 447
NGNP Alliance, 272
nickel-cadmium, 161
Nigeria, 85
Nixon, 199, 200
Nobel, 190, 297, 300, 434

noble gas, 232
noble gasses, 209, 233, 301
noble metals, 232, 233, 301, 410
North Korea, 336, 337, 343
Northeast Utilities, 135
NRC, 245, 249, 250, 253, 267, 275, 276, 279, 280, 281, 284, 290, 295, 307, 308, 316, 385, 391, 395, 396, 414, 415, 438, 445
Nstar, 135
Nuclear Regulatory Commission, 245, 249, 275, 307, 434, 438
NuScale, 281, 282, 414
Oak Ridge, 180, 186, 187, 188, 191, 192, 193, 200, 227, 248, 249, 250, 260, 262, 298, 410, 421, 450, 467
Oak Ridge National Laboratory, 186, 298, 421, 450, 467
Obama, 26, 396, 401, 434, 436
OCGT, 22, 119, 120, 137, 139, 143
OECD, 61, 69, 74, 83, 87, 88, 91, 92, 216, 403
Oklo, 331
Oman, 380
Opel, 378
Oppenheimer, 202
Oregon State University, 281
ORLY, 412
ORNL, 192, 196, 209, 210, 217, 225, 227, 228, 232, 233, 241, 250, 252, 254, 261, 262, 302, 393, 394, 408, 409, 410, 411, 414, 420, 421, 423, 467
Ottawa Valley Research Associates, 267
Pakistan, 305, 336, 337, 338, 342
particulate, 83, 92, 345, 393
PB-AHTR, 212, 214, 215, 393, 410
PBMR, 259, 347
PBR, 259
Pearl, 348, 417
Penumbra Energy, 267
Peterson, 210, 333, 410, 419, 481
petroleum, 28, 65, 85, 87, 108, 116, 155, 156, 158, 171, 174, 347, 348, 354, 355, 356, 358, 359, 363, 364, 370, 378, 381
Pew Charitable Trust, 386
Phiel, 302, 481
Phoenix, 149
pipeline, 124, 306, 347, 354
Pittinger, 129, 405
Pittsburg, 275
POET, 157
population, 27, 29, 59, 66, 69, 70, 71, 72, 73, 74, 80, 86, 87, 92, 93, 172, 324, 366, 380, 382, 393
Potential energy, 33, 45
primary energy, 102
production tax credit, 384
Production tax credits, 154, 388
Proliferation resistance, 252, 293
proliferation resistant, 252, 289, 337, 338, 342, 410

prosperity, 27, 30, 71, 72, 73, 74, 87, 88, 95, 100, 216, 224, 227, 343, 344, 393, 400, 440
proton accelerators, 298
Public Service of Colorado, 142
Pumped storage, 161
Qatar, 348, 359, 380
quads, 74, 102, 108, 125
Queens College, 231
Raccoon Mountain, 162
Radiophobia, 328, 329
radiotoxic, 98, 202, 206, 297, 333
Radkowsky, 190
Ramping, 140
Reed, 336, 416, 469
Regional Greenhouse Gas Initiative, 387
Reliance Industries, 292
Renewable energy certificates, 388
Renewable portfolio standards, 389
Restuccia, 368
Reverse osmosis, 381
Rhode Island, 135, 387
Rickover, 180, 181, 191, 199, 247, 399, 402, 424
Rokkasho, 161
Rolls Royce, 238
Rongbuk, 80, 403
Ropeik, 318, 469
RPS, 389
Rubbia, 190, 298, 300, 414
Rumford, 46
Russia, 202, 245, 248, 249, 258, 260, 265, 286, 290, 306, 328, 336, 337, 338, 390, 397

Ryegate, 153
Sachs, 91, 469
Sadoway, 160, 407
Safeguards, 242
Sammes, 368, 419
Sandia National Laboratory, 251, 333
sanitation, 73, 83, 344, 380
Sargent & Lundy, 217
Saudi Arabia, 85, 174, 380, 397
Savannah River National Laboratory, 251
Savannah River Site, 282, 284, 395
Scientific American, 91, 147
Scotland, 157
Seaborg, 176
Sellafield, 290
Senate, 88
sequestration, 110, 111, 112, 120, 404, 407
Shaheen, 252
shale, 121, 122, 123, 124, 126, 230, 235, 334, 347, 348, 350, 351, 353, 354, 405, 417, 418
Shanghai Institute of Applied Physics, 261, 262, 263
Shaw, 199, 275, 277, 283
Shell, 348, 351, 417, 418
Shenhua, 348
Shippingport, 181, 190, 247, 249, 275, 424
Shoreham, 201
Siemens, 119, 130, 166, 239, 408, 412
Sievert, 22, 313
single fluid LFTR, 207, 231
Slovic, 318, 415
SMRs, 251, 279, 280, 284

sodium-sulfur, 161
soot, 28, 78, 82, 270
SORCA, 307, 415
Sorensen, 25, 244, 248, 253, 399, 401, 411, 412, 447
Sorenson, 248
South Africa, 72, 93, 169, 238, 259, 273, 337, 338, 347, 358, 359
South Carolina, 282, 283
South Carolina Electric & Gas Company, 282
South Dakota, 157
Southern Company, 153
Spain, 36, 148, 149, 388
spallation, 298, 414
Springfield, 152, 153
S-PRISM, 289, 290
SSAS, 368, 369, 370
steam generator, 119, 236, 279, 290, 437
Steinhaus, 251
Stillman, 336, 416, 469
subsidies, 142, 154, 156, 385, 386, 392, 398, 420, 438
subsidy, 135, 147, 384, 385, 386, 392, 420
Subsidyscope, 386
sulfur-iodine, 357, 377
supercritical, 109, 110, 112, 114, 199, 215, 239, 394, 412
Sv, 22, 313, 321
SVBR-100, 290, 414
tar sands, 28, 65, 266, 350, 354, 394, 418
tax preferences, 142, 383, 384, 392, 398
Teller, 245, 248, 249, 251, 449, 467
Tennessee, 162, 250, 280, 421
TerraPower, 291, 292, 293, 294, 295, 414
Tesla's, 275
Texas, 142, 354, 391
The Atlantic, 305
The Economic Future of Nuclear Power, 220
thermal energy, 35, 40, 44, 45, 46, 47, 48, 49, 50, 64, 102, 114, 116, 117, 119, 144, 146, 148, 149, 158, 167, 183, 187, 192, 194, 237, 240, 271, 286, 293, 361, 369, 439
ThorEA, 299, 414
Thorenco, 255, 412
Thorium Energy Alliance, 4, 255
Thorium Energy Association, 4, 299, 414
Thorium One, 266, 413
Thorium Power, 190, 266, 413
Three Mile Island, 201, 240, 307, 421
Toshiba, 274, 275
trade deficit, 84, 87, 99
Transatomic Power, 254, 412
TRISO, 212, 213, 214, 215, 271, 272, 273
Tritium, 237, 238, 267, 312, 411, 464
Tshinghua University, 259
Tsinghua University, 215
Turkey, 204, 306
Uhlir, 265, 413
UK, 249, 286, 290, 299, 338, 388, 415
unconventional oil, 347
UNESCO, 73, 380, 419

United Kingdom, 59, 265, 332, 337
United Nations, 29, 88, 321, 328, 383, 398
United States, 24, 176, 204, 248, 252, 383, 391, 398
University of Chicago, 186, 220, 250, 411, 423
University of Kent, 319
University of Pittsburgh, 321
University of Tennessee, 250
uranium dioxide, 180, 431
Utah, 348
van der Veer, 348
Van Helmont, 151
Venezuela, 85, 348, 397
Vermont, 146, 149, 387, 388, 389, 420
Vermont Yankee, 319
Waste Isolation Pilot Plant, 331
Watt, 44, 47, 58, 62
weapons-grade plutonium, 202
Weinberg, 180, 181, 191, 199, 200, 247, 249, 262, 393, 399, 423, 425, 427, 447, 450
Wen Hui Bao, 261
Wen Jiabao, 263
Wen Wei Po, 261
Westinghouse, 258, 272, 274, 275, 277, 278, 279, 283, 284, 413, 414, 424, 442, 445
Wigner, 191
WIPP, 331, 334, 416
wood-chip, 152
Worcester, 153
World Bank, 73, 380
World Nuclear Association, 204, 258
Wyoming, 348
Yanch, 327, 416
Yemen, 338
Zhang Zuoyi, 259
zirconium, 180, 183, 185, 195, 210, 241, 243, 287, 293, 303, 431, 464

Acknowledgements

Ralph Moir encouraged me to write this book and then was punished by spending days sending me comments on the draft. Per Peterson let me incorporate some of his good ideas. Ed Phiel compiled a long list of LFTR advantages. Breaking my foot gave me time to write this, while my wife Ann waited on me hand and foot. I received extensive editorial review advice from Meredith Angwin and important corrections from Howard Shaffer. Cavan Stone, Jess Gehin, George Angwin, and Peter Roth also advised me.

Printed in Great Britain
by Amazon.co.uk, Ltd.,
Marston Gate.